CAMBRIDGE LIBRARY COLLECTION

Books of enduring scholarly value

Darwin

Two hundred years after his birth and 150 years after the publication of 'On the Origin of Species', Charles Darwin and his theories are still the focus of worldwide attention. This series offers not only works by Darwin, but also the writings of his mentors in Cambridge and elsewhere, and a survey of the impassioned scientific, philosophical and theological debates sparked by his 'dangerous idea'.

Principles of Geology

In 1833, the Scottish geologist Charles Lyell published the final volume of his groundbreaking trilogy, which Charles Darwin took with him on the Beagle. In it, Lyell describes the composition of the Earth's crust, examines shell fossils, and explains rock stratification, separating geological formations into three periods – Primary, Secondary and Tertiary. He chastises his fellow geologists for preferring to speculate on the possibilities of the past rather than exploring the realities of the present, and shows his readers the importance of testing the validity of scientific claims. Lyell expertly integrates this book with the two earlier volumes, extending his interpretation of his geological findings from his research in Europe, especially at Mount Etna. Volume 3 consists of 26 chapters, a comprehensive index and 93 woodcut illustrations of different rock formations. Lyell writes with infectious enthusiasm, conveying the excitement of his discoveries in this landmark book, which will interest geologists and historians of science alike.

Principles of Geology

*An Attempt to Explain the Former Changes
of the Earth's Surface, by Reference to Causes
now in Operation*

VOLUME 3

CHARLES LYELL

CAMBRIDGE
UNIVERSITY PRESS

CAMBRIDGE UNIVERSITY PRESS

Cambridge, New York, Melbourne, Madrid, Cape Town, Singapore,
São Paolo, Delhi, Dubai, Tokyo

Published in the United States of America by Cambridge University Press, New York

www.cambridge.org
Information on this title: www.cambridge.org/9781108001373

© in this compilation Cambridge University Press 2009

This edition first published 1833
This digitally printed version 2009

ISBN 978-1-108-00137-3 Paperback

PRINCIPLES

OF

GEOLOGY,

BEING

AN ATTEMPT TO EXPLAIN THE FORMER CHANGES OF THE EARTH'S SURFACE,

BY REFERENCE TO CAUSES NOW IN OPERATION.

BY

CHARLES LYELL, Esq., F.R.S.,

FOR. SEC. TO THE GEOL. SOC., PROF. OF GEOL. TO KING'S COLL., LONDON

IN THREE VOLUMES.

Vol. III.

LONDON:

JOHN MURRAY, ALBEMARLE-STREET.

MDCCCXXXIII.

RODERICK IMPEY MURCHISON, Esq., F.R.S.,

&c. &c. &c.

LATE PRESIDENT OF THE GEOLOGICAL SOCIETY.

MY DEAR MURCHISON,

I HAVE great pleasure in dedicating this volume to you, as it contains the results of some of our joint labours in the field, in Auvergne, Velay, and Piedmont—results which had not yet been communicated to the public through any other channel.

When we quitted England together for a tour on the continent, in May, 1828, the first sketch only of my ' Principles of Geology' was finished. Since that time you have watched the progress of the work with friendly interest, and, as President of the Geological Society, have twice expressed in your Anniversary Addresses, your participation in many of my views, which were warmly controverted by others. The eulogy which you have lately pronounced from the chair, on the last part of my work, (whether I attribute your approval to the exercise of an unbiassed judgment or to the partiality of a friend,) could not fail to be most gratifying to my feelings, and I trust that you will long enjoy health and energy to continue to promote with enthusiasm the advancement of your favourite science.

Believe me, my dear Murchison,

Yours, &c. &c.

CHARLES LYELL.

PREFACE.

THE original MS. of the 'Principles of Geology' was delivered to the publisher at the close of the year 1827, when it was proposed that it should appear in the course of the year following, in two volumes octavo. Since that time many causes have concurred to delay the completion of the work, and, in some degree, to modify the original plan. In May, 1828, when the preliminary chapters on the History of Geology, and some others which follow them in the first volume, were nearly finished, I became anxious to visit several parts of the continent, in order to acquire more information concerning the tertiary formations. Accordingly, I set out in May, 1828, in company with Mr. Murchison, on a tour through France and the north of Italy, where we examined together many districts which are particularly mentioned in the body of this work. We visited Auvergne, Velay, Cantal, and the Vivarais, and afterwards the environs of Aix, in Provence, and then passed by the Maritime Alps to Savona, thence crossing to Piedmont by the Valley of the Bormida.

At Turin we found Signor Bonelli engaged in the arrangement of a large collection of tertiary shells

obtained chiefly from the Italian strata; and as I had
already conceived the idea of classing the different
tertiary groups, by reference to the proportional
number of recent species found fossil in each, I was
at pains to learn what number Signor Bonelli had
identified with living species, and the degree of
precision with which such identifications could be
made. With a view of illustrating this point, he
showed us suites of shells common to the Subapennine beds and to the Mediterranean, pointing
out that in some instances not only the ordinary type
of the species, but even the different varieties had
their counterparts both in the fossil and recent series.
The same naturalist informed us that the fossil shells
of the hill of the Superga, at Turin, differed as a
group from those of Parma and other localities of
the Subapennine beds of northern Italy; and, on the
other hand, that the characteristic shells of the
Superga agreed with the species found at Bordeaux
and other parts of the South of France.

I was the more struck with this remark, as Mr. Murchison and myself had already inferred that the highly-
inclined strata of the Valley of the Bormida, which
agree with those of the Superga, were older than
the more horizontal Subapennine marls, by which the
plains of the Tanaro and the Po are skirted.

When we had explored some parts of the Vicentin
together, Mr. Murchison re-crossed the Alps, while I
directed my course to the south of Italy, first staying

at Parma, where I studied, in the cabinets of Signor
Guidotti, a beautiful collection of Italian tertiary
shells, consisting of more than 1000 species, many
of which had been identified with living testacea.
Signor Guidotti had not examined his fossils with
reference to their bearing on geological questions,
but computed, on a loose estimate, that there were
about 30 per cent. of living species in the Subapennine
beds. I then visited Florence, Sienna, and Rome,
and the results of my inquiries respecting the tertiary strata of those territories will be found partly in
the body of the work, and partly in the catalogues
given in Appendix II.

On my arrival at Naples I became acquainted with
Signor O. G. Costa, who had examined the fossil
shells of Otranto and Calabria, and had collected
many recent testacea from the seas surrounding the
Calabrian coasts. His comparison of the fossil and
living species had led him to a very different result
in regard to the southern extremity of Italy, from that
to which Signors Guidotti and Bonelli had arrived in
regard to the north, for he was of opinion that few of
the tertiary shells were of extinct species. In confirmation of this view, he showed me a suite of fossil
shells from the territory of Otranto, in which nearly
all the species were recent.

In October, 1828, I examined Ischia, and obtained
from the strata of that island the fossil shells named
in Appendix II., p. 57. They were all, with two or

three exceptions, recognized by Signor Costa as
species now inhabiting the Mediterranean, a circumstance which greatly astonished me, as I procured
some of them at the height of 2000 feet above the
level of the sea (Vol. iii. p. 126).

Early in November, 1828, I crossed from Naples to
Messina, and immediately afterwards examined Etna,
and collected on the flanks of that mountain, near
Trezza, the fossil shells alluded to in the third volume
(p. 79, and Appendix II., p. 53). The occurrence of
shells in this locality was not unknown to the naturalists of Catania, but having been recognized by
them as *recent* species, they were supposed to have
been carried up from the sea-shore to fertilize the
soil, and therefore disregarded. Their position is
well known to many of the peasants of the country, by
whom the fossils are called 'roba di diluvio.'

In the course of my tour I had been frequently led
to reflect on the precept of Descartes, 'that a philosopher should once in his life doubt every thing he
had been taught;' but I still retained so much faith
in my early geological creed as to feel the most lively
surprise, on visiting Sortino, Pentalica, Syracuse, and
other parts of the Val di Noto, at beholding a limestone of enormous thickness filled with recent shells,
or sometimes with the mere casts of shells, resting on
marl in which shells of Mediterranean species were
imbedded in a high state of preservation. All idea
of attaching a high antiquity to a regularly stratified

limestone, in which the casts and impressions of shells alone were discernible, vanished at once from my mind. At the same time, I was struck with the identity of the associated igneous rocks of the Val di Noto with well known varieties of 'trap' in Scotland and other parts of Europe, varieties, which I had also seen entering largely into the structure of Etna. I occasionally amused myself with speculating on the different rate of progress which Geology might have made, had it been first cultivated with success at Catania, where the phenomena above alluded to, and the great elevation of the modern tertiary beds in the Val di Noto, and the changes produced in the historical era by the Calabrian earthquakes, would have been familiarly known.

From Cape Passaro I passed on by Spaccaforno and Licata to Girgenti, where I abandoned my design of exploring the western part of Sicily, that I might return again to the Val di Noto and the neighbourhood of Etna, and verify the discoveries which I had made. With this view I travelled by Caltanisetta, Piazza, Caltagirone, Vizzini, Militello, Palagonia, Lago Naftia, and Radusa, to Castrogiovanni, and from thence to Palermo, at which last place I procured the shells named in Appendix II. p. 55. The sections on this new route confirmed me in my first opinions respecting the Val di Noto, as will appear by the 6th, 8th, and 9th chapters of the third Volume.

When I again reached Naples, in January, 1829, I

found that Signor O. G. Costa had examined the tertiary fossils which I had sent to him from different parts of Sicily, and declared them to be for the most part of recent species. I then bent my course homeward, seeing at Genoa, Professor Viviani and Dr. Sasso, the last of whom put into my hands his memoirs on the strata of Albenga (see vol. iii. p. 166), in which I found, that, according to his list of shells, the tertiary formations at the foot of the maritime Alps contained about 50 per cent. of recent species.

I next re-visited Turin, and communicated to Signor Bonelli the result of my inquiries respecting the tertiary beds of the south of Italy, and of Sicily, upon which he kindly offered to review his fossils, some of which had been obtained from those countries, and to compare them with the Subapennine shells of northern Italy. He also promised to draw up immediately a list of the shells characteristic of the greensand of the Superga, and common to that locality and Bordeaux, that I might publish it at the end of my second volume; but the death of this amiable and zealous naturalist soon afterwards deprived me of the benefit of his assistance.

I had now fully decided on attempting to establish four sub-divisions of the great tertiary epoch, the same which are fully illustrated in the present work. I considered the basin of Paris and London to be the type of the first division; the beds of the Superga, of the second; the Subapennine strata of northern Italy, of the third; and Ischia and the Val di Noto, of the fourth. I was also convinced that I had seen proofs, during my tour in Auvergne, Tuscany and Sicily, of volcanic rocks contemporaneous with the sedimentary strata of three of the above periods.

On my return to Paris, in February, 1829, I communicated to M. Desnoyers some of the new views to which my examination of Sicily had led me, and my intention to attempt a classification of the different tertiary formations in chronological order, by reference to the comparative proportion of living species of shells found fossil in each. He informed me, that during my tour he had been employed in printing the first part of his memoir, not yet published, ' on the Tertiary Formations more recent than the Paris basin,' in which he had insisted on the doctrine 'of the succession of tertiary formations of different ages.' At the end of the first part of his memoir, which was published before I left Paris*, he annexed a note on the accordance of many of my views with his own, and my intention of arranging the tertiary formations chronologically, according to the relative number of fossils in each group, which were identifiable with species now living.

At the same time I learned from M. Desnoyers, that M. Deshayes had, by the mere inspection of the fossil shells in his extensive museum, convinced himself that the different tertiary formations might be

* Ann. des Sci. Nat., tome xvi. p. 214.

arranged in a chronological series. I accordingly lost
no time in seeing M. Deshayes, who explained to me
the data on which he considered that the three tertiary periods mentioned in the Tables, Appendix I.,
might be established. I at once perceived that the
fossils obtained by me in my tour would form but an
inconsiderable contribution to so great a body of
zoological evidence as M. Deshayes had already in
his possession. I therefore requested him to examine
my shells when they arrived from Italy, and expressed
my great desire to obtain his co-operation in my work,
in which, as will appear in the sequel, I was fortunate
enough to succeed.

The preparation of my first volume had now been
suspended for nine months, and was not resumed until
my return to London in the beginning of March,
1829. Before the whole was printed another summer
arrived, and I again took the field to examine 'the
Crag,' on the coasts of Essex, Norfolk, and Suffolk.
The first volume appeared at length in January, 1830,
after which I applied myself to perfect what I had
written on 'the changes in the organic world,' a
subject which merely occupied four or five chapters
in my original sketch, but which was now expanded
into a small treatise. Before this part was completed
another summer overtook me, and I then set out on
a geological expedition to the south of France, the
Pyrenees, and Catalonia.

On my return to Paris, in September, 1830, I

studied for six weeks in the museum of M. Deshayes, examining his collection of fossil and recent shells, and profiting by his instructions in conchology. As he had not yet published any of the general results deducible from his valuable collection, I requested him to furnish me with lists of those species of shells which were common to two or more tertiary periods, as also the names of those known to occur both in some tertiary strata and in a living state. This he engaged to do, and we agreed that the information should be communicated in a tabular form. After several modifications of the plan first proposed for the Tables, we finally agreed upon the manner in which they should be constructed, and the execution was left entirely in the hands of M. Deshayes, in whose name they were to appear in my second volume.

The tables were sent to me in the course of the following spring (1831), and additions and corrections several months later. They contained not only the information which I had expected, but much more, for the names of several hundred species were added, as being common to two or more *formations* of the same *period*, whereas it was originally proposed to insert those only which were known to be common to two or more *distinct periods*. Thus, for example, more than 50 shells are now included in the tables, on the ground that they are common to the tertiary strata both of the London and Paris basins, although they

only occur in the *Eocene period* to which the strata
of those basins belong. The names thus added will
increase the value of the tables, and give a more complete view of the point to which fossil conchology has
now reached; at the same time, it must be admitted
that tables of shells cannot be perfected on this plan,
as the science advances from year to year, without
soon outgrowing the space which could reasonably be
allotted to fossil conchology in a work on geology, for
they would soon embrace the names of the greater
number of known shells, nearly all of these being
common to different groups of strata of the *same
period.* Some of the catalogues which I have given
in Appendix II., of fossil shells from the neighbourhood of the Red Sea, and from some other localities,
may illustrate this remark, as they lead us to anticipate that, at no distant time, we may find a large proportion of all the *Recent* species in a fossil state.

In *treatises on fossil conchology,* such as I trust
M. Deshayes will soon publish, we cannot have too
complete a catalogue of all the species which have
been found fossil in every locality, together with their
synonyms; but in geological works we can only illustrate the more important theoretical points by catalogues of those shells which are either characteristic
of particular periods, as being exclusively confined to
them, or which show the connexion of two periods,
by being common to each. For this purpose we

must select certain normal groups which do not approximate too closely to each other, and enumerate by name the species common to more than one of these. Thus, for example, we might omit in our tables the Newer Pliocene formations altogether, and enumerate the shells common to the Recent and Older Pliocene beds.

I have arranged the tertiary formations in four groups, as I had determined to do before I was acquainted with M. Deshayes; and in his tables he has referred the shells to three periods, according to which he had classed them before he had any communication with me. No confusion, however, will arise from this want of conformity between the tables and my classification, since I have named two of my periods (the Newer and Older Pliocene) as subdivisions of one of his; and by reference to the Synoptical Table, at p. 61, the reader will see which localities mentioned in M. Deshayes's Tables belong to the Newer and which to the Older Pliocene period.

In the summer of 1831 I made a geological excursion to the volcanic district of the Eifel, and on my return I determined to extend my work to three volumes, the second of which appeared in January, 1832. The last volume has been delayed till now by many interruptions, among which I may mention a tour, in the summer of 1832, up the valley of the Rhine, when I examined the loess (vol. iii. p. 151), and a visit, on my way home through Switzerland, to the Valorsine, where I had an opportunity of verifying the observations of M. Necker on the granite veins and altered stratified rocks of that district. I may also mention the time occupied in the correction of the second edition of the first and second volumes, and the delivery of a course of Lectures in May and June, 1832, at King's College, London, on which occasion I had an opportunity of communicating to the scientific world a great part of the views now explained in my last volume.

London, April, 1833.

CONTENTS.

Vol. III.

CHAPTER I.

PAGE

Connexion between the subjects treated of in the former parts of this work and those to be discussed in the present volume—Erroneous assumption of the earlier geologists respecting the discordance of the former and actual causes of change—Opposite system of inquiry adopted in this work Illustrations from the history of the progress of Geology of the respective merits of the two systems—Habit of indulging conjectures respecting irregular and extraordinary agents not yet abandoned—Necessity in the present state of science of prefixing to a work on Geology treatises respecting the changes now in progress in the animate and inanimate world . . 1

CHAPTER II.

Arrangement of the materials composing the earth's crust—The existing continents chiefly composed of subaqueous deposits—Distinction between sedimentary and volcanic rocks—Between primary, secondary, and tertiary—Origin of the primary—Transition formations—Difference between secondary and tertiary strata—Discovery of tertiary groups of successive periods—Paris basin—London and Hampshire basins—Tertiary strata of Bordeaux, Piedmont, Touraine, &c.—Subapennine beds—English crag—More recent deposits of Sicily, &c. 8

CHAPTER III.

Different circumstances under which the secondary and tertiary formations may have originated—Secondary series formed when the ocean prevailed: Tertiary during the conversion of sea into land, and the growth of a continent—Origin of interruption in the sequence of formations—The areas where new deposits take place are always varying—Causes which occasion this transference of the places of sedimentary deposition—Denudation augments the discordance in age of rocks in contact—Unconformability of overlying formations—In what manner the shifting of the areas of sedimentary deposition may combine with the gradual extinction and introduction of species to produce a series of deposits having distinct mineral and organic characters 23

CHAPTER IV.

Chronological relations of mineral masses the first object in geological
classification—Superposition, proof of more recent origin—Exceptions in regard to volcanic rocks—Relative age proved by included fragments of older
rocks—Proofs of contemporaneous origin derived from mineral characters—
Variations to which these characters are liable—Recurrence of distinct rocks
at successive periods—Proofs of contemporaneous origin derived from organic
remains—Zoological provinces are of limited extent, yet spread over wider
areas than homogeneous mineral deposits—Different modes whereby dissimilar mineral masses and distinct groups of species may be proved to have
been contemporaneous   35

CHAPTER V.

Classification of tertiary formations in chronological order—Comparative
value of different classes of organic remains—Fossil remains of testacea the
most important—Necessity of accurately determining species—Tables of shells
by M. Deshayes—Four subdivisions of the Tertiary epoch—Recent formations—Newer Pliocene period—Older Pliocene period—Miocene period
—Eocene period—The distinct zoological characters of these periods may not
imply sudden changes in the animate creation—The recent strata form a
common point of departure in distant regions—Numerical proportion of
recent species of shells in different tertiary periods—Mammiferous remains
of the successive tertiary eras—Synoptical Table of Recent and Tertiary
formations   45

CHAPTER VI.

Newer Pliocene formations—Reasons for considering in the first place the
more modern periods—Geological structure of Sicily—Formations of the
Val di Noto of newer Pliocene period—Divisible into three groups—Great
limestone—Schistose and arenaceous limestone—Blue marl with shells—
Strata subjacent to the above—Volcanic rocks of the Val di Noto—Dikes—
Tuffs and Peperinos—Volcanic conglomerates—Proofs of long intervals
between volcanic eruptions—Dip and direction of newer Pliocene strata
of Sicily   62

CHAPTER VII.

Marine and volcanic formations at the base of Etna—Their connexion
with the strata of the Val di Noto—Bay of Trezza—Cyclopian isles—Fossil
shells of recent species—Basalt and altered rocks in the Isle of Cyclops—

PAGE

Submarine lavas of the bay of Trezza not currents from Etna—Internal structure of the cone of Etna—Val di Calanna—Val del Bove not an ancient crater—Its precipices intersected by countless dikes—Scenery of the Val del Bove—Form, composition, and origin of the dikes—Lavas and breccias intersected by them 75

CHAPTER VIII.

Speculations on the origin of the Val del Bove on Etna—Subsidences—Antiquity of the cone of Etna—Mode of computing the age of volcanos—Their growth analogous to that of exogenous trees—Period required for the production of the lateral cones of Etna—Whether signs of Diluvial Waves are observable on Etna 95

CHAPTER IX.

Origin of the newer Pliocene strata of Sicily—Growth of submarine formations gradual—Rise of the same above the level of the sea probably caused by subterranean lava—Igneous newer Pliocene rocks formed at great depths, exceed in volume the lavas of Etna—Probable structure of these recent subterranean rocks—Changes which they may have superinduced upon strata in contact—Alterations of the surface during and since the emergence of the newer Pliocene strata—Forms of the Sicilian valleys—Sea cliffs—Proofs of successive elevation—Why the valleys in the newer Pliocene districts correspond in form to those in regions of higher antiquity—Migrations of animals and plants since the emergence of the newer Pliocene strata—Some species older than the stations they inhabit—Recapitulation 103

CHAPTER X.

Tertiary formations of Campania—Comparison of the recorded changes in this region with those commemorated by geological monuments—Differences in the composition of Somma and Vesuvius—Dikes of Somma, their origin —Cause of the parallelism of their opposite sides—Why coarser grained in the centre—Minor cones of the Phlegræan Fields—Age of the volcanic and associated rocks of Campania—Organic remains—External configuration of the country, how produced—No signs of diluvial waves—Marine Newer Pliocene strata visible only in countries of earthquakes—Illustrations from Chili — Peru — Parallel roads of Coquimbo — West-Indian archipelago — Honduras—East-Indian archipelago—Red Sea . . . 118

CHAPTER XI.

Newer Pliocene fresh-water formations—Valley of the Elsa—Travertins of Rome—Osseous breccias—Sicily—Caves near Palermo—Extinct animals

PAGE

in newer Pliocene breccias—Fossil bones of Marsupial animals in Australian caves—Formation of osseous breccias in the Morea—Newer Pliocene alluviums—Difference between alluviums and regular subaqueous strata—The former of various ages—Marine alluvium—Grooved surface of rocks—Erratic blocks of the Alps—Theory of. deluges caused by paroxysmal elevations untenable—How ice may have contributed to transport large blocks from the Alps—European alluviums chiefly tertiary—Newer Pliocene in Sicily—Löss of the Valley of the Rhine—Its origin—Contains recent shells   137

CHAPTER XII.

Geological monuments of the *older* Pliocene period—Subapennine formations—Opinions of Brocchi—Different groups termed by him Subapennine are not all of the same age—Mineral composition of the Subapennine formations—Marls—Yellow sand and gravel—Subapennine beds how formed—Illustration derived from the Upper Val d'Arno—Organic remains of Subapennine hills—Older Pliocene strata at the base of the Maritime Alps—Genoa—Savona—Albenga—Nice—Conglomerate of Valley of Magnan—Its origin—Tertiary strata at the eastern extremity of the Pyrenees   155

CHAPTER XIII.

Crag of Norfolk and Suffolk—Shown by its fossil contents to belong to the older Pliocene period—Heterogeneous in its composition—Superincumbent lacustrine deposits—Relative position of the crag—Forms of stratification—Strata composed of groups of oblique layers—Cause of this arrangement—Dislocations in the crag produced by subterranean movements—Protruded masses of chalk—Passage of marine crag into alluvium—Recent shells in a deposit at Sheppey, Ramsgate, and Brighton   .   .   .   171

CHAPTER XIV.

Volcanic rocks of the older Pliocene period—Italy—Volcanic region of Olot in Catalonia—Its extent and geological structure—Map—Number of cones—Scoriæ—Lava currents—Ravines in the latter cut by water—Ancient alluvium underlying lava—Jets of air called ' Bufadors '—Age of the Catalonian volcanos uncertain—Earthquake which destroyed Olot in 1421—Sardinian volcanos—District of the Eifel and Lower Rhine—Map—Geological structure of the country—Peculiar characteristics of the Eifel volcanos—Lake craters—Trass—Crater of the Roderberg—Age of the Eifel volcanic rocks uncertain—Brown coal formation   183

CHAPTER XV.

Miocene period—Marine formations—Faluns of Touraine—Comparison of the Faluns of the Loire and the English Crag—Basin of the Gironde and

PAGE

Landes—Fresh-water limestone of Saucats—Position of the limestone of Blaye—Eocene strata in the Bordeaux basin—Inland cliff near Dax—Strata of Piedmont—Superga—Valley of the Bormida—Molasse of Switzerland—Basin of Vienna—Styria—Hungary—Volhynia and Podolia—Montpellier 202

CHAPTER XVI.

Miocene alluviums—Auvergne—Mont Perrier—Extinct quadrupeds—Velay—Orleanais—Alluviums contemporaneous with Faluns of Touraine—Miocene fresh-water formations—Upper Val d'Arno—Extinct mammalia—Coal of Cadibona—Miocene volcanic rocks—Hungary—Transylvania—Styria—Auvergne—Velay   217

CHAPTER XVII.

Eocene period—Fresh-water formations—Central France—Map—Limagne d'Auvergne—Sandstone and conglomerate—Tertiary Red marl and sandstone like the secondary ' new red sandstone'—Green and white foliated marls—Indusial limestone—Gypseous marls—General arrangement and origin of the Travertin—Fresh-water formation of the Limagne—Puy en Velay—Analogy of the strata to those of Auvergne—Cantal—Resemblance of Aurillac limestone and its flints to our upper chalk—Proofs of the gradual deposition of marl—Concluding remarks   225

CHAPTER XVIII.

Marine formations of the Eocene period—Strata of the Paris basin how far analogous to the lacustrine deposits of Central France—Geographical connexion of the Limagne d'Auvergne and the Paris basin—Chain of lakes in the Eocene period—Classification of groups in the Paris basin—Observations of M. C. Prevost—Sketch of the different subdivisions of the Paris basin—Contemporaneous marine and fresh-water strata—Abundance of Cerithia in the Calcaire grossier—Upper marine formation indicates a subsidence—Part of the Calcaire grossier destroyed when the upper marine strata originated—All the Parisian groups belong to one great epoch—Microscopic shells—Bones of quadrupeds in gypsum—In what manner entombed—Number of species—All extinct—Strata with and without organic remains alternating—Our knowledge of the physical geography, fauna, and flora of the Eocene period considerable—Concluding remarks   .   .   .   241

CHAPTER XIX.

Volcanic rocks of the Eocene period—Auvergne—Igneous formations associated with lacustrine strata—Hill of Gergovia—Eruptions in Central

PAGE

France at successive periods—Mont Dor an extinct volcano—Velay—Plomb
du Cantal—Train of minor volcanos stretching from Auvergne to the Vivarais
—Monts Domes—Puy de Côme—Puy Rouge—Ravines excavated through
lava—Currents of lava at different heights—Subjacent alluviums of distinct
ages—The more modern lavas of Central France may belong to the Miocene
period—The integrity of the cones not inconsistent with this opinion—No
eruptions during the historical era—Division of volcanos into ante-diluvian
and post-diluvian inadmissible—Theories respecting the effects of the Flood
considered—Hypothesis of a partial flood—Of a universal deluge—Theory
of Dr. Buckland as controverted by Dr. Fleming—Recapitulation   257

CHAPTER XX.

Eocene formations, *continued*—Basin of the Cotentin, or Valognes—Rennes
—Basin of Belgium, or the Netherlands—Aix in Provence—Fossil insects—
Tertiary strata of England—Basins of London and Hampshire—Different
groups—Plastic clay and sand—London clay—Bagshot sand—Fresh-water
strata of the Isle of Wight—Palæotherium and other fossil mammalia of
Binstead—English Eocene strata conformable to chalk—Outliers on the
elevated parts of the chalk—Inferences drawn from their occurrence—Sketch
of a theory of the origin of the English tertiary strata   .   .   275

CHAPTER XXI.

Denudation of secondary strata during the deposition of the English
Eocene formations—Valley of the Weald between the North and South Downs
—Map—Secondary rocks of the Weald divisible into five groups—North and
South Downs—Section across the valley of the Weald—Anticlinal axis—
True scale of heights—Rise and denudation of the strata gradual—Chalk
escarpments once sea-cliffs—Lower terrace of ' firestone,' how caused—
Parallel ridges and valleys formed by harder and softer beds—No ruins of
the chalk on the central district of the Weald—Explanation of this pheno-
menon—Double system of valleys, the longitudinal and the transverse—
Transverse how formed—Gorges intersecting the chalk—Lewes Coomb—
Transverse valley of the Adur   285

CHAPTER XXII.

Denudation of the Valley of the Weald, *continued*—The alternative of the
proposition that the chalk of the North and South Downs were once continuous, considered—Dr. Buckland on the Valley of Kingsclere—Rise and
denudation of secondary rocks gradual—Concomitant deposition of tertiary
strata gradual—Composition of the latter such as would result from the
wreck of the secondary rocks—Valleys and furrows on the chalk how caused

PAGE

—Auvergne, the Paris basin, and south-east of England one region of earth-quakes during the Eocene period—Why the central parts of the London and Hampshire basins rise nearly as high as the denudation of the Weald— Effects of protruding force counteracted by the levelling operations of water —Thickness of masses removed from the central ridge of the Weald—Great escarpment of the chalk having a direction north-east and south-west— Curved and vertical strata in the Isle of Wight—These were convulsed after the deposition of the fresh-water beds of Headen Hill—Elevations of land posterior to the crag—Why no Eocene alluviums recognizable—Concluding remarks on the intermittent operations of earthquakes in the south-east of England, and the gradual formation of valleys—Recapitulation   .   303

CHAPTER XXIII.

Secondary formations—Brief enumeration of the principal groups—No species common to the secondary and tertiary rocks—Chasm between the Eocene and Maestricht beds—Duration of secondary periods—Former continents placed where it is now sea—Secondary fresh-water deposits why rare —Persistency of mineral composition why apparently greatest in older rocks —Supposed universality of red marl formations—Secondary rocks why more consolidated—Why more fractured and disturbed—Secondary volcanic rocks of many different ages   324

CHAPTER XXIV.

On the relative antiquity of different mountain-chains—Theory of M. Elie de Beaumont—His opinions controverted—His method of proving that different chains were raised at distinct periods—His proof that others were contemporaneous—His reasoning why not conclusive—His doctrine of the parallelism of contemporaneous lines of elevation—Objections—Theory of parallelism at variance with geological phenomena as exhibited in Great Britain—Objections of Mr. Conybeare—How far anticlinal lines formed at the same period are parallel—Difficulties in the way of determining the relative age of mountains   337

CHAPTER XXV.

On the rocks usually termed 'Primary'—Their relation to volcanic and sedimentary formations—The 'primary' class divisible into stratified and unstratified—Unstratified rocks called Plutonic—Granite veins—Their various forms and mineral composition—Proofs of their igneous origin—Granites of the same character produced at successive eras—Some of these newer than certain fossiliferous strata—Difficulty of determining the age of particular granites—Distinction between the volcanic and the plutonic rocks—Trappean rocks not separable from the volcanic—Passage from trap into granite— Theory of the origin of granite at every period from the earliest to the most recent   352

CHAPTER XXVI.

On the stratified rocks usually called 'primary'—Proofs from the disposition of their strata that they were originally deposited from water—Alternation of beds varying in composition and colour—Passage of gneiss into granite—Alteration of sedimentary strata by trappean and granitic dikes—Inference as to the origin of the strata called ' primary '—Conversion of argillaceous into hornblende schist—The term 'Hypogene' proposed as a substitute for primary—'Metamorphic ' for ' stratified primary ' rocks—No regular order of succession of hypogene formations—Passage from the metamorphic to the sedimentary strata—Cause of the high relative antiquity of the visible hypogene formations—That antiquity consistent with the hypothesis that they have been produced at each successive period in equal quantities—Great volume of hypogene rocks supposed to have been formed since the Eocene period—Concluding remarks 365

Table I. Showing the relations of the various classes of rocks, the Alluvial, the Aqueous, the Volcanic, and the Hypogene, of different periods 386

Table II. Showing the order of superposition of the principal European groups of sedimentary strata mentioned in this work . . 389

Notes in explanation of the Tables of fossil shells in Appendix I. . 395

Appendix I. Tables of fossil shells by Monsieur G. P. Deshayes . 1

Appendix II. Lists of fossil Shells chiefly collected by the author in Sicily and Italy, named by M. Deshayes . . . 53

Glossary, containing an explanation of geological and other scientific terms used in this work 61

Index 85

LIST OF PLATES AND WOOD-CUTS

IN THE THIRD VOLUME.

PLATES.

Frontispiece. View of the volcanos around Olot, in Catalonia. See p. 186. This view is taken from a sketch by the author; an attempt is made to represent by colours the different geological formations of which the country is composed. The blue line of mountains in the distance are the Pyrenees, which are to the north of the spectator, and consist of primary and ancient secondary rocks. In front of these are the secondary formations, described in chap. xiv., coloured purplish-grey of different tints, to express different distances. The flank of the hill, in the foreground, called Costa di Pujou, is composed partly of secondary rocks, which are seen to the left of a small bridle-road, and partly of volcanic, the red colour expressing lava and scoriæ.

Several very perfect volcanic cones, chiefly composed of red scoriæ, and having craters on their summits, are seen in the immediate neighbourhood of Olot, coloured red. The level plain on which that town stands has clearly been produced by the flowing down of many lava-streams from those hills into the bottom of a valley, probably once of considerable depth, like those of the surrounding country, but which has been in a great measure filled up by lava.

The reader should be informed, that in many impressions of this plate Montsacopa is mis-spelt ' Montescopa,' and Mount Garrinada is mis-spelt ' Gradenada.'

Plate I. The shells represented in this plate have been selected by M. Deshayes as characteristic of the Pliocene period of the Tables, Appendix I. The greater part of them are common both to the older and newer Pliocene periods of this work. Eight of the species, Nos. 1, 3, 5, 6, 7, 9, 13, and 14, are now living, but are given as being also found in the *Older* Pliocene formations. Fusus crispus is not found either *recent* or in the Miocene or Eocene formations, but occurs both in the Newer and Older Pliocene strata. Mitra plicatula has been found only in the older Pliocene deposits. The Turbo rugosus was considered as exclusively Pliocene when selected by M. Deshayes, but M. Boué has since found it in the Miocene strata at Vienna and Moravia (see Tables, Appendix I. p. 26). Buccinum semistriatum is also a Miocene shell, but was inserted as being peculiarly abundant in the Pliocene strata.

Plate II. All the shells figured in this plate, except Cardita Ajar, are very characteristic of the Miocene formations; that is to say, they are found in that period and no other. Cardita Ajar is also very common in the Miocene strata, but is also a Recent species. It has not yet been observed in any *Pliocene* deposit.

Plate III. The species of shells figured in this plate are characteristic of the Eocene period, as being exclusively confined to deposits of that period, and for the most part abundant in them.

Plate IV. The microscopic shells of the order Cephalopoda, figured in this plate, are characteristic of the Eocene period, and are distinct from the microscopic shells of the Older Pliocene formations of Italy. The figures are from unpublished drawings by M. Deshayes, who has selected some of the most remarkable types of form. The reader will observe, that the minute points, figures 4, 8, 11, 14, and 18, indicate the natural size of the species which are represented. (For observations on these shells see p. 251.)

Plate V. Geological Map of the south-east of England, exhibiting the Denudation of the Weald. This map has been compiled in great part from Mr. Greenough's Geological Map of England, and Mr. Mantell's Map of the south-east of England. (Illustrations of Geol. of Sussex, and fossils of Tilgate Forest, 1827.) The eastern extremity of the ' denudation ' is reduced from Mr. Murchison's Map of that district. (Geol. Trans., 2nd series, vol. ii. part i. plate 14.) The object of this map is fully explained in chapters xxi. and xxii. of this volume.

LIST OF WOOD-CUTS.

No.   Page

1. Diagram showing the order of succession of stratified masses   .   15
2. Diagram showing the relative position of the Primary, Secondary,
   and Tertiary strata   16
3. Diagram showing the relative age of the strata of the Paris basin, and
   those of the basin of the Loire, in Touraine   .   .   20
4. Diagram showing the same in the strata of Suffolk and Piedmont   21
5. Diagram containing sections in the Val di Noto, Sicily   .   .   64
6.⎫
7.⎭ Horizontal sections of dikes near Palagonia   .   .   69
8. Section of horizontal limestone in contact with inclined strata of Tuff
   in the hill of Novera, near Vizzini   .   .   .   70
9. Section of calcareous grit and peperino, east of Palagonia, south side
   of the pass   72
10. Section of the same beds on the north side of the pass   .   . 72
11. Outline view of the cone of Etna from the summit of the limestone
    platform of Primosole   75
12. Section from Paternò by Lago di Naftià to Palagonia   .   .   76
13. Section of beds of clay and sand capped by columnar basalt and con-
    glomerate at La Motta, near Catania   .   .   .   77
14. View of the Isle of Cyclops, in the Bay of Trezza   .   .   79
15. Diagram showing the contortions in the newer Pliocene strata of the
    Isle of Cyclops   80
16. Horizontal section showing the invasion of the newer Pliocene strata
    of the Isle of Cyclops by lava   81
17. Wood-cut showing the great valley on the east side of Etna   .   83
18. Diagram explanatory of the origin of the Valleys of Calanna and St.
    Giacomo, on Etna   86
19. View of dikes at the base of the Serre del Solfizio, Etna   .   90
20. View of tortuous dikes or veins of lava, Punto di Guimento, Etna . 91
21. View of the rocks Finochio, Capra, and Musara, in the Val del Bove 92
22. View from the summit of Etna into the Val del Bove   .   93
23. View of the Valley called Gozzo degli Martiri, below Melilli   . 110
24. Diagram showing the manner of obliteration of successive lines of sea-cliff 111
25. View of dikes or veins of lava at the Punto del Nasone, on Somma 122
26. Diagram showing the superposition of alluvium and cave deposits con-
    taining *extinct* quadrupeds to a limestone containing *recent* shells 139
27. Diagram showing the position of the Cave of San Ciro, near Palermo 141
28. Diagram showing the position of Tertiary strata at Genoa   .   166

No.   Page

29. Section from Monte Calvo to the sea by the Valley of Magnan, near
    Nice   167

30. Diagram showing the manner in which the Crag may be supposed to
    rest on the chalk   173

31. Section of shelly crag near Walton, Suffolk   .   .   .   174

32. Section at the light-house near Happisborough   .   .   174

33. Section of Little Cat Cliff, showing the inclination of the layers of
    quartzose sand in opposite directions   .   .   .   175

34. Lamination of shelly sand and loam, near the Signal-House, Walton   175

35. Diagram illustrative of the successive deposition of strata   .   .   176

36. Section of ripple marks caused by the wind on loose sand   .   176

37. Bent strata of loam in the cliffs between Cromer and Runton   .   178

38. Folding of the strata between East and West Runton   .   178

39. Section in the cliffs east of Sherringham   .   .   .   178

40. Section east of Sherringham, Norfolk   .   .   .   179

41. Side view of a promontory of chalk and crag, at Trimmingham, Norfolk 179

42. Northern protuberance of chalk, Trimmingham   .   .   180

43. Map of the volcanic district of Catalonia   .   .   .   184

44. Section of volcanic sand and ashes in a valley near Olot   .   .   187

45. Section above the bridge of Cellent   .   .   .   188

46. Section at Castell Follit   190

47. Superposition of rocks in the volcanic district of Catalonia   .   192

48. Map of the volcanic district of the Lower Rhine   .   .   194

49. View of the Gemunden Maar   195

50. Section of the same and other contiguous lake-craters   .   .   196

51. Section of tertiary strata overlying chalk near Dax   .   .   207

52. Section explaining the position of the Eocene strata in the Bordeaux basin 209

53. Section of Inland cliff near Dax   210

54. Position of the Miocene alluviums of Mont Perrier (or Boulade)   217

55. Section of the fresh-water formation of Cadibona   .   .   221

56. Map of Auvergne, Cantal, Velay, &c.   .   .   .   226

57. Section of Vertical marls near Clermont   .   .   .   231

58. }
59. } Superposition of the formations of the Paris basin   .   .   243

60. Section of the Hill of Gergovia near Clermont   .   .   259

61. Lavas of Auvergne resting on alluviums of different ages .   .   267

62. Map of the principal tertiary basins of the Eocene period   .   275

63. Section from the London to the Hampshire basin across the Valley of
    the Weald   288

64. Section of the country from the confines of the basin of London to that
    of Hants, with the principal heights above the level of the sea on a
    true scale   288

65. View of the chalk escarpment of the South Downs, taken from the
    Devil's Dike, looking towards the west and south-west .   .   290

No.   Page

66. Chalk escarpment as seen from the hill above Steyning, Sussex. The castle and village of Bramber in the foreground .   .   291

67. Section of lower terrace of firestone   .   .   .   292

68. Diagram explanatory of anticlinal and synclinal lines .   .   293

69.
70. } Sections illustrating the gradual denudation of the Weald Valley 294, 295

71. Section from the north escarpment of the South Downs to Barcombe 296

72. Section of cliffs west of Sherringham   .   .   .   297

73. View of the transverse valley of the Adur in the South Downs .   299

74. Supposed section of a transverse valley   .   .   .   300

75. View of Lewes Coomb   301

76. Section of a fault in the cliff-hills near Lewes   .   .   301

77. Hypothetical section to illustrate the question of the denudation of the Weald Valley   304

78. Ground plan of the Valley of Kingsclere .   .   .   305

79. Section across the Valley of Kingsclere from north to south .   305

80. Section of the Valley of Kingsclere with the heights on a true scale 306

81. Hypothetical section illustrating the denudation of the Weald Valley and the contemporaneous origin of the Eocene strata .   310

82.
83. } Diagrams illustrative of the relative antiquity of mountain-chains 340, 341

84. Diagram showing the relative position of the Hypogene sedimentary and volcanic rocks   353

85. Granite veins traversing stratified rocks   .   .   .   354

86. Granite veins traversing gneiss at Cape Wrath in Scotland .   354

87. Granite veins passing through hornblende slate, Carnsilver Cove, Cornwall   355

88. View of the junction of granite and limestone in Glen Tilt .   356

89. Lamination of clay-slate, Montagne de Seguinat, near Gavarnie, in the Pyrenees   366

90. Junction of granite with jurassic or oolite strata in the Alps .   371

91. Diagram showing the different order of position in the Plutonic and Sedimentary formations of different ages .   .   388

92. Diagram to explain the meaning of the term ' fault,' Glossary .   68

93. Diagram to explain the term ' salient angle,' Glossary .   79

ERRATA.

Page 89, line 11 from the top, *for* vivid, *read* livid.
—— 103, line 10 from the top, *for* newer, *read* older.
—— 104, line 9 from the top, *for* Colosseum, *read* Coliseum.
—— 110, No. of wood-cut, *for* No. 22, *read* No. 23.
—— 111,   Ditto,   *for* No. 23, *read* No. 24.
—— 192, line 10 from the bottom, *for* with, *read* without.
—— 193, line 2 from the bottom, *for* Von Oyenhausen, *read* Von Oeynhausen.
—— 193, line 3 from the bottom, *for* M. Nŏeggerath, *read* M. Noĕggerath.
—— 197, line 19 from the top, *for* Moseberg, *read* Mosenberg.

PRINCIPLES OF GEOLOGY.

CHAPTER I.

Connexion between the subjects treated of in the former parts of this work and those to be discussed in the present volume—Erroneous assumption of the earlier geologists respecting the discordance of the former and actual causes of change—Opposite system of inquiry adopted in this work—Illustrations from the history of the progress of Geology of the respective merits of the two systems—Habit of indulging conjectures respecting irregular and extraordinary agents not yet abandoned—Necessity in the present state of science of prefixing to a work on Geology treatises respecting the changes now in progress in the animate and inanimate world.

HAVING considered, in the preceding volumes, the actual operation of the causes of change which affect the earth's surface and its inhabitants, we are now about to enter upon a new division of our inquiry, and shall therefore offer a few preliminary observations, to fix in the reader's mind the connexion between two distinct parts of our work, and to explain in what manner the plan pursued by us differs from that more usually followed by preceding writers on Geology.

All naturalists, who have carefully examined the arrangement of the mineral masses composing the earth's crust, and who have studied their internal structure and fossil contents, have recognised therein the signs of a great succession of former changes ; and the causes of these changes have been the object of anxious inquiry. As the first theorists possessed but a scanty acquaintance with the present economy of the animate and inanimate world, and the vicissitudes to which these are subject, we find them in the situation of novices, who attempt to read a history written in a foreign language, doubting about the meaning of the most ordinary terms; disputing, for example, whether a shell was really a shell,—whether sand and pebbles were the result of aqueous trituration,—whether stratification was the effect of successive deposition from water ; and a thousand other elementary questions which now appear to us so easy and simple, that we can hardly conceive them to have once afforded matter for warm and tedious controversy.

In the first volume we enumerated many prepossessions which biassed the minds of the earlier inquirers, and checked an impartial desire of arriving at truth. But of all the causes to which we alluded, no one contributed so powerfully to give rise to a false method of philosophizing as the entire unconsciousness of the first geologists of the extent of their own ignorance respecting the operations of the existing agents of change.

They imagined themselves sufficiently acquainted with the mutations now in progress in the animate and inanimate world, to entitle them at once to affirm, whether the solution of certain problems in geology could ever be derived from the observation of the actual economy of nature, and having decided that they could not, they felt themselves at liberty to indulge their imaginations, in guessing at what *might be,* rather than in inquiring *what is ;* in other words, they employed themselves in conjecturing what might have been the course of nature at a remote period, rather than in the investigation of what was the course of nature in their own times.

It appeared to them more philosophical to speculate on the possibilities of the past, than patiently to explore the realities of the present, and having invented theories under the influence of such maxims, they were consistently unwilling to test their validity by the criterion of their accordance with the ordinary operations of nature. On the contrary, the claims of each new hypothesis to credibility appeared enhanced by the great contrast of the causes or forces introduced to those now developed in our terrestrial system during a period, as it has been termed, of *repose.*

Never was there a dogma more calculated to foster indolence and to blunt the keen edge of curiosity, than this assumption of the discordance between the former and the existing causes of change. It produced a state of mind unfavourable in the highest conceivable degree to the candid reception of the evidence of those minute, but incessant mutations, which every part of the earth's surface is undergoing, and by which the condition of its living inhabitants is continually made to vary. The student, instead of being encouraged with the hope of interpreting the enigmas presented to him in the earth's structure,—instead of being prompted to undertake laborious inquiries into the natural history of the organic world, and the complicated effects of the igneous and aqueous causes now in operation, was taught to despond from the first. Geology, it was affirmed, could never rise to the rank of an exact science, —the greater number of phenomena must for ever remain inexplicable, or only be partially elucidated by ingenious conjectures. Even the mystery which invested the subject was said to constitute one of its principal charms, affording, as it did, full scope to the fancy to indulge in a boundless field of speculation.

The course directly opposed to these theoretical views consists in an earnest and patient endeavour to reconcile the former indications of change with the evidence of gradual mutations now in progress; restricting us, in the first instance, to known causes, and then speculating on those which may be in activity in regions inaccessible to us. It seeks an interpretation of geological monuments by comparing the changes of which they give evidence with the vicissitudes now in progress, or *which may be* in progress.

We shall give a few examples in illustration of the practical results already derived from the two distinct methods of theorizing, for we have now the advantage of being enabled to judge by experience of their respective merits, and by the relative value of the fruits which they have produced.

In our historical sketch of the progress of geology, the reader has seen that a controversy was maintained for more than a century, respecting the origin of fossil shells and bones—were

they organic or inorganic substances ? That the latter opinion
should for a long time have prevailed, and that these bodies
should have been supposed to be fashioned into their present
form by a plastic virtue, or some other mysterious agency,
may appear absurd ; but it was, perhaps, as reasonable a conjecture as could be expected from those who did not appeal, in
the first instance, to the analogy of the living creation, as affording the only source of authentic information. It was only by
an accurate examination of living testacea, and by a comparison
of the osteology of the existing vertebrated animals with the
remains found entombed in ancient strata, that this favourite
dogma was exploded, and all were, at length, persuaded that
these substances were exclusively of organic origin.

In like manner, when a discussion had arisen as to the nature
of basalt and other mineral masses, evidently constituting a
particular class of rocks, the popular opinion inclined to a belief
that they were of aqueous, not of igneous origin.   These rocks,
it was said, might have been precipitated from an aqueous solution, from a chaotic fluid, or an ocean which rose over the continents, charged with the requisite mineral ingredients.   All
are now agreed that it would have been impossible for human
ingenuity to invent a theory more distant from the truth ; yet
we must cease to wonder, on that account, that it gained so
many proselytes, when we remember that its claims to probability arose partly from its confirming the assumed want of all
analogy between geological causes and those now in action.

By what train of investigation were all theorists brought round
at length to an opposite opinion, and induced to assent to the
igneous origin of these formations ? By an examination of
the structure of active volcanos, the mineral composition of
their lavas and ejections, and by comparing the undoubted products of fire with the ancient rocks in question.

We shall conclude with one more example. When the
organic origin of fossil shells had been conceded, their occurrence in strata forming some of the loftiest mountains in the
world, was admitted as a proof of a great alteration of the

relative level of sea and land, and doubts were then entertained whether this change might be accounted for by the partial drying up of the ocean, or by the elevation of the solid land. The former hypothesis, although afterwards abandoned by general consent, was at first embraced by a vast majority. A multitude of ingenious speculations were hazarded to show how the level of the ocean might have been depressed, and when these theories had all failed, the inquiry, as to what vicissitudes of this nature might now be taking place, was, as usual, resorted to in the last instance. The question was agitated, whether any changes in the level of sea and land had occurred during the historical period, and, by patient research, it was soon discovered that considerable tracts of land had been permanently elevated and depressed, while the level of the ocean remained unaltered. It was therefore necessary to reverse the doctrine which had acquired so much popularity, and the unexpected solution of a problem at first regarded as so enigmatical, gave, perhaps, the strongest stimulus ever yet afforded to investigate the ordinary operations of nature. For it must have appeared almost as improbable to the earlier geologists, that the laws of earthquakes should one day throw light on the origin of mountains, as it must to the first astronomers, that the fall of an apple should assist in explaining the motions of the moon.

Of late years the points of discussion in geology have been transferred to new questions, and those, for the most part, of a higher and more general nature ; but, notwithstanding the repeated warnings of experience, the ancient method of philosophising has not been materially modified.

We are now, for the most part, agreed as to what rocks are of igneous, and what of aqueous origin,—in what manner fossil shells, whether of the sea or of lakes, have been imbedded in strata,—how sand may have been converted into sandstone,— and are unanimous as to other propositions which are not of a complicated nature ; but when we ascend to those of a higher order, we find as little disposition, as formerly, to make a strenuous effort, in the first instance, to search out an explanation in the ordinary economy of Nature.   If, for example, we seek for the causes why mineral masses are associated together in certain groups ; why they are arranged in a certain order which is never inverted ; why there are many breaks in the continuity of the series ; why different organic remains are found in distinct sets of strata ; why there is often an abrupt passage from an assemblage of species contained in one formation to that in another immediately superimposed,—when these and other topics of an equally extensive kind are discussed, we find the habit of indulging conjectures, respecting irregular and extraordinary causes, to be still in full force.

We hear of sudden and violent revolutions of the globe, of the instantaneous elevation of mountain chains, of paroxysms of volcanic energy, declining according to some, and according to others increasing in violence, from the earliest to the latest ages.   We are also told of general catastrophes and a succession of deluges, of the alternation of periods of repose and disorder, of the refrigeration of the globe, of the sudden annihilation of whole races of animals and plants, and other hypotheses, in which we see the ancient spirit of speculation revived, and a desire manifested to cut, rather than patiently to untie, the Gordian knot.

In our attempt to unravel these difficult questions, we shall adopt a different course, restricting ourselves to the known or possible operations of existing causes ; feeling assured that we have not yet exhausted the resources which the study of the present course of nature may provide, and therefore that we are not authorized, in the infancy of our science, to recur to extraordinary agents.   We shall adhere to this plan, not only on the grounds explained in the first volume, but because, as we have above stated, history informs us that this method has always put geologists on the road that leads to truth,—suggesting views which, although imperfect at first, have been found capable of improvement, until at last adopted by universal consent.   On the other hand, the opposite method, that of speculating on a former distinct state of things, has led invariably to a multitude of contradictory systems, which have been overthrown one after the other,—which have been found quite incapable of modification,—and which are often required to be precisely reversed.

In regard to the subjects treated of in our first two volumes, if systematic treatises had been written on these topics, we should willingly have entered at once upon the description of geological monuments properly so called, referring to other authors for the elucidation of elementary and collateral questions, just as we shall appeal to the best authorities in conchology and comparative anatomy, in proof of many positions which, but for the labours of naturalists devoted to these departments, would have demanded long digressions. When we find it asserted, for example, that the bones of a fossil animal at Œningen were those of man, and the fact adduced as a proof of the deluge, we are now able at once to dismiss the argument as nugatory, and to affirm the skeleton to be that of a reptile, on the authority of an able anatomist ; and when we find among ancient writers the opinion of the gigantic stature of the human race in times of old, grounded on the magnitude of certain fossil teeth and bones, we are able to affirm these remains to belong to the elephant and rhinoceros, on the same authority.

But since, in our attempt to solve geological problems, we shall be called upon to refer to the operation of aqueous and igneous causes, the geographical distribution of animals and plants, the real existence of species, their successive extinction, and so forth, we were under the necessity of collecting together a variety of facts, and of entering into long trains of reasoning, which could only be accomplished in preliminary treatises.

These topics we regard as constituting the alphabet and grammar of geology ; not that we expect from such studies to obtain a key to the interpretation of all geological phenomena, but because they form the groundwork from which we must rise to the contemplation of more general questions relating to the complicated results to which, in an indefinite lapse of ages, the existing causes of change may give rise.

CHAPTER II.

Arrangement of the materials composing the earth's crust—The existing continents chiefly composed of subaqueous deposits—Distinction between sedimentary and volcanic rocks—Between primary, secondary, and tertiary—
Origin of the primary—Transition formations—Difference between secondary
and tertiary strata—Discovery of tertiary groups of successive periods—Paris
basin—London and Hampshire basins—Tertiary strata of Bordeaux, Piedmont, Touraine, &c.—Subapennine beds—English crag—More recent deposits
of Sicily, &c.

GENERAL ARRANGEMENT OF THE MATERIALS COMPOSING THE EARTH'S CRUST.

WHEN we examine into the structure of the earth's crust (by which we mean the small portion of the exterior of our planet accessible to human observation), whether we pursue our investigations by aid of mining operations, or by observing the sections laid open in the sea cliffs, or in the deep ravines of mountainous countries, we discover everywhere a series of mineral masses, which are not thrown together in a confused heap, but arranged with considerable order ; and even where their original position has undergone great subsequent disturbance, there still remain proofs of the order that once reigned.

We have already observed, that if we drain a lake, we frequently find at the bottom a series of recent deposits disposed with considerable regularity one above the other; the uppermost, perhaps, may be a stratum of peat, next below a more compact variety of the same, still lower a bed of laminated shell marl, alternating with peat, and then other beds of marl, divided by layers of clay. Now if a second pit be sunk through the same continuous lacustrine deposit, at some distance from the first, we often meet with nearly the same series of beds, yet with slight variations; some, for example, of the layers of sand, clay, or marl may be wanting, one or more of

them having thinned out and given place to others, or sometimes one of the masses, first examined, is observed to increase
in thickness to the exclusion of other beds. Besides this
limited continuity of particular strata, it is obvious that the
whole assemblage must terminate somewhere ; as, for example,
where they reach the boundary of the original lake-basin, and
where they will come in contact with the rocks which form the
boundary of, and; at the same time, pass under all the recent
accumulations.

In almost every estuary we may see, at low water, analogous
phenomena where the current has cut away part of some newly-
formed bank, consisting of a series of horizontal strata of peat,
sand, clay, and, sometimes, interposed beds of shells. Each of
these may often be traced over a considerable area, some extending farther than others, but all of necessity confined within
the basin of the estuary. Similar remarks are applicable, on a
much more extended scale, to the recent delta of a great river,
like the Ganges, after the periodical inundations have subsided,
and when sections are exposed of the river-banks and the cliffs
of numerous islands, in which horizontal beds of clay and sand
may be traced over an area many hundred miles in length, and
more than a hundred in breadth.

Subaqueous deposits. The greater part of our continents are
evidently composed of subaqueous deposits ; and in the manner
of their arrangement we discover many characters precisely similar to those above described ; but the different groups of strata
are, for the most part, on a greater scale, both in regard to depth
and area, than any observable in the new formations of lakes,
deltas, or estuaries. We find, for example, beds of limestone
several hundred feet in thickness, containing imbedded corals and
shells, stretching from one country to another, yet always giving
place, at length, to a distinct set of strata, which either rise up
from under it, like the rocks before alluded to as forming the
borders of a lake, or cover and conceal it. In other places,
we find beds of pebbles, and sand, or of clay of great thickness. The different formations composed of these materials

usually contain some peculiar organic remains ; as, for example, certain species of shells and corals, or certain plants.

Volcanic rocks. Besides these strata of aqueous origin, we find other rocks which are immediately recognized to be the products of fire, from their exact resemblance to those which have been produced in modern times by volcanos, and thus we immediately establish two distinct orders of mineral masses composing the crust of the globe—the sedimentary and the volcanic.

Primary rocks. But if we investigate a large portion of a continent which contains within it a lofty mountain range, we rarely fail to discover another class, very distinct from either of those above alluded to, and which we can neither assimilate to deposits such as are now accumulated in lakes or seas, nor to those generated by ordinary volcanic action. The class alluded to, consists of granite, granite schist, roofing slate, and many other rocks, of a much more compact and crystalline texture than the sedimentary and volcanic divisions before mentioned. In the unstratified portion of these crystalline rocks, as in the granite for example, no organic fossil remains have ever been discovered, and only a few faint traces of them in some of the *stratified* masses of the same class ; for we should state, that a considerable portion of these rocks are divided, not only into strata, but into laminæ, so closely imitating the internal arrangement of well-known aqueous deposits, as to leave scarcely any reasonable doubt that they owe this part of their texture to similar causes.

These remarkable formations have been called *primitive*, from being supposed to constitute the most ancient mineral productions known to us, and from a notion that they originated before the earth was inhabited by living beings, and while yet the planet was in a nascent state. Their high relative antiquity is indisputable; for in the oldest sedimentary strata, containing organic remains, we often meet with rounded pebbles, of the older crystalline rocks, which must therefore have been consolidated before the derivative strata were formed out of their ruins. They rise up from beneath the rocks of mechanical origin, entering into the structure of lofty mountains, so as to constitute, at the same time, the lowest and the most elevated portions of the crust of the globe.

Origin of primary rocks. Nothing strictly analogous to these ancient formations can now be seen in the progress of formation on the habitable surface of the earth, nothing, at least, within the range of human observation. The first speculators, however, in Geology, found no difficulty in explaining their origin, by supposing a former condition of the planet perfectly distinct from the present, when certain chemical processes were developed on a great scale, and whereby crystalline precipitates were formed, some more suddenly, in huge amorphous masses, such as granite; others by successive deposition and with a foliated and stratified structure, as in the rocks termed gneiss and mica-schist. A great part of these views have since been entirely abandoned, more especially with regard to the origin of granite, but it is interesting to trace the train of reasoning by which they were suggested. First, the stratified primitive rocks, exhibited, as we before mentioned, well-defined marks of successive accumulation, analogous to those so common in ordinary subaqueous deposits. As the latter formations were found divisible into natural groups, characterized by certain peculiarities of mineral composition, so also were the primitive. In the next place, there were discovered, in many districts, certain members of the so-called primitive series, either alternating with, or passing by intermediate gradations into rocks of a decidedly mechanical origin, containing traces of organic remains. From such gradual passage the aqueous origin of the stratified crystalline rocks was fairly inferred; and as we find in the different strata of subaqueous origin every gradation between a mechanical and a purely crystalline texture; between sand, for example, and saccharoid gypsum, the latter having, probably, been precipitated originally in a crystalline form, from water containing sulphate of lime in solution, so it was imagined that, in a former condition of the planet, the different degrees of crystallization in the older rocks might have been dependent on the varying state of the menstruum from which they were precipitated.

The presence of certain crystalline ingredients in the composition of many of the primary rocks, rendered it necessary to resort to many arbitrary hypotheses, in order to explain their precipitation from aqueous solution, and for this reason a difference in the condition of the planet, and of the pristine energy of chemical causes, was assumed. A train of speculation originally suggested by the observed effects of aqueous agents, was thus pushed beyond the limits of analogy, and it was not until a different and almost opposite course of induction was pursued, beginning with an examination of volcanic products, that more sound theoretical views were established.

Granite of igneous origin. As we are merely desirous, in this chapter, of fixing in the reader's mind the leading divisions of the rocks composing the earth's crust, we cannot enter, at present, into a detailed account of these researches, but shall only observe, that a passage was first traced from lava into other more crystalline igneous rocks, and from these again to granite, which last was found to send forth dykes and veins into the contiguous strata in a manner strictly analogous to that observed in volcanic rocks, and producing at the point of contact such changes as might be expected to result from the influence of a heated mass cooling down slowly under great pressure from a state of fusion. The want of stratification in granite supplied another point of analogy in confirmation of its igneous origin; and as some masses were found to send out veins through others, it was evident that there were granites of different ages, and that instead of forming in all cases the oldest part of the earth's crust, as had at first been supposed, the granites were often of comparatively recent origin, sometimes newer than the stratified rocks which covered them.

Stratified primary rocks. The theory of the origin of the other crystalline rocks was soon modified by these new views

respecting the nature of granite. First it was shown, by numerous examples, that ordinary volcanic dikes might produce great
alterations in the sedimentary strata which they traversed, causing them to assume a more crystalline texture, and obliterating
all traces of organic remains, without, at the same time, destroying either the lines of stratification, or even those which mark the
division into laminæ. It was also found, that granite dikes and
veins produced analogous, though somewhat different changes;
and hence it was suggested as highly probable, that the effects to
which small veins gave rise, to the distance of a few yards, might
be superinduced on a much grander scale where immense masses
of fused rock, intensely heated for ages, came in contact at great
depths from the surface with sedimentary formations, The slow
action of heat in such cases, it was thought, might occasion a state
of semi-fusion, so that, on the cooling down of the masses, the
different materials might be re-arranged in new forms, according
to their chemical affinities, and all traces of organic remains might
disappear, while the stratiform and lamellar texture remained.

May be of different ages. According to these views, the
primary strata may have assumed their crystalline structure at
as many successive periods as there have been distinct eras of
the formation of granite, and their difference of mineral composition may be attributed, not to an original difference of the
conditions under which they were deposited at the surface, but
to subsequent modifications superinduced by heat at great
depths below the surface.

The strict propriety of the term primitive, as applied to granite and to the granitiform and associated rocks, thus became
questionable, and the term primary was very generally substituted, as simply expressing the fact, that the crystalline
rocks, as a mass, were older than the *secondary*, or those which
are unequivocally of a mechanical origin and contain organic
remains.

Transition Formations. The reader may readily conceive,
even from the hasty sketch which we have thus given of the
supposed origin of the stratified primary rocks, that they may

occasionally graduate into the secondary; accordingly, an attempt was made, when the classification of rocks was chiefly derived from mineral structure, to institute an order called *transition*, the characters of which were intermediate between those of the primary and secondary formations. Some of the shales, for example, associated with these strata, often passed insensibly into clay slates, undistinguishable from those of the granitic series; and it was often difficult to determine whether some of the compound rocks of this transition series, called greywacke, were of mechanical or chemical origin. The imbedded organic remains were rare, and sometimes nearly obliterated; but by their aid the groups first called transition were at length identified with rocks, in other countries, which had undergone much less alteration, and wherein shells and zoophytes were abundant.

The term transition, however, was still retained, although no longer applicable in its original signification. It was now made to depend on the identity of certain species of organized fossils; yet reliance on mineral peculiarities was not fairly abandoned, as constituting part of the characters of the group. This circumstance became a fertile source of ambiguity and confusion; for although the species of the transition strata denoted a certain epoch, the intermediate state of mineral character gave no such indications, and ought never to have been made the basis of a chronological division of rocks.

Order of succession of stratified masses. All the subaqueous strata, which we before alluded to as overlying the primary, were at first called secondary; and when they had been found divisible into different groups, characterised by certain organic remains and mineral peculiarities, the relative position of these groups became a matter of high interest. It was soon found that the order of succession was never inverted, although the different formations were not coextensively distributed; so that, if there be four different formations, as *a, b, c, d,* in the annexed diagram (No. 1), which, in certain localities, may be seen in vertical superposition, the uppermost or newest of them, *a,* will in other places be in contact with *c,* or with the lowest

No. 1.

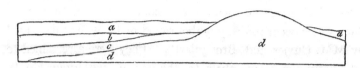

of the whole series, *d,* all the intermediate formations being absent.

Tertiary formations. After some progress had been made in classifying the secondary rocks, and in assigning to each its relative place in a chronological series, another division of sedimentary formations was established, called *tertiary,* as being of newer origin than the secondary. The fossil contents of the deposits belonging to this newly-instituted order are, upon the whole, very dissimilar from those of the secondary rocks, not only all the species, but many of the most remarkable animal and vegetable forms, being distinct. The tertiary formations were also found to consist very generally of detached and isolated masses, surrounded on all sides by primary and secondary rocks, and occupying a position, in reference to the latter, very like that of the waters of lakes, inland seas, and gulfs, in relation to a continent, and, like such waters, being often of great depth, though of limited area. The imbedded organic remains were chiefly those of marine animals, but with frequent intermixtures of terrestrial and fresh-water species so rarely found among the secondary fossils. Frequently there was evidence of the deposits having been purely lacustrine, a circumstance which has never yet been clearly ascertained in regard to any secondary group.

We shall consider more particularly, in the next chapter, how far this distinction of rocks into secondary and tertiary is founded in nature, and in what relation these two orders of mineral masses may be supposed to stand to each other. But before we offer any general views of this kind, it may be useful to present the reader with a succinct sketch of the principal points in the history of the discovery and classification of the tertiary strata.

Paris Basin. The first series of deposits belonging to this class, of which the characters were accurately determined, were those which occur in the neighbourhood of Paris, first described by MM. Cuvier and Brongniart*. They were ascertained to fill a depression in the chalk (as the beds *d*, in diagram No. 2, rest upon c), and to be composed of different materials, some-

No. 2.

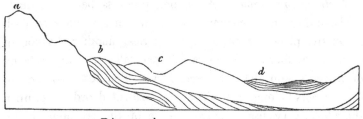

a, Primary rocks.
b, Older secondary formations.   c, Chalk.
d, Tertiary formation.

times including the remains of marine animals, and sometimes of fresh-water. By the aid of these fossils, several distinct alternations of marine and fresh-water formations were clearly shown to lie superimposed upon each other, and various speculations were hazarded respecting the manner in which the sea had successively abandoned and regained possession of tracts which had been occupied in the intervals by the waters of rivers or lakes. In one of the subordinate members of this Parisian series, a great number of scattered bones and skeletons of land animals were found entombed, the species being perfectly dissimilar from any known to exist, as indeed were those of almost all the animals and plants of which any portions were discovered in the associated deposits.

We shall defer, to another part of this work, a more detailed account of this interesting formation, and shall merely observe in this place, that the investigation of the fossil contents of these beds forms an era in the progress of the science. The

* Environs de Paris, 1811.

French naturalists brought to bear upon their geological researches so much skill and proficiency in comparative anatomy
and conchology, as to place in a strong light the importance of
the study of organic remains, and the comparatively subordinate
interest attached to the mere investigation of the structure and
mineral ingredients of rocks.

A variety of tertiary formations were soon afterwards found
in other parts of Europe, as in the south-east of England, in
Italy, Austria, and different parts of France, especially in the
basins of the Loire and Gironde, all strongly contrasted to
the secondary rocks. As in the latter class many different
divisions had been observed to preserve the same mineral characters and organic remains over wide areas, it was natural that
an attempt should first be made to trace the different subdivisions of the Parisian tertiary strata throughout Europe, for
some of these were not inferior in thickness to several of the
secondary formations that had a wide range.

But in this case the analogy, however probable, was not
found to hold good, and the error, though almost unavoidable,
retarded seriously the progress of geology. For as often as a
new tertiary group was discovered, as that of Italy, for example, an attempt was invariably made, in the first instance, to
discover in what characters it agreed with some one or more
subordinate members of the Parisian type. Every fancied point
of correspondence was magnified into undue importance, and
such trifling circumstances, as the colour of a bed of sand
or clay, were dwelt upon as proofs of identification, while the
difference in the mineral character and organic contents of the
group from the whole Parisian series was slurred over and
thrown into the shade.

By the influence of this illusion, the succession and chronological relations of different tertiary groups were kept out of
sight. The difficulty of clearly discerning these, arose from
the frequent isolation of the position of the tertiary formations before described, since, in proportion as the areas occupied
by them are limited, it is rare to discover a place where one

set of strata overlap another, in such a manner that the geologist might be enabled to determine the difference of age by direct superposition.

ORIGIN OF THE EUROPEAN TERTIARY STRATA AT SUCCESSIVE PERIODS.

We shall now very briefly enumerate some of the principal steps which eventually led to a conviction of the necessity of referring the European tertiary formations to distinct periods, and the leading data by which such a chronological series may be established.

London and Hampshire Basins.—Very soon after the investigation, before alluded to, of the Parisian strata, those of Hampshire and of the Basin of the Thames were examined in our own country. Mr. Webster found these English tertiary deposits to repose, like those in France, upon the chalk or newest rock of the secondary series. He identified a great variety of the shells occurring in the British and Parisian strata, and ascertained that, in the Isle of Wight, an alternation of marine and fresh-water beds occurred, very analogous to that observed in the basin of the Seine*. But no two sets of strata could well be more dissimilar in mineral composition, and they were only recognized to belong to the same era, by aid of the specific identity of their organic remains. The discordance, in other respects, was as complete as could well be imagined, for the principal marine formation in the one country consisted of blue clay, in the other of white limestone, and a variety of curious rocks in the neighbourhood of Paris had no representatives whatever in the south of England.

Subapennine Beds.—The next important discovery of tertiary strata was in Italy, where Brocchi traced them along the flanks of the Apennines, from one extremity of the peninsula to the other, usually forming a lower range of hills, called by him the Subapennines†. These formations, it is true, had

---

* Webster in Englefield's Isle of Wight and Geol. Trans., vol. ii. p. 161.

† Conch. Foss. Subap., 1814.

been pointed out by the older Italian writers, and some correct ideas, as we have seen, had been entertained respecting their recent origin, as compared to the inclined secondary rocks on which they rested*. But accurate data were now for the first time collected, for instituting a comparison between them and other members of the great European series of tertiary formations.

Brocchi came to the conclusion that nearly one-half of several hundred species of fossil shells procured by him from these Subapennine beds were identical with those now living in existing seas, an observation which did not hold true in respect to the organic remains of the Paris basin. It might have been supposed that this important point of discrepancy would at once have engendered great doubt as to the identity, in age, of any part of the Subapennine beds to any one member of the Parisian series; but, for reasons above alluded to, this objection was not thought of much weight, and it was supposed that a group of strata, called ' the upper marine formation,' in the basin of the Seine, might be represented by all the Subapennine clays and yellow sand.

English Crag.—Several years before, an English naturalist, Mr. Parkinson, had observed, that certain shelly strata, in Suffolk, which overlaid the blue clay of London, contained distinct fossil species of testacea, and that a considerable portion of these might be identified with species now inhabiting the neighbouring sea†. These overlying beds, which were provincially termed ' Crag,' were of small thickness, and were not regarded as of much geological importance. But when duly considered, they presented a fact worthy of great attention, viz., the superposition of a tertiary group, inclosing, like the Subapennine beds, a great intermixture of recent species of shells, upon beds wherein a very few remains of recent or living species were entombed.

Mr. Conybeare, in his excellent classification of the English

---

* See vol. i. p. 51, for opinions of Odoardi, in 1761.

† Geol. Trans., vol. i. p. 324. 1811.

strata*, placed the crag as the uppermost of the British series, and several geologists began soon to entertain an opinion that this newest of our tertiary formations might correspond in age to the Italian strata described by Brocchi.

Tertiary Strata of Touraine.—The next step towards esta- blishing a succession of tertiary periods was the evidence adduced to prove that certain formations, more recent than the uppermost members of the Parisian series, were also older than the Subapennine beds, so that they constituted deposits of an age intermediate between the two types above alluded to. Mr. Desnoyers, for example, ascertained that a group of marine strata in Touraine, in the basin of the Loire (*e*, dia- gram No. 3), rest upon the uppermost subdivision of the

No. 3.

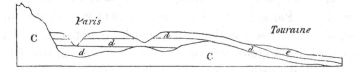

C, Chalk and other secondary formations.
d, Tertiary formation of Paris basin.
e, Superimposed marine tertiary beds of the Loire.

Parisian group *d*, which consists of a lacustrine formation, ex- tending continuously throughout a platform which intervenes between the basin of the Seine and that of the Loire. These overlaying marine strata, M. Desnoyers assimilated to the En- glish crag, to which they bear some analogy, although their or- ganic remains differ considerably, as will be afterwards shown.

A large tertiary deposit had already been observed in the south-west of France, around Bordeaux and Dax, and a de- scription of its fossils had been published by M. de Basterot†. Many of the species were peculiar, and differed from those of the strata now called Subapennine; yet these same peculiar and characteristic fossils reappeared in Piedmont, in a series of

---

* Outlines of the Geology of England and Wales, 1822.
† Mem. de la Soc. d'Hist. Nat. de Paris, tome ii., 1825.

strata inferior in position to the Subapennines (as *e* underlies
f, diagram No. 4.)

No. 4.

C, Chalk and older formations.
d, London clay. (old tertiary).
e, Tertiary strata of same age as beds of the Loire.
f, Crag and Subapennine tertiary deposits.

This inferior group, *e*, composed principally of green sand,
occurs in the hills of Mont Ferrat, and beds of the same age are
seen in the valley of the Bormida. They also form the hill of
the Superga, near Turin, where M. Bonelli formed a large collection of their fossils, and identified them with those discovered
near Bordeaux and in the basin of the Gironde.

But we are indebted to M. Deshayes for having proved, by
a careful comparison of the entire assemblage of shells found in
the above-mentioned localities, in Touraine, in the south-east
of France, and in Piedmont, that the whole of these three
groups possess the same zoological characters, and belong to
the same epoch, as also do the shells described by M. Constant
Prevost, as occurring in the basin of Vienna*.

Now the reader will perceive, by reference to the observations above made, and to the accompanying diagrams, that one
of the formations of this intervening period, *e*, has been found
superimposed upon the highest member of the Parisian series,
d; while another of the same set has been observed to underlie the Subapennine beds, *f*. Thus the chronological series,
d, e, f, is made out, in which the deposits, originally called
tertiary, those of the Paris and London basins, for example,
occupy the lowest position, and the beds called ' the Crag,'
and ' the Subapennines,' the highest.

Tertiary Strata newer than the Subapennine.—The fossil

* Sur la Constitution, &c. du bassin de Vienne, Journ. de Phys., Nov. 1820.

remains which characterize each of the three successive periods above alluded to, approximate more nearly to the assemblage of *species* now existing, in proportion as their origin is less remote from our own era, or, in other words, the recent species are always more numerous, and the extinct more rare, in proportion to the low antiquity of the formations. But the discordance between the state of the organic world indicated by the fossils of the Subapennine beds and the actual state of things is still considerable, and we naturally ask, are there no monuments of an intervening period?—no evidences of a gradual passage from one condition of the animate creation to that which now prevails, and which differs so widely?

It will appear, in the sequel, that such monuments are not wanting, and that there are marine strata entering into the composition of extensive districts, and of hills of no trifling height, which contain the exuviæ of testacea and zoophytes, hardly distinguishable, as a group, from those now peopling the neighbouring seas. Thus the line of demarcation between the actual period and that immediately antecedent, is quite evanescent, and the newest members of the tertiary series will be often found to blend with the formations of the historical era.

In Europe, these modern strata have been found in the district around Naples, in the territory of Otranto and Calabria, and more particularly in the Island of Sicily; and the bare enumeration of these localities cannot fail to remind the reader, that they belong to regions where the volcano and the earthquake are now active, and where we might have anticipated the discovery of emphatic proofs, that the conversion of sea into land had been of frequent occurrence at very modern periods.

CHAPTER III.

Different circumstances under which the secondary and tertiary formations may
have originated—Secondary series formed when the ocean prevailed: Tertiary
during the conversion of sea into land, and the growth of a continent—Origin
of interruption in the sequence of formations—The areas where new deposits
take place are always varying—Causes which occasion this transference of the
places of sedimentary deposition—Denudation augments the discordance in age
of rocks in contact—Unconformability of overlying formations—In what
manner the shifting of the areas of sedimentary deposition may combine with
the gradual extinction and introduction of species to produce a series of deposits
having distinct mineral and organic characters.

DIFFERENT CIRCUMSTANCES UNDER WHICH THE SECONDARY AND TERTIARY FORMATIONS MAY HAVE ORIGINATED.

WE have already glanced at the origin of some of the principal points of difference in the characters of the primary and
secondary rocks, and may now briefly consider the relation in
which the secondary stand to the tertiary, and the causes of
that succession of tertiary formations described in the last
chapter.

It is evident that large parts of Europe were simultaneously
submerged beneath the sea when different portions of the secondary series were formed, because we find homogeneous mineral
masses, including the remains of marine animals, referrible to
the secondary period, extending over great areas ; whereas the
detached and isolated position of tertiary groups, in basins
or depressions bounded by secondary and primary rocks,
favours the hypothesis of a sea interrupted by extensive tracts
of dry land.

State of the Surface when the Secondary Strata were formed.

Let us consider the changes that must be expected to accompany the gradual conversion of part of the bed of an ocean into
a continent, and the different characters that might be imparted

to subaqueous deposits formed during the period when the sea prevailed, as contrasted with those that might belong to the subsequent epoch when the land should predominate. First, we may suppose a vast submarine region, such as the bed of the western Atlantic, to receive for ages the turbid waters of several great rivers, like the Amazon, Orinoco, or Mississippi, each draining a considerable continent. The sediment thus introduced might be characterized by a peculiar colour and composition, and the same homogeneous mixture might be spread out over an immense area by the action of a powerful current, like the Gulf-stream. First one submarine basin, and then another, might be filled, or rendered shallow, by the influx of transported matter, the same species of animals and plants still continuing to inhabit the sea, so that the organic, as well as the mineral characters, might be constant throughout the whole series of deposits.

In another part of the same ocean, let us suppose masses of coralline and shelly limestone to grow, like those of the Pacific, simultaneously over a space several thousand miles in length, and thirty or forty degrees of latitude in breadth, while volcanic eruptions give rise, at different intervals, to igneous rocks, having a common subaqueous character in different parts of the vast area.

It is evident that, during such a state of a certain quarter of the globe, beds of limestone and other rocks might be formed, and retain a common character over spaces equal to a large portion of Europe.

State of the Surface when the Tertiary Groups were formed.

But when the area under consideration began to be converted into land, a very different condition of things would succeed. A series of subterranean movements might first give rise to small rocks and isles, and then, by subsequent elevations, to larger islands, by the junction of the former. These lands would consist partly of the mineral masses before described, whether coralline, sedimentary, or volcanic, and partly of the subjacent rocks, whatever they may have been, which constituted the original bed of the ocean.   Now the degradation of these lands would commence immediately upon their emergence, the waves of the sea undermining the cliffs, and torrents flowing from the surface, so that new strata would begin to form in different places ; and in proportion as the lands increased, these deposits would augment.

At length by the continued rising and sinking of different parts of the bed of the ocean, a number of distinct basins would be formed, wherein different kinds of sediment, each distinguished by some local character, might accumulate. Some of the groups of isles that had first risen would, in the course of ages, become the central mountain ranges of continents, and different lofty chains might thus be characterized by similar rocks of contemporaneous origin, the component strata having originated under analogous circumstances in the ocean before described.

Finally, when large tracts of land existed, there would be a variety of disconnected gulfs, inland seas, and lakes, each receiving the drainage of distinct hydrographical basins, and becoming the receptacles of strata distinguished by marked peculiarities of mineral composition.   The organic remains would also be more varied, for in one locality fresh-water species would be imbedded, as in deposits now forming in the lakes of Switzerland and the north of Italy ; in another, marine species, as in the Aral and Caspian ; in a third region, gulfs of brackish water would be converted into land, like those of Bothnia and Finland in the Baltic; in a fourth, there might be great fluviatile and marine formations along the borders of a chain of inland seas, like the deltas now growing at the mouths of the Don, Danube, Nile, Po, and Rhone, along the shores of the Azof, Euxine, and Mediterranean.   These deposits would each partake more or less of the peculiar mineral character of adjoining lands, the degradation of which would supply sediment to the different rivers.

Now if such be, in a great measure, the distinction between

the circumstances under which the secondary and tertiary series originated, it is quite natural that particular tertiary groups should occupy areas of comparatively small extent,—that they should frequently consist of littoral and lacustrine deposits, and that they should often contain those admixtures of terrestrial, fresh-water, and marine remains, which are so rare in secondary rocks. It might also be expected, that the tertiary volcanic formations should be much less exclusively submarine, and this we accordingly find to be the case.

CAUSES OF THE SUPERPOSITION OF SUCCESSIVE FORMATIONS HAVING DISTINCT MINERAL AND ORGANIC CHARACTERS.

But we have still to account for those remarkable breaks in the series of superimposed formations, which are common both to the secondary and tertiary rocks, but are more particularly frequent in the latter.

The elucidation of this curious point is the more important, because geologists of a certain school appeal to phenomena of this kind in support of their doctrine of great catastrophes, out of the ordinary course of nature, and sudden revolutions of the globe.

It is only by carefully considering the combined action of all the causes of change now in operation, whether in the animate or inanimate world, that we can hope to explain such complicated appearances as are exhibited in the general arrangement of mineral masses. In attempting, therefore, to trace the origin of these violations of continuity, we must re-consider many of the topics treated of in our two former volumes, such as the effects of the various agents of decay and reproduction, the imbedding of organic remains, and the extinction of species.

Shifting of the Areas of Sedimentary Deposition.—By reverting to our survey of the destroying and renovating agents, it will be seen that the surface of the terraqueous globe may be divided into two parts, one of which is undergoing repair, while the other, constituting, at any one period, by far the largest portion of the whole, is either suffering degradation, or remaining stationary without loss or increment. The reader will assent at once to this proposition, when he reflects that the dry land is, for the most part, wasting by the action of rain, rivers, and torrents, while the effects of vegetation have, as we have shown, only a conservative tendency, being very rarely instrumental in adding new masses of mineral matter to the surface of emerged lands; and when he also reflects that part of the bed of the sea is exposed to the excavating action of currents, while the greater part, remote from continents and islands, probably receives no new deposits whatever, being covered for ages with the clear blue waters uncharged with sediment. Here the relics of organic beings, lying in the ooze of the deep, may decompose like the leaves of the forest in autumn, and leave no wreck behind, but merely supply nourishment, by their decomposition, to succeeding races of marine animals and plants.

The other part of the terraqueous surface is the receptacle of new deposits, and in this portion alone, as we pointed out in the last volume, the remains of animals and plants become fossilized. Now the position of this area, where new formations are in progress, and where alone any memorials of the state of organic life are preserved, is always varying, and must for ever continue to vary; and, for the same reason, that portion of the terraqueous globe which is undergoing waste, also shifts its position, and these fluctuations depend partly on the action of aqueous, and partly of igneous causes.

In illustration of these positions we may observe, that the sediment of the Rhone, which is thrown into the lake of Geneva, is now conveyed to a spot a mile and a half distant from that where it accumulated in the tenth century, and six miles from the point where the delta began originally to form. We may look forward to the period when the lake will be filled up, and then a sudden change will take place in the distribution of the transported matter; for the mud and sand brought down from the Alps will thenceforth, instead of being deposited near Geneva, be carried nearly two hundred miles southwards, where the Rhone enters the Mediterranean.

The additional matter thus borne down to the lower delta of-the Rhone would not only accelerate its increase, but might affect the mineral character of the strata there deposited, and thus give rise to an upper group; or subdivision of beds, having a distinct character. But the filling up of a lake, and the consequent transfer of the sediment to a new place, may some-times give rise to a more abrupt transition from one group to another; as, for example, in a gulf like that of the St. Law-rence, where no deposits are now accumulated, the river being purged of all its impurities in its previous course through the Canadian lakes. Should the lowermost of these lakes be at any time filled up with sediment, or laid dry by earthquakes, the waters of the river would thenceforth become turbid, and strata would begin to be deposited in the gulf, where a new formation would immediately overlie the ancient rocks now constituting the bottom. In this case there would be an abrupt passage from the inferior and more ancient, to the newer superimposed formation.

The same sudden coming on of new sedimentary deposits, or the suspension of those which were in progress, must fre-quently occur in different submarine basins where there are currents which are always liable, in the course of ages, to change their direction. Suppose, for instance, a sea to be filling up in the same manner as the Adriatic, by the influx of the Po, Adige, and other rivers. The deltas, after advancing and converging, may at last come within the action of a trans-verse current, which may arrest the further deposition of matter, and sweep it away to a distant point. Such a current now appears to prey upon the delta of the Nile, and to carry eastward the annual accessions of sediment that once added rapidly to the plains of Egypt.

On the other hand, if a current charged with sediment vary its course, a circumstance which, as we have shown, must happen to all of them in the lapse of ages, the accumulation of transported matter will at once cease in one region, and com-mence in another.

Although the causes which occasion the transference of the places of sedimentary deposition are continually in action in every region, yet they are most frequent where subterranean movements alter, from time to time, the levels of land, and they must be immense during the successive elevations and depressions which must be supposed to accompany the rise of a great continent from the deep. A trifling change of level may sometimes throw a current into a new direction, or alter the course of a considerable river. Some tracts will be alter-nately submerged and laid dry by subterranean movements ; in one place a shoal will be formed, whereby the waters will drift matter over spaces where they once threw down their burden, and new cavities will elsewhere be produced, both marine and lacustrine, which will intercept the waters bearing sediment, and thereby stop the supply once carried to some distant basin.

We have before stated, that a few earthquakes of moderate power might cause a subsidence which would connect the sea of Azof with a large part of Asia now below the level of the ocean. This vast depression, recently shown by Humboldt to extend over an area of eighteen thousand square leagues, sur-rounds Lake Aral and the Caspian, on the shores of which seas it sinks in some parts to the depth of three hundred feet below the level of the ocean. The whole area might thus sud-denly become the receptacle of new beds of sand and shells, probably differing in mineral character from the masses pre-viously existing in that country, for an exact correspondence could only arise from a precise identity in the whole combina-tion of circumstances which should give rise to formations pro-duced at different periods in the same place.

Without entering into more detailed explanations, the reader will perceive that, according to the laws now governing the aqueous and igneous causes, distinct deposits must, at different periods, be thrown down on various parts of the earth's surface, and that, in the course of ages, the same area may become, again and again, the receptacle of such dissimilar sets of strata. During intervening periods, the space may either remain unaltered, or suffer what is termed *denudation*, in which case a superior set of strata are removed by the power of running water, and subjacent beds are laid bare, as happens wherever a sea encroaches upon a line of coast. By such means, it is obvious that the discordance in age of rocks in contact must often be greatly increased.

The frequent unconformability in the stratification of the inferior and overlying formation is another phenomenon in their arrangement, which may be considered as a natural consequence of those movements that accompany the gradual conversion of part of an ocean into land; for by such convulsions the older set of strata may become rent, shattered, inclined, and contorted to any amount. If the movement entirely cease before a new deposit is formed in the same tract, the superior strata may repose horizontally upon the dislocated series. But even if the subterranean convulsions continue with increasing violence, the more recent formations must remain comparatively undisturbed, because they cannot share in the immense derangement previously produced in the older beds, while the latter, on the contrary, cannot fail to participate in all the movements subsequently communicated to the newer.

Change of Species everywhere in progress.—If, then, it be conceded, that the combined action of the volcanic and the aqueous forces would give rise to a succession of distinct formations, and that these would be sometimes unconformable, let us next inquire in what manner these groups might become characterized by different assemblages of fossil remains.

We endeavoured to show, in the last volume, that the hypothesis of the gradual extinction of certain animals and plants, and the successive introduction of new species, was quite consistent with all that is known of the existing economy of the animate world; and if it be found the only hypothesis which is reconcilable with geological phenomena, we shall have

strong grounds for conceiving that such is the order of nature.

Fossilization of Plants and Animals partial.—We have
seen that the causes which limit the duration of species are not
confined, at any one time, to a particular part of the globe ;
and, for the same reason, if we suppose that their place is
supplied, from time to time, by new species, we may suppose their introduction to be no less generally in progress.
Hence, from all the foregoing premises, it would follow, that
the change of species would be in simultaneous operation everywhere throughout the habitable surface of sea and land ;
whereas the fossilization of plants and animals must always be
confined to those areas where new strata are produced.   These
areas, as we have proved, are always shifting their position, so
that the fossilizing process, whereby the commemoration of
the particular state of the organic world, at any given time,
is effected, may be said to move about, visiting and revisiting
different tracts in succession.

   In order more distinctly to elucidate our idea of the working
of this machinery, let us compare it to a somewhat analogous
case that might easily be imagined to occur in the history
of human affairs. Let the mortality of the population of a
large country represent the successive extinction of species,
and the births of new individuals the introduction of new
species. While these fluctuations are gradually taking place
everywhere, suppose commissioners to be appointed to visit
each province of the country in succession, taking an exact
account of the number, names, and individual peculiarities of
all the inhabitants, and leaving in each district a register containing a record of this information. If, after the completion
of one census, another is immediately made after the same
plan, and then another, there will, at last, be a series of
statistical documents in each province. When these are
arranged in chronological order, the contents of those which
stand next to each other will differ according to the length
of the intervals of time between the taking of each census.

If, for example, all the registers are made in a single year, the proportion of deaths and births will be so small during the interval between the compiling of two consecutive documents, that the individuals described in each will be nearly identical; whereas, if there are sixty provinces, and the survey of each requires a year, there will be an almost entire discordance between the persons enumerated in two consecutive registers.

There are undoubtedly some other causes besides the mere quantity of time which may augment or diminish the amount of discrepancy. Thus, for example, at some periods a pestilential disease may lessen the average duration of human life, or a variety of circumstances may cause the births to be unusually numerous, and the population to multiply, or, a province may be suddenly colonized by persons migrating from surrounding districts.

We must also remind the reader, that we do not propose the above case as an exact parallel to those geological phenomena which we desire to illustrate ; for the commissioners are supposed to visit the different provinces in rotation, whereas the commemorating processes by which organic remains become fossilized, although they are always shifting from one area to another, are yet very irregular in their movements. They may abandon and revisit many spaces again and again, before they once approach another district; and besides this source of irregularity, it may often happen, that while the depositing process is suspended, denudation may take place, which may be compared to the occasional destruction of some of the statistical documents before mentioned. It is evident, that where such accidents occur, the want of continuity in the series may become indefinitely great, and that the monuments which follow next in succession will by no means be equidistant from each other in point of time.

If this train of reasoning be admitted, the frequent distinctness of the fossil remains, in formations immediately in contact, would be a necessary consequence of the existing laws of sedimentary deposition, accompanied by the gradual birth and death of species.

We have already stated, that we should naturally look for a change in the mineral character in strata thrown down at distant intervals in the same place ; and, in like manner, we must also expect, for the reason last set forth, to meet occasionally with sudden transitions from one set of organic remains to another.   But the causes which have given rise to such differences in mineral characters have no necessary connexion with those which have produced a change in the species of imbedded plants and animals.

When the lowest of two sets of strata are much dislocated over a wide area, the upper being undisturbed, there is usually a considerable discordance in the organic remains of the two groups ; but this coincidence must not be ascribed to the agency of the disturbing forces, as if they had exterminated the living inhabitants of the surface.   The immense *lapse of time* required for the development of so great a series of subterranean movements, has in these cases allowed the species also throughout the globe to vary, and hence the two phenomena are usually concomitant.

Although these inferences appear to us very obvious, we are aware that they are directly opposed to many popular theories respecting catastrophes ; we shall, therefore, endeavour to place our views in a still clearer light before the reader.   Suppose we had discovered two buried cities at the foot of Vesuvius, immediately superimposed upon each other, with a great mass of tuff and lava intervening, just as Portici and Resina, if now covered with ashes, would overlie Herculaneum.   An antiquary might possibly be entitled to infer, from the inscriptions on public edifices, that the inhabitants of the inferior and older town were Greeks, and those of the modern, Italians.   But he would reason very hastily, if he also concluded from these data, that there had been a sudden change from the Greek to the Italian language in Campania.   Suppose he afterwards found *three* buried cities, one above the other, the intermediate one being Roman, while, as in the former example, the lowest was Greek, and the uppermost Italian, he would then perceive the fallacy of his former opinion, and would begin to suspect that the catastrophes, whereby the cities were inhumed, might have no relation whatever to the fluctuations in the language of the inhabitants ; and that, as the Roman tongue had evidently intervened between the Greek and Italian, so many other dialects may have been spoken in succession, and the passage from the Greek to the Italian may have been very gradual, some terms growing obsolete, while others were introduced from time to time.

If this antiquary could have shown that the volcanic paroxysms of Vesuvius were so governed as that cities should be buried one above the other, just as often as any variation occurred in the language of the inhabitants, then, indeed, the abrupt passage from a Greek to a Roman, and from a Roman to an Italian city, would afford proof of fluctuations no less sudden in the language of the people.

So in Geology, if we could assume that it is part of the plan of nature to preserve, in every region of the globe, an unbroken series of monuments to commemorate the vicissitudes of the organic creation, we might infer the sudden extirpation of species, and the simultaneous introduction of others, as often as two formations in contact include dissimilar organic fossils. But we must shut our eyes to the whole economy of the existing causes, aqueous, igneous, and organic, if we fail to perceive *that such is not the plan of Nature.*

CHAPTER IV.

Chronological relations of mineral masses the first object in geological classification—Superposition, proof of more recent origin—Exceptions in regard to
volcanic rocks—Relative age proved by included fragments of older rocks—
Proofs of contemporaneous origin derived from mineral characters—Variations
to which these characters are liable—Recurrence of distinct rocks at successive
periods—Proofs of contemporaneous origin derived from organic remains—
Zoological provinces are of limited extent, yet spread over wider areas than
homogeneous mineral deposits—Different modes whereby dissimilar mineral
masses and distinct groups of species may be proved to have been contemporaneous.

DETERMINATION OF THE RELATIVE AGES OF ROCKS.

IN attempting to classify the mineral masses which compose
the crust of the earth, the principal object which the geologist
must keep in view, is to determine with accuracy their chronological relations, for it is abundantly clear, that different rocks
have been formed in succession ; and in order thoroughly to
comprehend the manner in which they enter into the structure
of our continents, we should study them with reference to the
time and mode of their formation.

We shall now, therefore, consider by what characters the
relative ages of different rocks may be established, whereby we
may be supplied at once with sound information of the greatest
practical utility, and which may throw, at the same time, the
fullest light on the ancient history of the globe.

Proofs of relative age by superposition.

It is evident that where we find a series of horizontal strata,
of sedimentary origin, the uppermost bed must be newer than
those which it overlies, and that when we observe one distinct
set of strata reposing upon another, the inferior is the older of
the two. In countries where the original position of mineral

masses has been disturbed, at different periods, by convulsions of extraordinary violence, as in the Alps and other mountainous districts, there are instances where the original position of strata has been reversed ; but such exceptions are rare, and are usually on a small scale, and an experienced observer can generally ascertain the true relations of the rocks in question, by examining some adjoining districts where the derangement has been less extensive.

In regard to volcanic formations, if we find a stratum of tuff or ejected matter, or a stream of lava covering sedimentary strata, we may infer, with confidence, that the igneous rock is the more recent ; but, on the other hand, the superposition of aqueous deposits to a volcanic mass does not always prove the former to be of newer origin.   If, indeed, we discover strata of tuff with imbedded shells, or, as in the Vicentine and other places, rolled blocks of lava with adhering shells and corals, we may then be sure that these masses of volcanic origin covered the bottom of the sea, before the superincumbent strata were thrown down.   But as lava rises from below, and does not always reach the surface, it may sometimes penetrate a certain number of strata, and then cool down, so as to constitute a solid mass of newer origin, although inferior in position.   It is, for the most part, by the passage of veins proceeding from such igneous rocks through contiguous sedimentary strata, or by such hardening and other alteration of the overlying bed, as might be expected to result from contact with a heated mass, that we are enabled to decide whether the volcanic matter was previously consolidated, or subsequently introduced.

Proofs by included fragments of older rocks.

A Geologist is sometimes at a loss, after investigating a district composed of two distinct formations, to determine the relative ages of each, from want of sections exhibiting their superposition.   In such cases, another kind of evidence, of a character no less conclusive, can sometimes be obtained.   One group of strata has frequently been derived from the degradation of another in the immediate neighbourhood, and may be observed to include within it fragments of such older rocks. Thus, for example, we may find chalk with flints, and in another part of the same country, a distinct series, consisting of alternations of clay, sand, and pebbles. If some of these pebbles consist of flints, with silicified fossil-shells of the same species as those in the chalk, we may confidently infer, that the chalk is the oldest of the two formations.

We remarked in the second chapter, that some granite must have existed before the most ancient of our secondary rocks, because some of the latter contain rounded pebbles of granite. But for the existence of such evidence, we might not have felt assured that all the granite which we see had not been protruded from below in a state of fusion, subsequently to the origin of the secondary strata.

Proofs of contemporaneous origin derived from mineral characters.

When we have established the relative age of two formations in a given place, by direct superposition, or by other evidence, a far more difficult task remains, to trace the continuity of the same formation, or, in other cases, to find means of referring detached groups of rocks to a contemporaneous origin. Such identifications in age are chiefly derivable from two sources— mineral character and organic contents ; but the utmost skill and caution are required in the application of such tests, for scarcely any general rules can be laid down respecting either, that do not admit of important exceptions.

If, at certain periods of the past, rocks of peculiar mineral composition had been precipitated simultaneously upon the floor of an ' universal ocean,' so as to invest the whole earth in a succession of concentric coats, the determination of relative dates in geology might have been a matter of the greatest simplicity. To explain, indeed, the phenomenon would have been difficult, or rather impossible, as such appearances would have implied a former state of the globe, without any analogy to that now prevailing. Suppose, for example, there were three masses extending over every continent,—the upper of chalk and chloritic sand; the next below, of blue argillaceous limestone; and the third and lowest, of red marl and sandstone; we must imagine that all the rivers and currents of the world had been charged, at the first period, with red mud and sand; at the second, with blue calcareo-argillaceous mud; and at a subsequent epoch, with chalky sediment and chloritic sand.

But if the ocean were universal, there could have been no land to waste away by the action of the sea and rivers, and, therefore, no known source whence the homogeneous sedimentary matter could have been derived. Few, perhaps, of the earlier geologists went so far as to believe implicitly in such universality of formations, but they inclined to an opinion, that they were continuous over areas almost indefinite; and since such a disposition of mineral masses would, if true, have been the least complex and most convenient for the purposes of classification, it is probable that a belief in its reality was often promoted by the hope that it might prove true. As to the objection, that such an arrangement of mineral masses could never result from any combination of causes now in action, it never weighed with the earlier cultivators of the science, since they indulged no expectation of being ever able to account for geological phenomena by reference to the known economy of nature. On the contrary, they set out, as we have already seen, with the assumption that the past and present conditions of the planet were too dissimilar to admit of exact comparison.

But if we inquire into the true composition of any stratum, or set of strata, and endeavour to pursue these continuously through a country, we often find that the character of the mass changes gradually, and becomes at length so different, that we should never have suspected its identity, if we had not been enabled to trace its passage from one form to another.

We soon discover that rocks dissimilar in mineral composition have originated simultaneously; we find, moreover, evidence in certain districts, of the recurrence of rocks of precisely the same mineral character at very different periods; as, for example, two formations of red sandstone, with a great series of other strata intervening between them. Such repetitions might have been anticipated, since these red sandstones are produced by the decomposition of granite, gneiss, and mica-schist; and districts composed exclusively of these, must again and again be exposed to decomposition, and to the erosive action of running water.

But notwithstanding the variations before alluded to in the composition of one continuous set of strata, many rocks retain the same homogeneous structure and composition, throughout considerable areas, and frequently, after a change of mineral character, preserve their new peculiarities throughout another tract of great extent. Thus, for example, we may trace a limestone for a hundred miles, and then observe that it becomes more arenaceous, until it finally passes into sand or sandstone. We may then follow the last-mentioned formation throughout another district as extensive as that occupied by the limestone first examined.

Proofs of contemporaneous origin derived from organic remains.

We devoted several chapters, in the last volume, to show that the habitable surface of the sea and land may be divided into a considerable number of distinct provinces, each peopled by a peculiar assemblage of animals and plants, and we endeavoured to point out the origin of these separate divisions. It was shown that climate is only one of many causes on which they depend, and that difference of longitude, as well as latitude, is generally accompanied by a dissimilarity of indigenous species of organic beings.

As different seas, therefore, and lakes are inhabited at the same period, by different species of aquatic animals and plants, and as the lands adjoining these may be peopled by distinct terrestrial species, it follows that distinct organic remains are imbedded in contemporaneous deposits. If it were otherwise

—if the same species abounded in every climate, or even in every part of the globe where a corresponding temperature, and other conditions favourable to their existence were found, the identification of mineral masses of the same age, by means of their included organic contents, would be a matter of ,much greater facility.

But, fortunately, the extent of the same zoological provinces, especially those of marine animals, is very great, so that we are entitled to expect, from analogy, that the identity of fossil species, throughout large areas, will often enable us to connect together a great variety of detached and dissimilar formations.

Thus, for example, it will be seen, by reference to our first volume, that deposits now forming in different parts of the Mediterranean, as in the deltas of the Rhone and the Nile, are distinct in mineral composition ; for calcareous rocks are precipitated from the waters of the former river, while pebbles are carried into its delta, and there cemented, by carbonate of lime, into a conglomerate; whereas strata of soft mud and fine sand are formed exclusively in the Nilotic delta. The Po, again, carries down fine sand and mud into the Adriatic ; but since this sediment is derived from the degradation of a different assemblage of mountains from those drained by the Rhone or the Nile, we may safely assume that there will never be an exact identity in their respective deposits.

If we pass to another quarter of the Mediterranean, as, for example, to the sea on the coast of Campania, or near the base of Etna in Sicily, or to the Grecian archipelago, we find in all these localities that distinct combinations of rocks are in progress. Occasional showers of volcanic ashes are falling into the sea, and streams of lava are flowing along its bottom ; and in the intervals between volcanic eruptions, beds of sand and clay are frequently derived from the waste of cliffs, or the turbid waters of rivers. Limestones, moreover, such as the Italian travertins, are here and there precipitated from the waters of mineral springs, while shells and corals accumulate in various localities. Yet the entire Mediterranean, where the above-mentioned formations are simultaneously in progress, may be considered as one zoological province; for, although certain species of testacea and zoophytes may be very local, and each region may probably have some species peculiar to it, still a considerable number are common to the whole sea. If, therefore, at some future period, the bed of this inland sea should be converted into land, the geologist might be enabled, by reference to organic remains, to prove the contemporaneous origin of various mineral masses throughout a space equal in area to a great portion of Europe. The Black Sea, moreover, is inhabited by so many identical species, that the delta of the Danube and the Don might, by the same evidence, be shown to have originated simultaneously.

Such identity of fossils, we may remark, not only enables us to refer to the same era, distinct rocks widely separated from each other in the horizontal plane, but also others which may be considerably distant in the vertical series. Thus, for example, we may find alternating beds of clay, sand, and lava, two thousand feet in thickness, the whole of which may be proved to belong to the same epoch, by the specific identity of the fossil shells dispersed throughout the whole series. It may be objected, that different species would, during the same zoological period, inhabit the sea at different depths, and that the case above supposed could never occur; but, for reasons explained in the last volume*, we believe that rivers and tidal currents often act upon the banks of littoral shells, so that a sea of great depth may be filled with strata, containing throughout a considerable number of the same fossils.

The reader, however, will perceive, by referring to what we have said of zoological provinces, that they are sometimes separated from each other by very narrow barriers, and for this reason contiguous rocks may be formed at the same time, differing widely both in mineral contents and organic remains. Thus, for example, the testacea, zoophytes, and fish of the Red Sea, may be considered, as a group, to be very distinct from

* Chap. xvii. p. 280.

those inhabiting the adjoining parts of the Mediterranean, although the two seas are only separated by the narrow isthmus of Suez. We shall show, in a subsequent chapter, that calcareous formations have accumulated, on a great scale, in the Red Sea, in modern times, and that fossil shells of existing species are well preserved therein; while we know that, at the mouth of the Nile, large deposits of mud are amassed, including the remains of Mediterranean species. Hence it follows, that if, at some future period, the bed of the Red Sea should be laid dry, the geologist might experience great difficulties in endeavouring to ascertain the relative age of these formations, which, although dissimilar both in organic and mineral characters, were of synchronous origin.

There might, perhaps, be no means of clearing up the obscurity of such a question, yet we must not forget that the north-western shores of the Arabian Gulf, the plains of Egypt, and the isthmus of Suez, are all parts of one province of *terrestrial* species. Small streams, therefore, occasional land-floods, and those winds which drift clouds of sand along the deserts, might carry down into the Red Sea the same shells of fluviatile and land testacea, which the Nile is sweeping into its delta, together with some remains of terrestrial plants, whereby the groups of strata, before alluded to, might, notwithstanding the discrepancy of their mineral composition, and *marine* organic fossils, be shown to have belonged to the same epoch.

In like manner, the rivers which descend into the Caribbean Sea and Gulf of Mexico on one side, and into the Pacific on the other, carry down the same fluviatile and terrestrial spoils into seas which are inhabited by different groups of marine species.

But it will much more frequently happen, that the coexistence of *terrestrial* species, of distinct zoological and botanical provinces, will be proved by the specific identity of the *marine* organic remains which inhabited the intervening space. Thus, for example, the distinct terrestrial species of the south of Europe, north of Africa, and north-west of Asia, might all be shown to have been contemporaneous, if we suppose the rivers flowing from those three countries to carry the remains of different species of the animal and vegetable kingdoms into the Mediterranean.

In like manner, the sea intervening between the northern shores of Australia and the islands of the Indian ocean, contains a great proportion of the same species of corallines and testacea, yet the *land animals and plants* of the two regions are very dissimilar, even the islands nearest to Australia, as Java, New Guinea, and others, being inhabited by a distinct assemblage of terrestrial species. It is well known that there are calcareous rocks, volcanic tuff, and other strata in progress, in different parts of these intermediate seas, wherein marine organic remains might be preserved and associated with the terrestrial fossils above alluded to.

As it frequently happens that the barriers between different provinces of animals and plants are not very strongly marked, especially where they are determined by differences of temperature, there will usually be a passage from one set of species to another, as in a sea extending from the temperate to the tropical zone. In such cases, we may be enabled to prove, by the fossils of intermediate deposits, the connexion between the distinct provinces, since these intervening spaces will be inhabited by many species, common both to the temperate and equatorial seas.

On the other hand, we may be sometimes able, by aid of a peculiar homogeneous deposit, to prove the former coexistence of distinct animals and plants in distant regions. Suppose, for example, that in the course of ages the sediment of a river, like that of the Red River in Louisiana, is dispersed over an area several hundred leagues in length, so as to pass from the tropics into the temperate zone, the fossil remains imbedded in red mud might indicate the different forms which inhabited, at the same period, those remote regions of the earth.

It appears, then, that mineral and organic characters, although often inconstant, may, nevertheless, enable us to establish the contemporaneous origin of formations in distant countries. As the same species of organic beings usually extend over wider areas than deposits of a homogeneous composition, they are more valuable in geological classification than mineral peculiarities; but it fortunately happens, that where the one criterion fails, we can often avail ourselves of the other. Thus, for example, sedimentary strata are as likely to preserve the same colour and composition in a part of the ocean reaching from the borders of the tropics to the temperate zone, as in any other quarter of the globe; but in such spaces the variation of species is always most considerable.

In regard to the habitations of species, the marine tribes are of more importance than the terrestrial, not only because they are liable to be fossilized in subaqueous deposits in the greatest abundance, but because they have, for the most part, a wider geographical range. Sometimes, however, it may happen, as we have shown, that the remains of species of some one province of terrestrial plants and animals may be carried down into two seas inhabited by distinct marine species; and here again we have an illustration of the principle, that when one means of identification fails, another is often at hand to assist us.

In conclusion, we may observe, that in endeavouring to prove the contemporaneous origin of strata in remote countries by organic remains, we must form our conclusions from a great number of species, since a single species may be enabled to survive vicissitudes in the earth's surface, whereby thousands of others are exterminated. When a change of climate takes place, some may migrate and become denizens of other latitudes, and so abound there, as to characterize strata of a subsequent era. In the last volume we have stated our reasons for inferring that such migrations are never sufficiently general to interfere seriously with geological conclusions, provided we do not found our theories on the occurrence of a small number of fossil species.

CHAPTER V.

Classification of tertiary formations in chronological order—Comparative value of different classes of organic remains—Fossil remains of testacea the most important—Necessity of accurately determining species—Tables of shells by M. Deshayes—Four subdivisions of the Tertiary epoch—Recent Formations— Newer Pliocene period—Older Pliocene period—Miocene period—Eocene period —The distinct zoological characters of these periods may not imply sudden changes in the animate creation—The recent strata form a common point of departure in distant regions—Numerical proportion of recent species of shells in different tertiary periods—Mammiferous remains of the successive tertiary eras—Synoptical Table of Recent and Tertiary formations.

CLASSIFICATION OF TERTIARY FORMATIONS IN CHRONOLOGICAL ORDER.

WE explained in the last chapter the principles on which the relative ages of different formations may be ascertained, and we found the character to be chiefly derivable from superposition, mineral structure, and organic remains. It is by combining the evidence deducible from all these sources, that we determine the chronological succession of distinct formations, and this principle is well illustrated by the investigation of those European tertiary strata to the discovery of which we have already alluded.

It will be seen, that in proportion as we have extended our inquiries over a larger area, it has become necessary to intercalate new groups of an age intermediate between those first examined, and we have every reason to expect that, as the science advances, new links in the chain will be supplied, and that the passage from one period to another will become less abrupt. We may even hope, without travelling to distant regions,— without even transgressing the limits of western Europe, to render the series far more complete. The fossil shells, for example, of many of the Subalpine formations, on the northern limits of the plain of the Po, have not yet been carefully collected and compared with those of other countries, and we are almost entirely ignorant of many deposits known to exist in Spain and Portugal.

The theoretical views developed in the last chapter, respecting breaks in the sequence of geological monuments, will explain our reasons for anticipating the discovery of intermediate gradations as often as new regions of great extent are explored.

Comparative value of different classes of organic remains.

In the mean time, we must endeavour to make the most systematic arrangement in our power of those formations which are already known, and in attempting to classify these in chronological order, we have already stated that we must chiefly depend on the evidence afforded by their fossil organic contents. In the execution of this task, we have first to consider what class of remains are most useful, for although every kind of fossil animal and plant is interesting, and cannot fail to throw light on the former history of the globe at a certain period, yet those classes of remains which are of rare and casual occurrence, are absolutely of no use for the purposes of general classification. If we have nothing but plants in one assemblage of strata, and the bones of mammalia in another, we can obviously draw no conclusion respecting the number of species of organic beings common to two epochs; or if we have a great variety, both of vertebrated animals and plants, in one series, and only shells in another, we can form no opinion respecting the remoteness or proximity of the two eras. We might, perhaps, draw some conclusions as to relative antiquity, if we could compare each of these monuments to a third ; as, for example, if the species of shells should be almost all identical with those now living, while the plants and vertebrated animals were all extinct; for we might then infer that the shelly deposit was the most recent of the two. But in this case it will be seen that the information flows from a direct comparison of the species of corresponding orders of the animal and vegetable kingdoms,— of plants with plants, and shells with shells ; the only mode of making a systematic arrangement by reference to organic remains.

Although the bones of mammalia in the tertiary strata, and those of reptiles in the secondary, afford us instruction of the most interesting kind, yet the species are too few, and confined to too small a number of localities, to be of great importance in characterizing the minor subdivisions of geological formations. Skeletons of fish are by no means frequent in a good state of preservation, and the science of ichthyology must be farther advanced, before we can hope to determine their specific character with sufficient precision. The same may be said of fossil botany, notwithstanding the great progress that has recently been made in that department; and even in regard to zoophytes, which are so much more abundant in a fossil state than any of the classes above enumerated, we are still greatly impeded in our endeavour to classify strata by their aid, in consequence of the smallness of the number of recent species which have been examined in those tropical seas where they occur in the greatest profusion.

Fossil remains of testacea of chief importance. The testacea are by far the most important of all classes of organic beings which have left their spoils in the subaqueous deposits; they are the medals which nature has chiefly selected to record the history of the former changes of the globe. There is scarcely any great series of strata that does not contain some marine or fresh-water shells, and these fossils are often found so entire, especially in the tertiary formations, that when disengaged from the matrix, they have all the appearance of having been just procured from the sea. Their colour, indeed, is usually wanting, but the parts whereon specific characters are founded remain unimpaired; and although the animals themselves are gone, yet their form and habits can generally be inferred from the shell which covered them.

The utility of the testacea, in geological classification, is greatly enhanced by the circumstance, that some forms are proper to the sea, others to the land, and others to fresh-water.

Rivers scarcely ever fail to carry down into their deltas some
land shells, together-with species which are at once fluviatile
and lacustrine. The Rhone, for example, receives annually,
from the Durance, many shells which are drifted down in an
entire state from the higher Alps of Dauphiny, and these
species, such as *Bulimus montanus,* are carried down into the
delta of the Rhone to a climate far different from that of their
native habitation. The young hermit crabs may often be seen
on the shores of the Mediterranean, near the mouth of the
Rhone, inhabiting these univalves, brought down to them from
so great a distance*. At the same time that some fresh-water
and land species are carried into the sea, other individuals of
the same become fossil in inland lakes, and by this means we
learn what species of fresh-water and marine testacea coexisted
at particular eras; and from this again we are able to make out
the connexion between various plants and mammifers imbedded
in those lacustrine deposits, and the testacea which lived in the
ocean at the same time.

There are two other characters of the molluscous animals
which render them extremely valuable in settling chronological
questions in geology. The first of these is a wide geographical
range, and the second (probably a consequence of the former),
is the superior duration of species in this class. It is evident
that if the habitation of a species be very local, it cannot aid us
greatly in establishing the contemporaneous origin of distant
groups of strata, in the manner pointed out in the last chapter;
and if a wide geographical range be useful in connecting formations far separated in space, the longevity of species is no
less serviceable in establishing the relations of strata considerably distant from each other in point of time.

We shall revert in the sequel to the curious fact, that in
tracing back these series of tertiary deposits, many of the existing species of testacea accompany us after the disappearance of
all the recent mammalia, as well as the fossil remains of living

* M. Marcel de Serres pointed out this fact to me when I visited Montpellier,
July, 1828.

species of several other classes.   We even find the skeletons of extinct quadrupeds in deposits wherein all the land and fresh-water shells are of recent species *.

Necessity of accurately determining species.—The reader will already perceive that the systematic arrangement of strata, so far as it rests on organic remains, must depend essentially on the accurate determination of *species*, and the geologist must therefore have recourse to the ablest naturalists, who have de-voted their lives to the study of certain departments of organic nature.   It is scarcely possible that they who are continually employed in laborious investigations in the field, and in ascer-taining the relative position and characters of mineral masses, should have leisure to acquire a profound knowledge of fossil osteology, conchology, and other branches; but it is desirable that, in the latter science at least, they should become acquainted with the principles on which the specific characters are deter-mined, and on which the habits of species are inferred from their peculiar forms.   When the specimens are in an imperfect state of preservation, or the shells happen to belong to genera in which it is difficult to decide on the species, except when the inhabitant itself is present, or when any other grounds of ambi-guity arise, we must reject, or lay small stress upon, the evidence, lest we vitiate our general results by false identifications and analogies.   We cannot do better than consider the steps by which the science of botanical geography has reached its pre-sent stage of advancement, and endeavour to introduce the same severe comparison of the specific characters, in drawing all our geological inferences.

Tables of shells by M. Deshayes.—In the Appendix the reader will find a tabular view of the results obtained by the comparison of more than three thousand tertiary shells, with nearly five thousand living species, all of which, with few ex-ceptions, are contained in the rich collection of M. Deshayes. Having enjoyed an opportunity of examining, again and again, the specimens on which this eminent conchologist has founded

* See vol. i. chap. vi.

his identifications, and having been witness to the great time and labour devoted by him to this arduous work, I feel confidence in the results, so far as the data given in his list will carry us.   It was necessary to compare nearly forty thousand specimens, in order to construct these tables, since not only the varieties of every species required examination, but the different individuals, also, belonging to each which had been found fossil in various localities.   The correctness of the localities themselves was ascertained with scrupulous exactness, together with the relative position of the strata; and if any doubts existed on these questions, the specimens were discarded as of no geological value.   A large proportion of the shells were procured, by M. Deshayes himself, from the Paris basin, many were contributed by different French geologists, and some were collected by myself from different parts of Europe.

It would have been impossible to give lists of more than three thousand fossil-shells in a work not devoted exclusively to conchology ; but we were desirous of presenting the reader with a catalogue of those fossils which M. Deshayes has been able to identify with living species, as also of those which are common to two distinct tertiary eras.   By this means a comparison may be made of the testacea of each geological epoch, with the actual state of the organic creation, and, at the same time, the relations of different tertiary deposits to each other exhibited.   The number of shells mentioned by name in the tables, in order to convey this information, is seven hundred and eighty-two, of which four hundred and twenty-six have been found both living and fossil, and three hundred and fifty-six fossil only, but in the deposits of more than one era.   An exception, however, to the strictness of this rule has been made in regard to the fossil-shells common to the London and Paris basins, fifty-one of which have been enumerated by name, though these formations do not belong to different eras.

It has been more usual for geologists to give tables of characteristic shells; that is to say, of those found in the strata of one period and not common to any other.   These typical species are certainly of the first importance, and some of them

Plate I

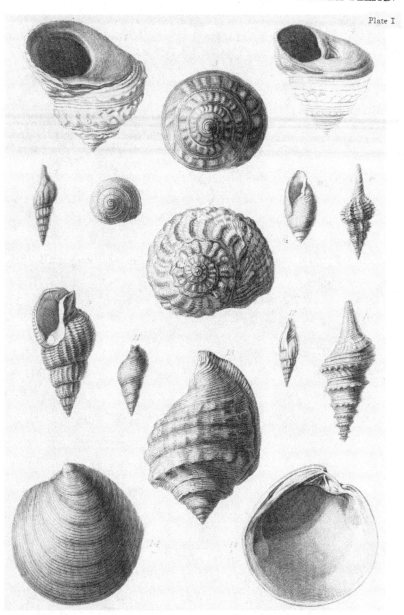

1.2. *Turbo rugosus. Lin:_ 3.4. Trochus magus. Lin:_ 5. Solarium variegatum. Lam*.
6. *Tornatella fasciata. Lam*_ 7. Pleurotoma vulpecula. Broc:_ 8. Fusus crispus. Bors:
9. Buccinum prismaticum. Bors:_ 10. Pleurotoma rotata. Broc:_ 11. Buccinum semistriatum Broc:_ 12. Mitra plicatula. Broc:_ 13. Cassidaria echinophora Lam*_ 14. Cytherea
exoleta. Lam* var.*

P. Oudart del.   T. Bradley sc

Plate II

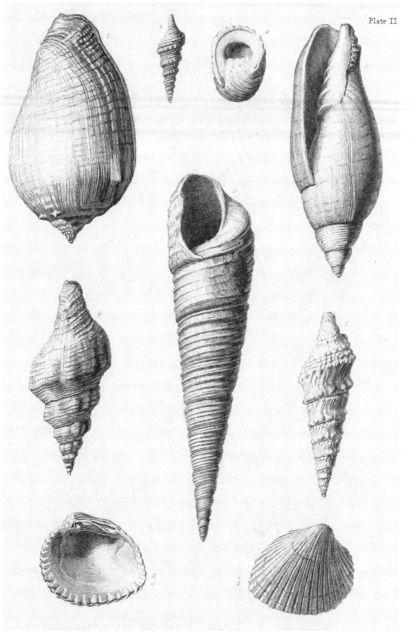

1 Voluta rarispina, Lamᵏ _ 2. Mitra, Dufrenei, Bast. _ 3. Pleurotoma denticula, Bast.
4. Nerita Plutonis, Brong. _ 5. Turritella Proto, Bast. _ 6 Fasciolaria turbinelloïdes, Dcsh.
7. Pleurotoma tuberculosa, Bast. _ 8. a.b. Cardita Ajar. Brug.

P. Oudart del.

T. Bradley sculp.

Plate III

1. *Voluta costaria. Lam.* — 2. *Pleurotoma clavicularis.* — 3. *Cassidaria carinata. Lam.*
4. *Nerita tricarinata. Lam.* — 5. *Calyptræa trochiformis. Lam.* — 6. *Turritella imbricataria. Lam.* — 7. *Voluta digitalina. Lam.* — 8. *Natica epiglottina Lam.* — 9 *Solarium canaliculatum. Lam.* — 10. *Cardita planicosta. Desh.*

P.Oudart del.   T.Bradley sc.

Plate IV.

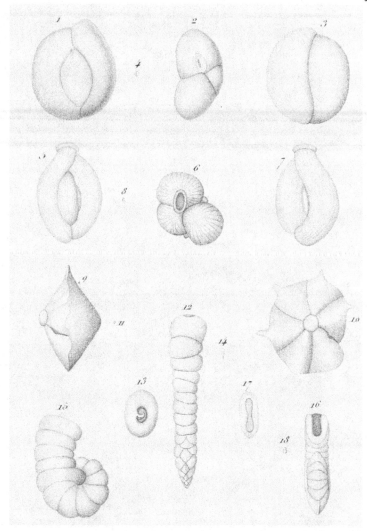

MICROSCOPIC FOSSIL SHELLS.   EOCENE TERTIARY PERIOD.

PARIS BASIN

1. 2. 3. 4. ._ Triloculina   inflata . Desh._ 5. 6. 7 8. Quinqueloculina striata . Desh.
9. 10. 11 Calcarina rarispina . Desh. __ 12. 13. 14. Clavulina corrugata . Desh.
16. 16. 17. 18. Spirolina stenostoma . Desh.

London, Published by John Murray, Dec.r 1831.

will be seen figured in the plates illustrative of the different
tertiary eras ; but we were more anxious, in this work, to place
in a clear light a point of the greatest theoretical interest, which
has been often overlooked or controverted, viz., the identity of
many living and fossil species, as also the connexion of the
zoological remains of deposits formed at successive periods.

The value of such extensive comparisons, as those of which
the annexed tables of M. Deshayes give the results, depends
greatly on the circumstance, that all the identifications have
been made by the same naturalist.   The amount of variation
which ought to determine a species is, in cases where they
approach near to each other, a question of the nicest discrimina-
tion, and requires a degree of judgment and tact that can hardly
be possessed by different zoologists in exactly the same degree.
The standard, therefore, by which differences are to be measured, can scarcely ever be perfectly invariable, and one great
object to be sought for is, that, at least, it should be uniform.
If the distinctions are all made by the same naturalist, and his
knowledge and skill be considerable, the results may be relied
on with sufficient confidence, as far as regards our geological
conclusions.

If one conchologist should inform us that out of 1122 species
of fossil testacea, discovered in the Paris basin, he has only
been enabled to identify thirty-eight with recent species, while
another should declare, that out of two hundred and twenty-
six Sicilian fossil shells, no less than two hundred and sixteen
belonged to living species, we might suspect that one of these
observers allowed a greater degree of latitude to the variability
of the specific character than the other; but when, in both
instances, the conclusions are drawn by the same eminent conchologist, we are immediately satisfied that the relations of these
two groups, to the existing state of the animate creation, are as
distinct as are indicated by the numerical results.

It is not pretended that the tables, to which we refer, com-
prise all the known tertiary shells.   In the museums of Italy
there are magnificent collections, to which M. Deshayes had no

access, and the additions to the recent species in the cabinets
of conchologists in London have been so great of late years,
that in many extensive genera the number of species has been
more than doubled. But as the greater part of these newly-
discovered shells have been brought from the Pacific and other
distant seas, it is probable that these accessions would not materially alter the results given in the tables, and it must, at all
events, be remembered, that the only effect of such additional
information would be, to increase the number of identifications
of recent with fossil species, while the proportional number of
analogues in the different periods might probably remain
nearly the same.

SUBDIVISIONS OF THE TERTIARY EPOCH.

Recent formations.—We shall now proceed to consider the
subdivisions of tertiary strata which may be founded on the
results of a comparison of their respective fossils, and to give
names to the periods to which they each belong. The tertiary
epoch has been divided into three periods in the tables; we
shall, however, endeavour to establish *four*, all distinct from the
actual period, or that which has elapsed since the earth has been
tenanted by man. To the events of this latter era, which we
shall term the *recent*, we have exclusively confined ourselves in
the two preceding volumes. All sedimentary deposits, all volcanic rocks, in a word, every geological monument, whether
belonging to the animate or inanimate world, which appertains
to this epoch, may be termed *recent*. Some *recent* species, therefore, are found *fossil* in various tertiary periods, and, on the
other hand, others, like the Dodo, may be *extinct*, for it is sufficient that they should once have coexisted with man, to make
them referrible to this era.

Some authors apply the term *contemporaneous* to all the
formations which have originated during the human epoch ;
but as the word is so frequently in use to express the synchronous origin of distinct formations, it would be a source of great
inconvenience and ambiguity, if we were to attach to it a technical sense.

We may sometimes prove, that certain strata belong to the recent period by aid of historical evidence, as parts of the delta of the Po, Rhone, and Nile, for example; at other times, by discovering imbedded remains of man or his works; but when we have no evidence of this kind, and we hesitate whether to ascribe a particular deposit to the recent era, or that immediately preceding, we must generally incline to refer it to the latter, for it will appear in the sequel, that the changes of the historical era are quite insignificant when contrasted with those even of the newest tertiary period.

Newer Pliocene period.—This most modern of the four subdivisions of the whole tertiary epoch, we propose to call the *Newer Pliocene*, which, together with the *Older* Pliocene, constitute one group in the annexed tables of M. Deshayes.

We derive the term Pliocene from πλειων, major, and καινος, recens, as the major part of the fossil testacea of this epoch are referrible to recent species*. Whether in all cases there may hereafter prove to be an absolute preponderance of recent species, in every group of strata assigned to this period in the tables, is very doubtful; but the proportion of living species, where least considerable, usually approaches to one-half of the total number, and appears always to exceed a third; and as our acquaintance with the testacea of the Mediterranean, and some other seas, increases, it is probable that a greater proportion will be identified.

* In the terms Pliocene, Miocene, and Eocene, the Greek diphthongs *ei* and *ai* are changed into the vowels *i* and *e*, in conformity with the idiom of our language. Thus we have Encenia, an inaugural ceremony, derived from εν and καινος, recens; and as examples of the conversion of *ei* into *i*, we have icosahedron.

I have been much indebted to my friend, the Rev. W. Whewell, for assisting me in inventing and anglicizing these terms, and I sincerely wish that the numerous foreign diphthongs, barbarous terminations, and Latin plurals, which have been so plentifully introduced of late years into our scientific language, had been avoided as successfully as they are by French naturalists, and as they were by the earlier English writers, when our language was more flexible than it is now. But while I commend the French for accommodating foreign terms to the structure of their own language, I must confess that no naturalists have been more unscholar-like in their mode of fabricating Greek derivatives and compounds, many of the latter being a bastard offspring of Greek and Latin.

The newer Pliocene formations, before alluded to, pass insensibly into those of the *Recent* epoch, and contain an immense preponderance of recent species. It will be seen that of two hundred and twenty-six species, found in the Sicilian beds, only ten are of extinct or unknown species, although the antiquity of these tertiary deposits, as contrasted with our most remote historical eras, is immensely great. In the volcanic and sedimentary strata of the district round Naples, the proportion appears to be even still smaller.

Older Pliocene period.—These formations, therefore, and others wherein the plurality of living species is so very decided, we shall term the *Newer* Pliocene, while those of the tertiary period immediately preceding may be called the *Older* Pliocene. To the latter belong the formations of Tuscany, and of the Subapennine hills in the north of Italy, as also the English Crag.

It appears that in the period last mentioned, the proportion of recent species varies from upwards of a third to somewhat more than half of the entire number; but it must be recollected, that this relation to the recent epoch is only *one* of its zoological characters, and that certain *peculiar species* of testacea also distinguish its deposits from all other strata. The relative position of the beds referrible to this era has been explained in diagrams Nos. 3 and 4, letter *f*, chapter II.

Miocene period.—The next antecedent tertiary epoch we shall name Miocene, from μειων, minor, and καινος, recens, a minority only of fossil shells imbedded in the formations of this period being of recent species. The total number of Miocene shells, referred to in the annexed tables, amounts to 1021, of which one hundred and seventy-six only are recent, being in the proportion of rather less than eighteen in one hundred. Of species common to this period, and to the two divisions of the Pliocene epoch before alluded to, there are one hundred and ninety-six, whereof one hundred and fourteen are living, and the remaining eighty-two extinct, or only known as fossil.

As there are a certain number of fossil species which are characteristic of the Pliocene strata before described, so also there are many shells exclusively confined to the Miocene period. We have already stated, that in Touraine and in the South of France near Bordeaux, in Piedmont, in the basin of Vienna, and other localities, these Miocene formations are largely developed, and their relative position has been shown in diagrams Nos. 3 and 4, letter *e*, chapter II.

Eocene period.—The period next antecedent we shall call Eocene, from ἠως, aurora, and καινος, recens, because the extremely small proportion of living species contained in these strata, indicates what may be considered the first commencement, or *dawn*, of the existing state of the animate creation. To this era the formations first called tertiary, of the Paris and London basins, are referrible. Their position is shown in the diagrams Nos. 3 and 4, letter *d*, in the second chapter.

The total number of fossil shells of this period already known, is one thousand two hundred and thirty-eight, of which number forty-two only are living species, being nearly in the proportion of three and a half in one hundred. Of fossil species, not known as recent, forty-two are common to the Eocene and Miocene epochs. In the Paris basin alone, 1122 species have been found fossil, of which thirty-eight only are still living.

The geographical distribution of those recent species which are found fossil in formations of such high antiquity as those of the Paris and London basins, is a subject of the highest interest.

It will be seen by reference to the tables, that in the more modern formations, where so large a proportion of the fossil shells belong to species still living, they also belong, for the most part, to species now inhabiting the seas immediately adjoining the countries where they occur fossil; whereas the recent species, found in the older tertiary strata, are frequently inhabitants of distant latitudes, and usually of warmer climates. Of the forty-two Eocene species, which occur fossil in England, France, and Belgium, and which are still living, about half now inhabit within, or near the tropics, and almost all the rest are denizens of the more southern parts of Europe. If some

Eocene species still flourish in the same latitudes where they are found fossil, they are species which, like *Lucina divaricata*, are now found in many seas, even those of different quarters of the globe, and this wide geographical range indicates a capacity of enduring a variety of external circumstances, which may enable a species to survive considerable changes of climate and other re-volutions of the earth's surface. One fluviatile species (*Melania inquinata*), fossil in the Paris basin, is now only known in the Philippine islands, and during the lowering of the temperature of the earth's surface, may perhaps have escaped destruction by transportation to the south. We have pointed out in the second volume (chap. vii.), how rapidly the eggs of fresh-water species might, by the instrumentality of water-fowl, be transported from one region to another. Other Eocene species, which still survive and range from the temperate zone to the equator, may formerly have extended from the pole to the temperate zone, and what was once the southern limit of their range may now be the most northern.

Even if we had not established several remarkable facts in attestation of the longevity of certain tertiary species, we might still have anticipated that the duration of the living species of aquatic and terrestrial testacea would be very unequal. For it is clear that those which now inhabit many different regions and climates, may survive the influence of destroying causes, which might extirpate the greater part of the species now living. We might expect, therefore, some species to survive several successive states of the organic world, just as Nestor was said to have outlived three generations of men.

The distinctness of periods may indicate our imperfect infor-mation.—In regard to distinct zoological periods, the reader will understand, from our observations in the third chapter, that we consider the wide lines of demarcation that sometimes separate different tertiary epochs, as quite unconnected with extraordinary revolutions of the surface of the globe, and as arising, partly, like chasms in the history of nations, out of the present imperfect state of our information, and partly from

the irregular manner in which geological memorials are preserved, as already explained. We have little doubt that it
will be necessary hereafter to intercalate other periods, and
that many of the deposits, now referred to a single era, will be
found to have been formed at very distinct periods of time, so
that, notwithstanding our separation of tertiary strata into four
groups, we shall continue to use the term *contemporaneous*
with a great deal of latitude.

We throw out these hints, because we are apprehensive lest
zoological periods in geology, like artificial divisions in other
branches of natural history, should acquire too much importance, from being supposed to be founded on some great interruptions in the regular series of events in the organic world,
whereas, like the genera and orders in zoology and botany,
we ought to regard them as invented for the convenience of
systematic arrangement, always expecting to discover intermediate gradations between the boundary lines that we have first
drawn.

In natural history we select a certain species as a generic
type, and then arrange all its congeners in a series, according
to the degrees of their deviation from that type, or according as they approach to the characters of the genus which
precedes or follows. In like manner, we may select certain
geological formations as typical of particular epochs; and
having accomplished this step, we may then arrange the groups
referred to the same period in chronological order, according as
they deviate in their organic contents from the *normal* groups,
or according as they approximate to the type of an antecedent
or subsequent epoch.

If intermediate formations shall hereafter be found between
the Eocene and Miocene, and between those of the last period
and the Pliocene, we may still find an appropriate place for all,
by forming subdivisions on the same principle as that which
has determined us to separate the lower from the upper Pliocene groups. Thus, for example, we might have three divisions
of the Eocene epoch,—the older, middle, and newer; and

three similar subdivisions, both of the Miocene and Pliocene epochs. In that case, the formations of the middle period must be considered as the types from which the assemblage of organic remains in the groups immediately antecedent or sub-sequent will diverge.

The recent strata form a common point of departure in all countries.—We derive one great advantage from beginning our classification of formations by a comparison of the fossils of the more recent strata with the species now living, namely, the ac-quisition of a common point of departure in every region of the globe. Thus, for example, if strata should be discovered in India or South America, containing the same small proportion of recent shells as are found in the Paris basin, *they* also might be termed Eocene, and, on analogous data, an approximation might be made to the relative dates of strata placed in the arctic and tropical regions, or the comparative age ascertained of European deposits, and those which are trodden by our anti-podes.

There might be no species common to the two groups; yet we might infer their synchronous origin from the common relation which they bear to the existing state of the animate creation. We may afterwards avail ourselves of the dates thus established, as eras to which the monuments of preceding periods may be referred.

Numerical proportion of recent shells in the different Ter-tiary periods.—There are seventeen species of shells discovered, which are common to all the tertiary periods, thirteen of which are still living, while four are extinct, or only known as fossil*. These seventeen species show a connexion between all these geological epochs, whilst we have seen that a much greater number are common to the Eocene and Miocene periods, and a still greater to the Miocene and Pliocene.

We have already stated, that in the older tertiary formations, we find a very small proportion of fossil species identical with

---

* See the Tables of M. Deshayes in Appendix I.

those now living, and that, as we approach the superior and newer sets of strata, we find the remains of existing animals and plants in greater abundance. It is almost as difficult to find an unknown species in some of the newer Pliocene deposits, although very ancient, and elevated at great heights above the level of the sea, as to meet with recent species in the Eocene strata.

This increase of existing species, and gradual disappearance of the extinct, as we trace the series of formations from the older to the newer, is strictly analogous, as we before observed, to the fluctuations of a population such as might be recorded at successive periods, from the time when the oldest of the individuals now living was born to the present moment. The disappearance of persons who never were contemporaries of the greater part of the present generation, would be seen to have kept pace with the birth of those who now rank amongst the oldest men living, just as the Eocene and Miocene species are observed to have given place to those Pliocene testacea which are now contemporary with man.

In reference to the organic remains of the different groups which we have named, we may say that about a thirtieth part of the Eocene shells are of recent species, about one-fifth of the Miocene, more than a third, and often more than half, of the older Pliocene, and nine-tenths of the newer Pliocene.

Mammiferous remains of the successive tertiary eras.—But although a thirtieth part of the Eocene testacea have been identified with species now living, none of the associated mammiferous remains belong to species which now exist, either in Europe or elsewhere. Some of these equalled the horse, and others the rhinoceros, in size, and they could not possibly have escaped observation, had they survived down to our time. More than forty of these Eocene mammifers are referrible to a division of the order Pachydermata, which has now only four living representatives on the globe. Of these, not only the species but the genera are distinct from any of those which have been established for the classification of living animals.

In the Miocene mammalia we find a few of the generic forms most frequent in the Eocene strata associated with some of those now existing, and in the Pliocene we find an intermixture of extinct and recent species of quadrupeds. There is, therefore, a considerable degree of accordance between the results deducible from an examination of the fossil testacea, and those derived from the mammiferous fossils. But although the latter are more important in respect to the unequivocal evidence afforded by them of the extinction of species, yet, for reasons before explained, they are of comparatively small value in the general classification of strata in geology.

It will appear evident, from what we have said in the last volume respecting the fossilization of terrestrial species, that the imbedding of their remains depends on rare casualties, and that they are, for the most part, preserved in detached alluvions covering the emerged land, or in osseous breccias and stalagmites formed in caverns and fissures, or in isolated lacustrine formations. These fissures and caves may sometimes remain open during successive geological periods, and the alluvions, spread over the surface, may be disturbed, again and again, until the mammalia of successive epochs are mingled and confounded together. Hence we must be careful, when we endeavour to refer the remains of mammalia to certain tertiary periods, that we ascertain, not only their association with testacea of which the date is known, but also that the remains were intermixed in such a manner as to leave no doubt of the former coexistence of the species.

In the next page will be found a Synoptical Table of the Recent and Tertiary formations alluded to in this chapter.

N.B. By aid of this table, the reader will be able to refer almost all the localities of the Pliocene formations enumerated in the Tables of M. Deshayes (Appendix I) to the newer or older division of the Pliocene period established in the foregoing chapter.

Synoptical Table of Recent and Tertiary Formations.

PERIODS.		Character of Formations.	Localities of the different Formations.
I. Recent.		Marine.	Coral Formations of Pacific. Delta of Po, Ganges, &c.
		Freshwater.	Modern deposits in Lake Superior— Lake of Geneva—Marl lakes of Scotland—Italian travertin, &c.
		Volcanic.	Jorullo — Monte Nuovo — Modern lavas of Iceland, Etna, Vesuvius, &c.
II. Tertiary.	1. Newer Pliocene.	Marine.	Strata of the Val di Noto in Sicily, Ischia, Morea? Uddevalla.
		Freshwater.	Valley of the Elsa around Colle in Tuscany.
		Volcanic.	Older parts of Vesuvius, Etna, and Ischia—Volcanic rocks of the Val di Noto in Sicily.
	2. Older Pliocene.	Marine.	Northern Subapennine formations, as at Parma, Asti, Sienna, Perpignan, Nice—English Crag.
		Freshwater.	Alternating with marine beds near the town of Sienna.
		Volcanic.	Volcanos of Tuscany and Campagna di Roma.
	3. Miocene.	Marine.	Strata of Touraine, Bordeaux, Valley of the Bormida, and the Superga near Turin—Basin of Vienna.
		Freshwater.	Alternating with marine at Saucats, twelve miles south of Bordeaux.
		Volcanic.	Hungarian and Transylvanian volcanic rocks. Part of the volcanos of Auvergne, Cantal, and Velay?
	4. Eocene.	Marine.	Paris and London Basins.
		Freshwater.	Alternating with marine in Paris basin—Isle of Wight—purely lacustrine in Auvergne, Cantal, and Velay.
		Volcanic.	Oldest part of volcanic rocks of Auvergne.

CHAPTER VI.

Newer Pliocene formations—Reasons for considering in the first place the more modern periods—Geological structure of Sicily—Formations of the Val di Noto of newer Pliocene period—Divisible into three groups—Great limestone—Schistose and arenaceous limestone—Blue marl with shells—Strata subjacent to the above—Volcanic rocks of the Val di Noto—Dikes—Tuffs and Peperinos —Volcanic conglomerates—Proofs of long intervals between volcanic eruptions —Dip and direction of newer Pliocene strata of Sicily.

NEWER PLIOCENE FORMATIONS.

HAVING endeavoured, in the last chapter, to explain the principles on which the different tertiary formations may be arranged in chronological order, we shall now proceed to consider the newest division of formations, or that which we have named the newer Pliocene.

It may appear to some of our readers, that we reverse the natural order of historical research by thus describing, in the first place, the monuments of a period which immediately preceded our own era, and passing afterwards to the events of antecedent ages. But, in the present state of our science, this retrospective order of inquiry is the only one which can conduct us gradually from the known to the unknown, from the simple to the more complex phenomena. We have already explained our reasons for beginning this work with an examination, in the first two volumes, of the events of the *recent* epoch, from which the greater number of rules of interpretation in geology may be derived. The formations of the newer Pliocene period will be considered next in order, because these have undergone the least degree of alteration, both in position and internal structure, subsequently to their origin. They are monuments of which the characters are more easily deciphered than those belonging to more remote periods, for they have been less mutilated by the hand of time. The organic remains, more especially of this era, are most important, not only as being in a more perfect state of preservation, but also as being chiefly referrible to species now living ; so that their habits are known to us by direct comparison, and not merely by inference from analogy, as in the case of extinct species.

Geological structure of Sicily.—We shall first describe an extensive district in Sicily, where the newer Pliocene strata are largely developed, and where they are raised to considerable heights above the level of the sea.   After presenting the reader with a view of these formations, we shall endeavour to explain the manner in which they originated, and speculate on the subterranean changes of which their present position affords evidence.

The island of Sicily consists partly of primary and secondary rocks, which occupy, perhaps, about two-thirds of its superficial area *, and the remaining part is covered by tertiary formations, which are of great extent in the southern and central parts of the island, while portions are found bordering nearly the whole of the coasts.

Formations of the Val di Noto.—If we first turn our attention to the Val di Noto, a district which intervenes between Etna and the southern promontory of Sicily, we find a considerable tract, containing within it hills which are from one to two thousand feet in height, entirely composed of limestone, marl, sandstone, and associated volcanic rocks, which belong to the newer Pliocene era.   The recent shells of the Mediterranean abound throughout the sedimentary strata, and there are abundant proofs that the igneous rocks were the produce of successive submarine eruptions, repeated at intervals during the time when the subaqueous formations were in progress.

These rising grounds of the Val di Noto are separated from the cone of Etna, and the marine strata whereon it rests, by the low level plain of Catania, just elevated above the level of the sea, and watered by the Simeto.   The traveller who passes

* We may shortly expect a full account of the Geology of this island from Professor Hoffmann, who has devoted more than a year to its examination.

from Catania to Syracuse, has an opportunity of observing, on the sides of the valley, many deep sections of the modern formations above described, especially if he makes a slight detour by Sortino and the Valley of Pentalica.

The whole series of strata, in the Val di Noto, is divisible into three principal groups, exclusive of the associated volcanic rocks. The uppermost mass consists of limestone, which sometimes acquires the enormous thickness of seven or eight hundred feet, below which is a series much inferior in thickness, consisting of a calcareous sandstone, conglomerate and schistose limestone, and beneath this again, blue marl. The whole of the above groups contain shells and zoophytes, nearly all of which are referrible to species now inhabiting the contiguous sea.

Castrogiovanni.   No. 5.

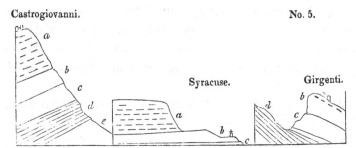

a, Great limestone of Val di Noto.
b, Schistose and arenaceous limestone of Florida, &c.
c, Blue marl with shells.
d, White laminated marl.
e, Blue clay and gypsum, &c. without shells.

Great limestone formation (*a*, diagram No. 5).—In mineral character this rock often corresponds to the yellowish white building-stone of Paris, well known by the name of *Calcaire grossier*, but it often passes into a much more compact stone. In the deep ravine-like valleys of Sortino and Pentalica, it is seen in nearly horizontal strata, as solid and as regularly bedded as the greater part of our ancient secondary formations. It abounds in natural caverns, which, in many places, as in the valley of Pentalica, have been enlarged and multiplied by artificial excavations.

The shells in the limestone are often very indistinct, sometimes nothing but casts remaining, but in many localities,
especially where there is a slight intermixture of volcanic sand,
they are more entire, and, as we have already stated, can almost
all be identified with recent Mediterranean testacea. Several
species of the genus Pecten are exceedingly numerous, particularly the large scallop (*P. Jacobæus*), now so common on
the coasts of Sicily. The shells which I collected from this
limestone at Syracuse, Villasmonde, Militello (V di Noto), and
Girgenti, have been examined by M. Deshayes, and found
to be all referrible to species now living, with three or four
exceptions*.

The mineral characters of this great calcareous formation
vary considerably in different parts of the island. In the south,
near the town of Noto, the rock puts on the compactness,
together with the spheroidal concretionary structure of some of
the Italian travertins. At the same place, also, it contains the
leaves of plants and reeds, as if a stream of fresh-water, charged
with carbonate of lime and terrestrial vegetable remains, had
entered the sea in the neighbourhood. At Spaccaforno, and
other places in the south of Sicily, a similar compact variety of
the limestone occurs, where it is for the most part pure white,
often very thick bedded, and occasionally without any lines of
stratification. This hard white rock is often four or five hundred feet in thickness, and appears to contain no fossil shells.
It has much the appearance of having been precipitated from
the waters of mineral springs, such as frequently rise up at the
bottom of the sea in the volcanic regions of the Mediterranean.
As these springs give out an equal quantity of mineral matter
at all seasons, they are much more likely to give rise to unstratified masses, than a river which is swoln and charged with

---

* For lists of these see Appendix II. I procured at Villasmonde, seven species;
at Militello, ten ; in the limestone of Girgenti, of which the ancient temples are
built, ten species ; from the limestone and subjacent clay at Syracuse, twenty-six
species; in the limestone and clay near Palermo, also belonging to the newer
Pliocene formation, one hundred shells.

sedimentary matter of different kinds, and in unequal quantities, at particular seasons of the year.

The great limestone above mentioned prevails not only in the Val di Noto, but re-appears in the centre of the island, capping the hill of Castrogiovanni, at the height of three thousand feet above the level of the sea. It is cavernous there, as at Sortino and Syracuse, and contains fossil shells and casts of shells of the same species*.

Schistose and arenaceous limestone, &c. (b, diagram No. 5.)—The limestone above-mentioned passes downwards into a white calcareous sand, which has sometimes a tendency to an oolitic and pisolitic structure, analogous to that which we have described when speaking of the travertin of Tivoli†. At Florida, near Syracuse, it contains a sufficient number of small calcareous pebbles to constitute a conglomerate, where also beds of sandy limestone are associated, replete with numerous fragments of shells, and much resembling, in structure, the English cornbrash. A diagonal lamination is often observable in the calcareous sandy beds analogous to that represented in the first volume (chap. xiv. diagram No. 6), and to that exhibited in many sections of the English crag, to which we shall afterwards allude.

In some parts of the island this sandy calcareous division b, seems to be represented by yellow sand, exactly resembling that so frequently superimposed on the blue shelly marl of the Subapennines in the Italian peninsula. Thus, near Grammichele, on the road to Caltagirone, beds of incoherent yellow sand, several hundred feet in thickness, with occasional layers of shells, repose upon the blue shelly marl of Caltagirone.

When we consider the arenaceous character of this formation, the disposition of the laminæ, and the broken shells sometimes imbedded in it, it is difficult not to suspect that it was

---

* Dr. Daubeny correctly identified the Val di Noto limestone of Syracuse with that of the summit of Castrogiovanni.—Jameson, Ed. Phil. Journ., No. xxv. p. 107, July, 1825.

† Vol. i. chap. xii.

formed in shallower water, and nearer the action of superficial currents, than the superincumbent limestone, which was evidently accumulated in a sea of considerable depth. If we adopt this view, we must suppose a considerable subsidence of the bed of the sea, subsequent to the deposition of the arenaceous beds in the Val di Noto.

Blue marl with shells (*c*, diagram No. 5).—Under the sandy beds last mentioned is found an argillaceous deposit of variable thickness, called *Creta* in Sicily. It resembles the blue marl of the Subapennine hills, and, like it, encloses fossil shells and corals in a beautiful state of preservation. Of these I collected a great abundance from the clay, on the south side of the harbour of Syracuse, and twenty species in the environs of Caltanisetta, all of which, with three exceptions, M. Deshayes was able to identify with recent species*. From similar blue marl, alternating with yellow sand, at Caltagirone, at an elevation of about five hundred feet above the level of the sea, I obtained forty species of shells, of which all but six were recognized as identical with recent species†. The position of this argillaceous formation is well seen at Castrogiovanni and Girgenti, as represented in the sections, diagram No. 5. In both of these localities, the limestone of the Val di Noto re-appears, passing downwards into a calcareous sandstone, below which is a shelly blue clay.

Strata beneath the blue marl.—The clay rests, in both localities, on an older series of white and blue marls, probably belonging to the tertiary period, but of which I was unable to determine the age, having procured from it no organic remains save the skeletons of fish which I found in the white thinly-laminated marls‡.

---

* See list of these shells, Appendix II.

† See Appendix II.

‡ I found these fossil fish in great abundance on the road, half a mile north-west of Radusa, on the road to Castrogiovanni, where the marls are fetid, and near Castrogiovanni in gypseous marls, at the mile-stone No. 88, and between that and No. 89. Lord Northampton has since presented to the Geological Society

The marls are sometimes gypseous, and belong to a great argillaceous formation which stretches over a considerable part of Sicily, and contains sulphur and salt in great abundance. The strata of this group have been in some places contorted in the most extraordinary manner, their convolutions often resembling those seen in the most disturbed districts of primary clay slate.

But we wish, at present, to direct the reader's exclusive attention to strata decidedly referrible to the newer Pliocene era, and we have yet to mention the igneous rocks associated with the sedimentary formations already alluded to.

Volcanic Rocks of the Val di Noto.—The volcanic rocks occasionally associated with the limestones, sands, and marls already described, constitute a very prominent feature throughout the Val di Noto. Great confusion might have been expected to prevail, where lava and ejected sand and scoriæ are intermixed with the marine strata, and, accordingly, we find it often impossible to recognize the exact part of the series to which the beds thus interfered with belong.

Sometimes there are proofs of the posterior origin of the lava, and sometimes of the newer date of the stratified rock, for we find dikes of lava intersecting both the marl and limestone, while, in other places, calcareous beds repose upon lava, and are unaltered at the point of contact. Thus the shelly limestone of Capo Santa Croce rests in horizontal strata upon a mass of lava, which had evidently been long exposed to the action of the waves, so that the surface has been worn perfectly smooth. The limestone is unchanged at its junction with the igneous rock, and incloses within it pebbles of the lava *.

The volcanic formations of the Val di Noto usually consist of the most ordinary variety of basalt with or without olivine. The rock is sometimes compact, often very vesicular. The

some which he obtained from the same localities, but I have met with no zoologists who could name the species.

* This locality is described by Professor Hoffman, Archiv für Mineralogie, &c. Berlin, 1831.

vesicles are occasionally empty, both in dikes and currents, and are in some localities filled with calcareous spar, arragonite, and zeolites. The structure is; in some places, spheroidal, in others, though rarely, columnar. I found dykes of amygdaloid, wacke, and prismatic basalt, intersecting the limestone at the bottom of the hollow, called Gozzo degli Martiri, below Melilli.

Dikes.—Dikes of vesicular and amygdaloidal lava are also seen traversing peperino, west of Palagonia, near a mill by the road side.

No. 6.   No. 7.

Horizontal section of Dikes near Palagonia.

a, Lava.
b, Peperino, consisting of volcanic sand, mixed with fragments of lava and
    of limestone.

In this case we may suppose the peperino to have resulted from showers of volcanic sand and scoriæ, together with fragments of limestone thrown out by a submarine explosion, similar to that which lately gave rise to the volcanic island off Sciacca. When the mass was, to a certain degree, consolidated, it may have been rent open, so that the lava ascended through fissures, the walls of which were perfectly even and parallel. After the melted matter that filled the rent had cooled down, it must have been fractured and shifted horizontally by a lateral movement.

In the second figure, No. 7, the lava has more the appearance of a vein which forced its way through the peperino, availing itself, perhaps, of a slight passage opened by rents caused by earthquakes. Some of the pores of the lava, in these dikes, are empty, while others are filled with carbonate of lime.

The annexed diagrams (Nos. 6 and 7) represent a ground plan of the rocks as they are exposed to view on a horizontal surface. We think it highly probable that similar appearances would be seen, if we could examine the floor of the sea in that part of the Mediterranean where the waves have recently washed away the new volcanic island, for when a superincumbent mass of ejected fragments has been removed by denudation, we may expect to see sections of dikes traversing tuff, or, in other words, sections of the channels of communication by which the subterranean lavas reached the surface.

On the summit of the limestone platform of the Val di Noto, I more than once saw analogous dikes, not only of lava but of volcanic tuff, rising vertically through the horizontal strata, and having no connexion with any igneous masses now apparent on the surface. In regard to the *dikes of tuff or peperino*, we may suppose them to have been open fissures at the bottom of the sea, into which volcanic sand and scoriæ were drifted by a current.

Tuffs and Peperinos.—In the hill of Novera, between Vizzini and Militelli, a mass of limestone, horizontally stratified, comes in contact with inclined strata of tuff (see diagram No. 8),

No. 8.

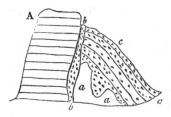

A, Limestone.
aa, Calcareous breccia with fragments of lava.
b, Black tuff.
c, Tuff.

while a mixed calcareous and volcanic breccia, *a a*, supports the inclined layers of tuff, *c*. The vertical fissure, *b b*, is filled with volcanic sand of a different colour. An inspection of this section will convince the reader that the limestone must have been greatly dislocated during the time that the submarine eruptions were taking place.

At the town of Vizzini, a dike of lava intersects the argillaceous strata, and converts them into siliceous schist, which has been contorted and shivered into an immense number of fragments.

We have stated that the beds of limestone, clay, and sand, in the Val di Noto, are often partially intermixed with volcanic ejections, such as may have been showered down into the sea during eruptions, or may have been swept by rivers from the land. When the volcanic matter predominates, these compound rocks constitute the peperinos of the Italian mineralogists, some of which are highly calcareous, full of shells, and extremely hard, being capable of a high polish like marble. In some parts of the Val di Noto they are variously mottled with spots of red and yellow, and contain small angular fragments, similar to the lapilli thrown from volcanos.

It is recorded that, during the late eruption off the southern coast of Sicily, opposite Sciacca, the sea was in a state of violent ebullition, and filled, for several weeks continuously, with red or chocolate-coloured mud, consisting of finely-comminuted scoriæ. During this period, it is clear that the waves and currents that have since had power to sweep away the island, and disperse its materials far and wide over the bed of the sea, must with still greater ease have carried to vast distances the fine red mud, which was seen boiling up from the bottom, so that it may have entered largely into the composition of modern peperinos.

Professor Hoffman relates that, during the eruption (June, 1831), the surface of the sea was strewed over, at the distance of thirty miles from the new volcano, with so dense a covering of scoriæ, that the fishermen were obliged to part it with their oars, in order to propel their boats through the water. It is, therefore, quite consistent with analogy, that we should find the ancient tuffs and peperinos so much more generally distributed than the submarine lavas.

In the road which leads from Palagonia to Lago Naftia, and at the distance of about a mile and a half from the former place, there is a small pass where the hills, on both sides, consist of a calcareous grit, intermixed with some grains of volcanic sand.

No. 9.

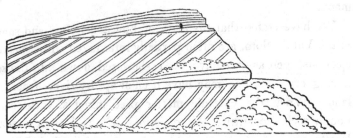

Section of calcareous grit and peperino, east of Patagonia.   South side of pass.
Vertical height about thirty feet

No. 10.

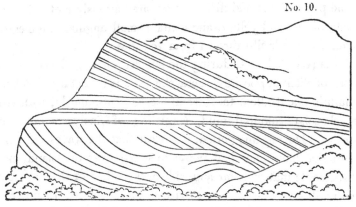

Section of the same beds on the north side of the pass.

The disposition of the strata, on both sides of the pass, is most singular, and remarkably well exposed, as the harder layers have resisted the weathering of the atmosphere and project in relief. The sections exhibited on both sides of the pass are nearly vertical, and do not exactly correspond, as will be seen in the annexed diagrams (Nos. 9 and 10). It is somewhat difficult to conceive in what manner this arrangement of the layers was occasioned, but we may, perhaps, suppose it to have arisen from the throwing down of calcareous sand and volcanic matter, upon steep slanting banks at the bottom of the sea, in which case they might have accumulated at various angles of between thirty and fifty degrees, as may be frequently seen in the sections of volcanic cones in Ischia and elsewhere. The denuding power of the waves may, then, have cut off the upper portion of these banks, so that nearly horizontal layers may have been superimposed unconformably, after which another bank may have been formed in a similar manner to the first.

Volcanic conglomerates.—In the Val di Noto we sometimes meet with conglomerates entirely composed of volcanic pebbles. They usually occur in the neighbourhood of masses of lava, and may, perhaps, have been the shingle produced by the wasting cliffs of small islands in a volcanic archipelago. The formation of similar beds of volcanic pebbles may now be seen in progress on the beach north of Catania, where the waves are undermining one of the modern lavas of Etna; and the same may also be seen on the shores of Ischia.

Proofs of gradual accumulation.—In one part of the great limestone formation near Lentini, I found some imbedded volcanic pebbles, covered with full-grown serpulæ, supplying a beautiful proof of a considerable interval of time having elapsed between the rounding of these pebbles and their inclosure in a solid stratum. I also observed, not far from Vizzini, a very striking illustration of the length of the intervals which occasionally separated the flows of distinct lava-currents. A bed of oysters, perfectly identifiable with our common eatable species, no less than *twenty feet in thickness*, is there seen resting upon a current of basaltic lava; upon the oyster-bed again is superimposed a second mass of lava, together with tuff or peperino. Near Galieri, not far from the same locality, a horizontal bed, about a foot and a half in thickness, composed entirely of a common Mediterranean coral (*Caryophyllia cespitosa*, Lam.), is also seen in the midst of the same series of alternating igneous and aqueous formations. These corals stand erect as they grew, and after being traced for hundreds of yards, are again found at a corresponding height on the opposite side of the valley.

Dip and direction.—The disturbance which the newer Pliocene strata have undergone in Sicily, subsequent to their deposition, differs greatly in different places; in general, however, the beds are nearly horizontal, and are not often highly inclined. The calcareous schists, on which part of the town of Lentini is built, are much fractured, and dip at an angle of twenty-five degrees to the north-west. In some of the valleys in the neighbourhood an anticlinal dip is seen, the beds on one side being inclined to the north-west, and on the other to the south-east.

Throughout a considerable part of Sicily which I examined, the dips of the tertiary strata were north-east and south-west; as, for example, in the district included between Terranuova, Girgenti, Caltanisetta, and Piazza, where there are several parallel lines, or ridges of elevation, which run north-west and south-east.

CHAPTER VII.

Marine and volcanic formations at the base of Etna—Their connexion with the strata of the Val di Noto—Bay of Trezza—Cyclopian isles—Fossil shells of recent species—Basalt and altered rocks in the Isle of Cyclops—Submarine lavas of the Bay of Trezza not currents from Etna—Internal structure of the cone of Etna—Val di Calanna—Val del Bove not an ancient crater—Its precipices intersected by countless dykes—Scenery of the Val del Bove—Form, composition, and origin of the dykes—Lavas and breccias intersected by them.

MARINE AND VOLCANIC FORMATIONS AT THE BASE OF ETNA.

The phenomena considered in the last chapter suggest many theoretical views of the highest interest in Geology ; but before we enter upon these topics we are desirous of describing some formations in Valdemone, which are analogous to those of the Val di Noto, and to point out the relation of such rocks to the modern lavas of Etna.

If the traveller passes along the table-land, formed by the great limestone of the Val di Noto, until it terminates suddenly near Primosole, he there sees the plain of Catania at his feet,

No. 11.

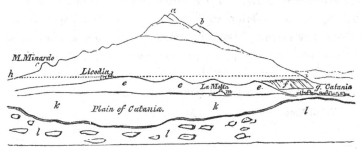

View of Etna from the summit of the limestone platform of Primosole.

a, Highest cone. b, Montagnuola. c, Monte Minardo, with smaller lateral cones above. d, Town of Licodia dei Monaci. e, Marine formation called creta, argillaceous and sandy beds with a few shells, and associated volcanic rocks. f, Escarpment of stratified subaqueous volcanic tuff, &c., north-west of Catania. g, Town of Catania. h, i, Dotted line expressing the highest boundary along which the marine strata are occasionally seen. k, Plain of Catania. l, Limestone platform of Primosole of the newer Pliocene. m, La Motta di Catania.

and before him, to the north, the cone of Etna (see diagram No. 11). At the base of the cone he beholds a low line of hills *e, e* (No. 11), formed of clays and marls, associated with yellowish sand, similar to the formation provincially termed ' Creta,' in various parts of Sicily.

This marine formation, which is composed partly of volcanic and partly of sedimentary rocks, is seen to underlie the modern lavas of Etna. To what extent it forms the base of the mountain cannot be observed, for want of sections of the lower part of the cone, but the marine sub-Etnean beds are not observed to rise to a greater elevation than eight hundred, or, at the utmost, one thousand feet above the level of the sea. We should remind the reader, that the annexed drawing is not a section, but an outline view of Etna, as seen from Primosole, so that the proportional height of the volcanic cone, which is, in reality, ten times greater than that of the hills of ' Creta,' at its base, is not represented, the summit of the cone being ten or twelve miles more distant from the plain of Catania, than Licodia.

Connexion of the sub-Etnean strata with those of the Val di Noto.—These marine strata are found both on the southern and eastern foot of Etna, and it is impossible not to infer that they belong to the inferior argillaceous series of the Val di Noto, which they resemble both in mineral and organic characters. In one locality they appear on the opposite sides of the Valley of the Simeto, covered on the north by the lavas of Etna, and on the south by the Val di Noto limestone.

Val di Noto.   No. 12.   Etna.

Section from Paternò by Lago di Naftia to Palagonia.

a, Plain of the Simeto.   *b*, Base of the cone of Etna, composed of modern lavas.   *c*, Limestone of the Val di Noto.   *d*, Clay, sand, and associated submarine volcanic rocks.

If in the country adjacent to the Lago di Naftia, through

which the annexed section is drawn, and in several other districts where the 'creta' prevails, together with associated
submarine lavas, and where there is no limestone capping, a
volcano should now burst forth, and give rise to a great cone,
the position of such a cone would exactly correspond to that of
the modern Etna, with relation to the rocks on which it rests.

Southern base of Etna.—The marine strata of clay and sand
already alluded to, alternate in thin layers at the southern base
of Etna, sometimes attaining a thickness of three hundred feet,
or more, without any intermixture of volcanic matter. Crystals
of selenite are dispersed through the clay, accompanied by a
few shells, almost entirely of recent Mediterranean species.
This formation of blue marl and yellow sand greatly resembles
in character that of the Italian Subapennine beds, and, like
them, often presents a surface denuded of vegetation, in consequence of the action of the rains on soft incoherent materials.

In travelling by Paternò, Misterbianco, and La Motta, we
pass through deep narrow valleys excavated through these
beds, which are sometimes capped, as at La Motta, by columnar basalt, accompanied by strata of tuff and volcanic conglomerate. (Diagram No. 13.)

No. 13.

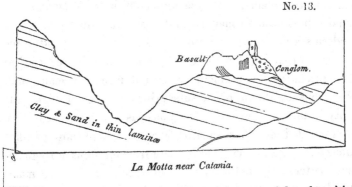

La Motta near Catania.

The latter rock is composed of rolled masses of basalt, which
may either have originated when first the lava was produced
in a volcanic archipelago, or subsequently when the whole
country was rising from beneath the level of the sea. Its
occurrence in this situation is striking, as not a single pebble

can be observed in the entire thickness of subjacent beds of sand and clay.

The dip of the marine strata, at the base of Etna, is by no means uniform; on the eastern side, for example, they are sometimes inclined towards the sea, and at others towards the mountain. Near the aqueduct at Aderno, on the southern side, I observed two sections, in quarries not far distant from each other, where beds of clay and yellow sand dipped, in one locality, at an angle of forty-five degrees to the east-south-east, and in the other at a much higher inclination in the opposite direction. These facts would be of small interest, if an attempt had not been made to represent these mixed marine and volcanic deposits which encircle part of the base of Etna, as the outer margin of a so-called ' elevation crater *.'

Near Catania the marine formation, consisting chiefly of volcanic tuff thinly laminated, terminates in a steep inland cliff, or escarpment, which is from six hundred to eight hundred feet in height. A low flat, composed of recent lava and volcanic sand, intervenes between the sea and the base of this escarpment, which may be well seen at Fasano. (*f*, diagram No. 11.)

Eastern side of Etna—Bay of Trezza.—Proceeding northwards from Catania, we have opportunities of examining the same sub-Etnean formations laid open more distinctly in the modern sea-cliffs, especially in the Bay of Trezza and in the Cyclopian islands (Dei Faraglioni), which may be regarded as the extremity of a promontory severed from the main land. Numerous are the proofs of submarine eruptions of high antiquity in this spot, where the argillaceous and sandy beds have been invaded and intersected by lava, and where those peculiar tufaceous breccias occur which result from ejections of fragmentary matter, projected from a volcanic vent. I observed many angular and hardened fragments of laminated clay (creta), in different states of alteration, between La Trezza and Nizzitta, and in the hills above Aci Castello, a town on the main land contiguous to the Cyclopian isles, which could not be mistaken

* See vol. i. chap. xxii.

by one familiar with Somma and the minor cones of Ischia, for anything but masses thrown out by volcanic explosions. From the tuffs and marls of this district I collected a great variety of marine shells *, almost all of which have been identified with species now inhabiting the Mediterranean, and, for the most part, now frequent on the coast immediately adjacent. Some few of these fossil shells retain part of their colour, which is the same as in their living analogues.

The largest of the Cyclopian islets, or rather rocks, is distant two hundred yards from the land, and is only three hundred yards in circumference, and about two hundred feet in height. The summit and northern sides are formed of a mass of stratified marl (creta), the laminæ of which are occasionally subdivided by thin arenaceous layers. These strata rest on a mass of columnar lava (see wood-cut, No. 14) †, which appears to have forced itself into, and to have heaved up the stratified mass.

No. 14.

View of the Isle of Cyclops in the Bay of Trezza.

* See, in Appendix No. II., a list, by M. Deshayes, of sixty-five species, which I procured from the hills called Monte Cavalaccio, Rocca di Ferro, and Rocca di Bempolere (or Borgia).

† This cut is from an original drawing by my friend Capt. W. H. Smyth. R. N.

This theory of the intrusion of the basalt is confirmed by the fact, that in some places the clay has been greatly altered, and hardened by the action of heat, and occasionally contorted in the most extraordinary manner, the lamination not having been obliterated, but, on the contrary, rendered much more conspicuous by the indurating process.

The annexed wood-cut (No. 15) is a careful representation of a portion of the altered rock, a few feet square, where the alternate thin laminæ of sand and clay have put on the appearance which we often observe in some of the most contorted of the primary schists.

No. 15.

Contortions in the newer Pliocene strata, Isle of Cyclops.

A great fissure, running from east to west, nearly divides the island into two parts, and lays open its internal structure. In the section thus exhibited, a dike of lava is seen, first cutting through an older mass of lava, and then penetrating the superincumbent tertiary strata. In one locality, the lava ramifies and terminates in thin veins, from a few feet to a few inches in thickness (see diagram No. 16).

No. 16.

Newer Pliocene strata invaded by lava. Isle of Cyclops (horizontal section).

a, Lava.   b, laminated clay and sand.   c, the same altered.

The arenaceous laminæ are much hardened at the point of contact, and the clays are converted into siliceous schist.   In this
island the altered rocks assume a honeycombed structure on
their weathered surface, singularly contrasted with the smooth
and even outline which the same beds present in their usual
soft and yielding state.

The pores of the lava are sometimes coated, or entirely filled,
with carbonate of lime, and with a zeolite resembling analcime,
which has been called cyclopite.   The latter mineral has also
been found in small fissures traversing the altered marl, showing that the same cause which introduced the minerals into the
cavities of the lava, whether we suppose sublimation or aqueous
infiltration, conveyed it also into the open rents of the contiguous sedimentary strata.

Lavas of the Cyclopian Isles not currents from Etna.—The
phenomena of the Bay of Trezza are very important, for it is
evident that the submarine lavas were produced by eruptions
on the spot, an inference which follows not only from the presence of dikes and veins, but from those tuffs above Castello
d'Aci, which contain angular fragments of hardened marl, evidently thrown up, together with the sand and scoriæ, by volcanic

explosions. We may, therefore, suppose this volcanic action to have been as independent of the modern vents of Etna, as that which gave rise to the analogous formations in the Val di Noto. It is quite evident that the lavas of the Cyclopian isles are not the lower extremities of currents which flowed down from the highest crater of Etna, or from the region where lateral eruptions are now frequent,—lavas which, after entering the sea, were afterwards upraised into their present position. It is more probable that the basalts of the Bay of Trezza, and those along the southern foot of Etna, at La Motta, Adernò, Paternò, Licodia, and other places, originated in the same sea in which the eruptions of the Val di Noto took place.

There are, however, as we have observed, no sections to prove that the central and oldest parts of Etna repose on similar submarine formations. The modern lavas of the volcano are continually extending their area, and covering, from time to time, a larger portion of the marine strata; but we know not where this operation commenced, so that we cannot demonstrate the posteriority of the whole cone to these newer Pliocene strata.

We might imagine that when the volcanos of the Val di Noto were in activity, and when the eruptions of the Bay of Trezza were taking place, Etna already existed as a volcano, the upper part only of the cone projecting above the level of the waters, as in the case of Stromboli at present. By such an hypothesis, we might refer the origin of the older part of Etna to the same period as that of the sedimentary strata and volcanic rocks of the Val di Noto.

But, for our own part, we see no grounds for inclining to such a theory, since we must admit that a sufficient series of ages has elapsed since the limestone of the Val di Noto was deposited, to allow the same to be elevated to the height of from two thousand to three thousand feet, in which case there may also have been sufficient time for the growth of a volcanic pile like Etna, since the newer Pliocene strata now seen at the base of the volcano originated.

INTERNAL STRUCTURE OF THE CONE OF ETNA.

In our first volume we merely described that part of Etna which has been formed during the historical era; an insignificant portion of the whole mass. Nearly all the remainder may be referred to the tertiary period immediately antecedent to the *recent* epoch. We before stated, that the great cone is, in general, of a very symmetrical form, but is broken, on its eastern side, by a deep valley, called the Val del Bove*, which,

No. 17.

Great valley on the east side of Etna.

a, highest cone. *b*, Montagnuola. *c*, Head of Val del Bove. *d, d*, Serre del Solfizio. *e*, Zaffarana. *f*, One of the lateral cones. *g*, Monti Rossi.

commencing near the summit of the mountain, descends into the woody region, and is then continued, on one side, by a second and narrower valley, called the Val di Calanna. Below the latter another, named the Val di St. Giacomo, begins,—a long narrow ravine, which is prolonged to the neighbourhood of Zaffarana (*e*, No. 17), on the confines of the fertile region. These natural incisions, into the side of the volcano, are of such

* In the provincial dialect of the peasants called ' Val del Bué,' for here the herdsman

———— 'in reductâ valle *mugientium*
Prospectat errantes greges.—'

Dr. Buckland was, I believe, the first English geologist who examined this valley with attention, and I am indebted to him for having described it to me, before my visit to Sicily, as more worthy of attention than any single spot in that island, or perhaps in Europe. I have already stated, that the view of this valley, which I have given in the frontispiece of the second volume, does not pretend to convey any idea of the grandeur of the scene.

depth, that they expose to view a great part of the structure of the entire mass, which, in the Val del Bove, is laid open to the depth of from four thousand to five thousand feet from the summit of Etna. The geologist thus enjoys an opportunity of ascertaining how far the internal conformation of the cone corresponds with what he might have anticipated as the result of that mode of increase which has been witnessed during the historical era.

It is clear, from what we before said of the gradual manner in which the principal cone increases, partly by streams of lava and showers of volcanic ashes ejected from the summit, partly by the throwing up of minor hills and the issuing of lava-currents on the flanks of the mountain, that the whole cone must consist of a series of cones enveloping others, the regularity of each being only interrupted by the interference of the lateral volcanos.

We might, therefore, have anticipated that a section of Etna, as exposed in a ravine which should begin near the summit and extend nearly to the sea, would correspond very closely to the section of the ancient Vesuvius, commencing with the escarpment of Somma, and ending with the Fossa Grande ; but with this difference, that where the ravine intersects the woody region of Etna, indications must appear of changes brought about by lateral eruptions. Now the section before alluded to, which can be traced from the head of the Val del Bove to the inferior borders of the woody region, fully answers such expectations. We find, almost everywhere, a series of layers of tuff and breccia interstratified with lavas, which slope gently to the sea, at an angle of from twenty to thirty degrees ; and as we rise to the parallel of the zone of lateral eruptions, and still more as we approach the summit, we discover indications of disturbances, occasioned by the passage of lava from below, and the successive inhumation of lateral cones.

Val di Calanna.—On leaving Zaffarana, on the borders of the fertile region, we enter the ravine-like valley of St. Giacomo, and see on the north side, or on our right as we ascend, rising ground composed of the modern lavas of Etna. On our left, a lofty cliff, wherein a regular series of beds is exhibited, composed of tuffs and lavas, descending with a gentle inclination towards the sea.   In this lower part of the section there are no intersecting dikes, nor any signs of minor cones interfering with the regular slope of the alternating volcanic products.   If we then pass upwards through a defile, called the ' Portello di Calanna,' we enter a second valley, that of Calanna, resembling the ravine before mentioned, but wider and much deeper.   Here again we find, on our right, many currents of modern lava, piled one upon the other, and on our left a continuation of our former section, in a perpendicular cliff from four hundred to five hundred feet high.   As this lofty wall sweeps in a curve, it has very much the appearance of the escarpment which Somma presents towards Vesuvius, and this resemblance is increased by the occurrence of two or three vertical dikes which traverse the gently-inclined volcanic beds.   When I first beheld this precipice, I fancied that I had entered a lateral crater, but was soon undeceived, by discovering that on all sides, both at the head of the valley, in the hill of Zocolaro, and at its side and lower extremity, the dip of the beds was always in the same direction, all slanting to the east, or towards the sea, instead of sloping to the north, east, and south, as would have been the case had they constituted three walls of an ancient crater.

It is not difficult to explain how the valleys of St. Giacomo and Calanna originated, when once the line of lofty precipices on the north side of them had been formed.   Many lava-currents flowing down successively from the higher regions of Etna, along the foot of a great escarpment of volcanic rock, have at length been turned by a promontory at the head of the valley of Calanna, which runs out at right angles, to the great line of precipices.   This promontory consists of the hills called Zocolaro and Calanna, and of a ridge of inferior height which connects them.   (See diagram No. 18.)

No. 18.

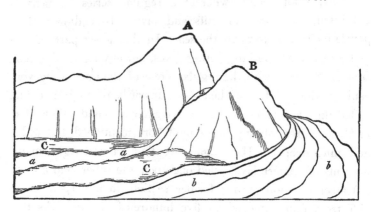

A, Zocolaro.   B, Monte di Calanna.
C, Plain at the head of the Valley of Calanna.
a, Lava of 1819 descending the precipice and flowing through the valley.
b, Lavas of 1811 and 1819 flowing round the hill of Calanna.

The flows of melted matter have been deflected from their course by this projecting mass, just as a tidal current, after setting against a line of sea-cliffs, is often thrown off into a new direction by some rocky headland.

Lava-streams, it is well known, become solid externally, even while yet in motion, and their sides may be compared to two rocky walls, which are sometimes inclined at an angle of forty-five degrees. When such streams descend a considerable slope at the base of a line of precipices, and are turned from their course by a projecting rock, they move right onwards in a new direction, so as to leave a considerable space (as in the Valley of Calanna) between them and the cliffs which may be continuous below the point of deflection.

It happened in 1811 and 1819, that the flows of lava overtopped the ridge intervening between the hills of Zocolaro and Calanna, so that they fell in a cascade over a lofty precipice, and began to fill up the valley. (See letter a, diagram No. 18.)

The narrow cavity of St. Giacomo will admit of an explanation precisely similar to that already offered for Calanna.

Val del Bove.—After passing up through the defile, called
the ' Rocca di Calanna," we enter a third valley of truly
magnificent dimensions—the Val del Bove—a vast amphitheatre four or five miles in diameter, surrounded by nearly
vertical precipices, varying from one thousand to above three
thousand feet in height, the loftiest being at the upper end,
and the height gradually diminishing on both sides. The
feature which first strikes the geologist as distinguishing this
valley from those before mentioned, is the prodigious mul-
titudes of vertical dikes, which are seen in all directions traversing the volcanic beds. The circular form of this great
chasm, and the occurrence of these countless dikes, amounting
perhaps to several thousands in number, so forcibly recalled
to my mind the phenomena of the Atrio del Cavallo, on
Vesuvius, that I imagined once more that I had entered a
vast crater, on a scale as far exceeding that of Somma, as Etna
surpasses Vesuvius in magnitude.

But having already been deceived in regard to the crescent-
shaped precipice of the valley of Calanna, I began attentively
to explore the different sides of the great amphitheatre, in order
to satisfy myself whether the semicircular wall of the Val del
Bove had ever formed the boundary of a crater, and whether
the beds had the same quâquâ-versal dip which is so beautifully exhibited in the escarpment of Somma. If the supposed
analogy between Somma and the Val del Bove should hold true,
the tuffs and lavas, at the head of the valley, would dip to the
west, those on the north side towards the north, and those on
the southern side to the south. But such I did not find to be
the inclination of the beds ; they all dip towards the sea, or
nearly east, as was before seen to be the case in the Valley of
Calanna.

There are undoubtedly exceptions to this general rule, which
might deceive a geologist who was strongly prepossessed with
a belief that he had discovered the hollow of an ancient crater.
It is evident that, wherever lateral cones are intersected in the
precipices, a series of tuffs and lavas, very similar to those which

enter into the structure of the great cone, will be seen dipping at a much more rapid angle.

The lavas and tuffs, which have conformed to the sides of Etna, dip at angles of from fifteen to twenty-five degrees, while the slope of the lateral cones is from thirty-five to fifty degrees. Now, wherever we meet with sections of these buried cones in the precipices bordering the Val del Bove, (and they are frequent in the cliffs called the Serre del Solfizio, and in those near the head of the valley not far from the rock of Musara,) we find the beds dipping at high angles and inclined in various directions*.

Scenery of the Val del Bove.—Without entering at present into any further discussions respecting the origin of the Val del Bove, we shall proceed to describe some of its most remarkable features. Let the reader picture to himself a large amphitheatre, five miles in diameter, and surrounded on three sides by precipices from two thousand to three thousand feet in height. If he has beheld that most picturesque scene in the chain of the Pyrenees, the celebrated 'cirque of Gavarnie,' he may form some conception of the magnificent circle of precipitous rocks which inclose, on three sides, the great plain of the Val del Bove. This plain has been deluged by repeated streams of lava, and although it appears almost level when viewed from a distance, it is, in fact, more uneven than the surface of the most tempestuous sea. Besides the minor irregularities of the lava, the valley is in one part interrupted by a ridge of rocks, two of which, Musara and Capra, are very prominent. It can hardly be said that they

> ———— 'like giants stand
> To sentinel enchanted land; '

for although, like the Trosachs, they are of gigantic dimen-

* I perceive that Professor Hoffmann, who visited the Val del Bove after me (in January, 1831), has speculated on its structure as corresponding to that of the so-called elevation craters, which hypothesis would require that there should be a quâquâ-versal dip, such as I have above alluded to. I can only account for this difference of opinion, by supposing the Professor to have overlooked the phenomena of the buried cones.—Archiv. für Mineralogie, &c. Berlin, 1831.

sions, and appear almost isolated as seen from many points, yet
the stern and severe grandeur of the scenery which they adorn
is not such as would be selected by a poet for a vale of enchant-
ment.   The character of the scene would accord far better with
Milton's picture of the infernal world ; and if we imagine ourselves to behold in motion, in the darkness of the night, one of
those fiery currents, which have so often traversed the great
valley, we may well recall

> ———— ' yon dreary plain, forlorn and wild,
> The seat of desolation, void of light
> Save what the glimmering of these livid flames
> Cast pale and dreadful.'

The face of the precipices already mentioned is broken in the
most picturesque manner by the vertical walls of lava which
traverse them.   These masses usually stand out in relief, are
exceedingly diversified in form, and often of immense altitude.
In the autumn, their black outline may often be seen relieved
by clouds of fleecy vapour which settle behind them, and do
not disperse until midday, continuing to fill the valley while the
sun is shining on every other part of Sicily, and on the higher
regions of Etna.

As soon as the vapours begin to rise, the changes of scene
are varied in the highest degree, different rocks being unveiled
and hidden by turns, and the summit of Etna often breaking
through the clouds for a moment with its dazzling snows, and
being then as suddenly withdrawn from the view.

An unusual silence prevails, for there are no torrents dash-
ing from the rocks, nor any movement of running water in
this valley, such as may almost invariably be heard in mountainous regions.   Every drop of water that falls from the
heavens, or flows from the melting ice and snow, is instantly
absorbed by the porous lava ; and such is the dearth of springs,
that the herdsman is compelled to supply his flocks, during the
hot season, from stores of snow laid up in hollows of the mountain during winter.

The strips of green herbage and forest-land, which have

here and there escaped the burning lavas, serve, by contrast, to
heighten the desolation of the scene.   When I visited the valley,
nine years after the eruption of 1819, I saw hundreds of trees,
or rather the white skeletons of trees, on the borders of the
black lava, the trunks and branches being all leafless, and
deprived of their bark by the scorching heat emitted from the
melted rock ; an image recalling those beautiful lines—

> ——— ' As when heaven's fire
> Hath scath'd the forest oaks, or mountain pines,
> With singed top their stately growth, though bare,
> Stands on the blasted heath.'

Form, composition, and origin of the Dikes.—But without
indulging the imagination any longer in descriptions of scenery,
we may observe, that the dikes before mentioned form unques-
tionably the most interesting geological phenomenon in the Val
del Bove.

No. 19.

Dikes at the base of the Serre del Solfizio, Etna.

Some of these are composed of trachyte, others of compact

blue basalt with olivine.　They vary in breadth from two to twenty feet and upwards, and usually project from the face of the cliffs, as represented in the annexed drawing (No. 19). They consist of harder materials than the strata which they traverse, and therefore waste away less rapidly under the influence of that repeated congelation and thawing to which the rocks in this zone of Etna are exposed.　The dikes are, for the most part, vertical, but sometimes they run in a tortuous course through the tuffs and breccias, as represented in diagram, No. 20.　In the escarpment of Somma where, as we be-

No. 20.

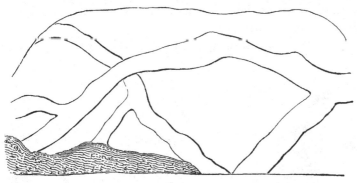

Veins of Lava.　　　*Punto di Guimento.*

fore observed, similar walls of lava cut through alternating beds of sand and scoriæ, a coating of coal-black rock, approaching in its nature and appearance to pitch-stone, is seen at the contact of the dike with the intersected beds.　I did not observe such parting layers at the junction of the Etnean dikes which I examined, but they may perhaps be discoverable.

The geographical position of these dikes is most interesting, as they occur in that zone of the mountain where lateral eruptions are frequent ; whereas, in the valley of Calanna, which is below that parallel, and in a region where lateral eruptions are extremely rare, scarcely any dikes are seen, and none whatever still lower in the valley of St. Giacomo.　This is precisely what we should have expected, if we consider the vertical fissures now filled with rock to have been the feeders of lateral

cones, or, in other words, the channels which gave passage to
the lava-currents and scoriæ that have issued from vents in the
forest-zone.

Some fissures may have been filled from above, but I did not
see any which, by terminating downwards, gave proof of such
an origin. Almost all the isolated masses in the Val del Bove,
such as Capra, Musara, and others, are traversed by dikes, and
may, perhaps, have partly owed their preservation to that circumstance, if at least the action of occasional floods has been
one of the destroying causes in the Val del Bove; for there is
nothing which affords so much protection to a mass of strata
against the undermining action of running water, as a perpendicular dike of hard rock.

In the accompanying drawing (No. 21) the flowing of the

No. 21.

View of the rocks Finochio, Capra, and Musara, Val del Bove.

lavas of 1811 and 1819, between the rocks Finochio, Capra,
and Musara, is represented. The height of the two last-mentioned isolated masses has been much diminished by the elevation of their base, caused by these currents. They may,
perhaps, be the remnants of cones, which existed before the Val
del Bove was formed, and may hereafter be once more buried
by the lavas that are now accumulating in the valley.

From no point of view are the dikes more conspicuous than from the summit of the highest cone of Etna; a view of some of them is given in the annexed drawing *.

No. 22.

View from the summit of Etna into the Val del Bove.

The small cone and crater immediately below were among those formed during the eruptions of 1810 *and* 1811.

Lavas and breccias.—In regard to the volcanic masses which are intersected by dikes in the Val del Bove, they consist, in great part, of graystone lavas, of an intermediate character between basalt and trachyte, and partly of the trachytic varieties of lava. Beds of scoriæ and sand, also, are very numerous, alternating with breccias formed of angular blocks of igneous rock. It is possible that some of the breccias may be referred to aqueous causes, as we have before seen that great floods do

* This drawing is part of a panoramic sketch which I made from the summit of the cone, December 1st, 1828, when every part of Etna was free from clouds except the Val del Bove.

occasionally sweep down the flanks of Etna when eruptions take place in winter, and when the snows are melted by lava.

Many of the angular fragments may have been thrown out by volcanic explosions, which, falling on the hardened surface of moving lava-currents, may have been carried to a considerable distance. It may also happen, that when lava advances very slowly, in the manner of the flow of 1819, described in the first volume *, the angular masses resulting from the frequent breaking of the mass as it rolls over upon itself, may produce these breccias. It is at least certain, that the upper portion of the lava-currents of 1811 and 1819, now consist of angular masses, to the depth of many yards.

D'Aubuisson has compared the surface of one of the ancient lavas of Auvergne to that of a river suddenly frozen over by the stoppage of immense fragments of drift-ice, a description perfectly applicable to these modern Etnean flows.

* Chap. xxi.

(95)

CHAPTER VIII.

Speculations on the origin of the Val del Bove on Etna—Subsidences—Antiquity
of the cone of Etna—Mode of computing the age of volcanos—Their growth
analogous to that of exogenous trees—Period required for the production of
the lateral cones of Etna—Whether signs of Diluvial Waves are observable
on Etna.

ORIGIN OF THE VAL DEL BOVE.

BEFORE concluding our observations on the cone of Etna, the
structure of which was considered in the last chapter, we desire
to call the reader's attention to several questions :—first, in
regard to the probable origin of the great valley already described ; secondly, whether any estimate can be made of the
length of the period required for the accumulation of the great
cone ; and, thirdly, whether there are any signs on the surface
of the older parts of the mountain, of those devastating waves
which, according to the theories of some geologists, have swept
again and again over our continents.

Origin of the Val del Bove.—We explained our reasons in
the last chapter for not assenting to the opinion, that the great
cavity on the eastern side of Etna was the hollow of an immense
crater, from which the volcanic masses of the surrounding walls
were produced. On the other hand, we think it impossible to
ascribe the valley to the action of running water alone; for if
it had been excavated exclusively by that power, its depth
would have increased in the descent; whereas, on the contrary,
the precipices are most lofty at the upper extremity, and diminish gradually on approaching the lower region of the volcano.

The structure of the surrounding walls is such as we should
expect to see exhibited on any other side of Etna, if a cavity
of equal depth should be caused, whether by subsidence, or by
the blowing up of part of the flanks of the volcano, or by either
of these causes co-operating with the removing action of running water.

It is recorded, as we have already seen in our history of earthquakes, that in the year 1772 an immense subsidence took place on Papandayang, the largest volcano in the island of Java, and that, during the catastrophe, an extent of ground, *fifteen miles in length and six in breadth*, gave way, so that no less than forty villages were engulphed, and the cone lost no less than four thousand feet of its height *.

Now we might imagine a similar event, or a series of subsidences to have formerly occurred on the eastern side of Etna, although such catastrophes have not been witnessed in modern times, or only on a very trifling scale. A narrow ravine, about a mile long, twenty feet wide, and from twenty to thirty-six in depth, has been formed, within the historical era, on the flanks of the volcano, near the town of Mascalucia; and a small circular tract, called the Cisterna, near the summit, sank down in the year 1792, to the depth of about forty feet, and left on all sides of the chasm a vertical section of the beds, exactly resembling those which are seen in the precipices of the Val del Bove. At some remote periods, therefore, we might suppose more extensive portions of the mountain to have fallen in during great earthquakes.

But some geologists will, perhaps, incline to the opinion, that the removed mass was blown up by paroxysmal explosions, such as that which, in the year 79, destroyed the ancient cone of Vesuvius, and gave rise to the escarpment of Somma. The Val del Bove, it will be remembered, lies within the zone of lateral eruptions, so that a repetition of volcanic explosions might have taken place, after which the action of running water may have contributed powerfully to degrade the rocks, and to transport the materials to the sea. We have before alluded to the effects of a violent flood, which swept through the Val del Bove in the year 1755, when a fiery torrent of lava had suddenly overflowed a great depth of snow in winter. †.

In the present imperfect state of our knowledge of the his-

* Vol. i. chap. xxv.
† See vol. i. chap. xxi.

tory of volcanos, we have some difficulty in deciding on the relative probability of these hypotheses ; but if we embrace the theory of explosions from below, the cavity would not constitute a *crater* in the ordinary acceptation of that term, still less would it accord with the notion of the so-called ' elevation craters.'

ANTIQUITY OF THE CONE OF ETNA.

We have stated in a former volume, that confined notions in regard to the quantity of past time, have tended, more than any other prepossessions, to retard the progress of sound theoretical views in Geology ; the inadequacy of our conceptions of the earth's antiquity having cramped the freedom of our speculations in this science, very much in the same way as a belief in the existence of a vaulted firmament once retarded the progress of astronomy. It was not until Descartes assumed the indefinite extent of the celestial spaces, and removed the supposed boundaries of the universe, that just opinions began to be entertained of the relative distances of the heavenly bodies ; and until we habituate ourselves to contemplate the possibility of an indefinite lapse of ages having been comprised within each of the more modern periods of the earth's history, we shall be in danger of forming most erroneous and partial views in Geology.

Mode of computing the age of volcanos.—If history had bequeathed to us a faithful record of the eruptions of Etna, and a hundred other of the principal active volcanos of the globe, during the last three thousand years,—if we had an exact account of the volume of lava and matter ejected during that period, and the times of their production,—we might, perhaps, be able to form a correct estimate of the average rate of the growth of a volcanic cone. For we might obtain a mean result from the comparison of the eruptions of so great a number of vents, however irregular might be the development of the igneous action in any one of them, if contemplated singly during a brief period.

It would be necessary to balance protracted periods of in-

H

action against the occasional outburst of paroxysmal explosions. Sometimes we should have evidence of a repose of seventeen centuries, like that which was interposed in Ischia, between the end of the fourth century, B. C., and the beginning of the fourteenth century of our era *. Occasionally a tremendous eruption, like that of Jorullo, would be recorded, giving rise, at once, to a considerable mountain.

If we desire to approximate to the age of a cone such as Etna, we ought first to obtain some data in regard to the thickness of matter which has been added during the historical era, and then endeavour to estimate the time required for the accumulation of such alternating lavas and beds of sand and scoriæ as are superimposed upon each other in the Val del Bove; afterwards we should try to deduce, from observations on other volcanos, the more or less rapid increase of burning mountains in all the different stages of their growth.

Mode of increase of volcanos analogous to that of exogenous trees.—There is a considerable analogy between the mode of increase of a volcanic cone and that of trees of *exogenous* growth. These trees augment, both in height and diameter, by the successive application externally of cone upon cone of new ligneous matter, so that if we make a transverse section near the base of the trunk, we intersect a much greater number of layers than nearer to the summit. When branches occasionally shoot out from the trunk they first pierce the bark, and then, after growing to a certain size, if they chance to be broken off, they may become inclosed in the body of the tree, as it augments in size, forming knots in the wood, which are themselves composed of layers of ligneous matter, cone within cone.

In like manner a volcanic mountain, as we have seen, consists of a succession of conical masses enveloping others, while lateral cones, having a similar internal structure, often project, in the first instance, like branches from the surface of the main cone, and then becoming buried again, are hidden like the knots of a tree.

---

* See vol. i. chap. xix.

We can ascertain the age of an oak or pine, by counting the number of concentric rings of annual growth, seen in a transverse section near the base, so that we may know the date at which the seedling began to vegetate. The Baobab-tree of Senegal (*Adansonia digitata*) is supposed to exceed almost any other in longevity; Adanson inferred that one which he measured, and found to be thirty feet in diameter, had attained the age of 5150 years. Having made an incision to a certain depth, he first counted three hundred rings of annual growth, and observed what thickness the tree had gained in that period. The average rate of growth of younger trees, of the same species, was then ascertained, and the calculation made according to a supposed mean rate of increase. De Candolle considers it not improbable, that the celebrated Taxodium of Chapultepec, in Mexico (*Cupressus disticha*, Linn.), which is one hundred and seventeen feet in circumference, may be still more aged *.

It is, however, impossible, until more data are collected respecting the average intensity of the volcanic action, to make anything like an approximation to the age of a cone like Etna, because, in this case, the successive envelopes of lava and scoriæ are not continuous, like the layers of wood in a tree, and afford us no definite measure of time. Each conical envelope is made up of a great number of distinct lava-currents and showers of sand and scoriæ, differing in quantity, and which may have been accumulated in unequal periods of time. Yet we cannot fail to form the most exalted conception of the antiquity of this mountain, when we consider that its base is about ninety miles in circumference; so that it would require ninety flows of lava, each a mile in breadth at their termination, to raise the present foot of the volcano as much as the average height of one lava-current.

There are no records within the historical era which lead to the opinion, that the altitude of Etna has materially varied within the last two thousand years. Of the eighty most con-

* On the Longevity of Trees, Bibliot. Univ., May, 1831.

spicuous minor cones which adorn its flanks, only one of the largest, Monti Rossi, has been produced within the times of authentic history.   Even this hill, thrown up in the year 1669, although 450 feet in height, only ranks as a cone of second magnitude.   Monte Minardo, near Bronte, rises, even now, to the height of 750 feet, although its base has been elevated by more modern lavas and ejections.   The dimensions of these larger cones appear to bear testimony to *paroxysms* of volcanic activity, after which we may conclude, from analogy, that the fires of Etna remained dormant for many years—since nearly a century of rest has sometimes followed a violent eruption in the historical era.   It must also be remembered, that of the small number of eruptions which occur in a century, one only is estimated to issue from the summit of Etna for every two, that proceed from the sides.   Nor do all the lateral eruptions give rise to such cones as would be enumerated amongst the smallest of the eighty hills above enumerated ; some produce merely insignificant monticules, soon destined to be buried, as we before explained.

How many years then must we not suppose to have been expended in the formation of the eighty cones ?   It is difficult to imagine that a fourth part of them have originated during the last thirty centuries.   But if we conjecture the whole of them to have been formed in twelve thousand years, how inconsiderable an era would this portion of time constitute in the history of the volcano !   If we could strip off from Etna all the lateral monticules now visible, together with the lavas and scoriæ that have been poured out from them, and from the highest crater, during the period of their growth, the diminution of the entire mass would be extremely slight !   Etna might lose, perhaps, several miles in diameter at its base, and some hundreds of feet in elevation, but it would still be the loftiest of Sicilian mountains, studded with other cones, which would be recalled, as it were, into existence by the removal of the rocks under which they are now buried.

There seems nothing in the deep sections of the Val del

Bove, to indicate that the lava currents of remote periods were greater in volume than those of modern times ; and there are abundant proofs that the countless beds of solid rock and scoriæ were accumulated, as now, in succession.   On the grounds, therefore, already explained, we must infer that a mass, eight thousand or nine thousand feet in thickness, must have required an immense series of ages anterior to our historical periods, for its growth ; yet the whole must be regarded as the product of a modern portion of the newer Pliocene epoch. Such, at least, is the conclusion that we draw from the geological data already detailed, which show that the oldest parts of the mountain, if not of posterior date to the marine strata which are visible around its base, were at least of coeval origin.

Whether signs of Diluvial Waves are observable on Etna.— Some geologists contend, that the sudden elevation of large continents from beneath the waters of the sea, have again and again produced waves which have swept over vast regions of the earth, and left enormous rolled blocks strewed over the surface *.   That there are signs of local floods of extreme violence, on various parts of the surface of the dry land, is incontrovertible, and in the former volumes we have pointed out causes which must for ever continue to give rise to such phenomena ; but for the proofs of these general cataclysms we have searched in vain.   It is clear that no devastating wave has passed over the forest zone of Etna, since any of the lateral cones before mentioned were thrown up ; for none of these heaps of loose sand and scoriæ could have resisted for a moment the denuding action of a violent flood.

To some, perhaps, it may appear that hills of such incoherent materials cannot be of immense antiquity, because the mere action of the atmosphere must, in the course of several thousand years, have obliterated their original forms.   But there is no weight in this objection, for the older hills are covered with trees and herbage, which protect them from waste ; and in

* Sedgwick, Anniv. Address to the Geol. Soc., p. 35.   Feb. 1831.

regard to the newer ones, such is the porosity of their component materials, that the rain which falls upon them is instantly
absorbed, and, for the same reason that the rivers on Etna have
a subterranean course, there are none descending the sides of
the minor cones.

No sensible alteration has been observed in the form of these
cones since the earliest periods of which there are memorials;
and we see no reason for anticipating, that in the course of the
next ten thousand or twenty thousand years they will undergo
any great alteration in their appearance, unless they should be
shattered by earthquakes, or covered by volcanic ejections.

We shall afterwards point out, that, in other parts of Europe,
similar loose cones of scoriæ, which we believe to be of higher
antiquity than the whole mass of Etna, stand uninjured at
inferior elevations above the level of the sea.

Origin of the newer Pliocene strata of Sicily—Growth of submarine formations gradual—Rise of the same above the level of the sea probably caused by subterranean lava—Igneous newer Pliocene rocks, formed at great depths, exceed in volume the lavas of Etna—Probable structure of these recent subterranean rocks—Changes which they may have superinduced upon strata in contact—Alterations of the surface during and since the emergence of the newer Pliocene strata—Forms of the Sicilian valleys—Sea cliffs—Proofs of successive elevation—Why the valleys in the newer Pliocene districts correspond in form to those in regions of higher antiquity—Migrations of animals and plants since the emergence of the newer Pliocene strata—Some species older than the stations they inhabit—Recapitulation.

ORIGIN OF THE NEWER PLIOCENE STRATA OF SICILY.

HAVING in the last two chapters described the tertiary formations of the Val di Noto and Valdemone, both igneous and aqueous, we shall now proceed more fully to consider their origin, and the manner in which they may be supposed to have assumed their present position. The consideration of this subject may be naturally divided into three parts: first, we shall inquire in what manner the submarine formations were accumulated beneath the waters; secondly, whether they emerged slowly or suddenly, and what modifications in the earth's crust, at considerable depths below the surface, may be indicated by their rise; thirdly, the mutations which the surface and its inhabitants have undergone during and since the period of emergence.

Growth of Submarine formations.—First, then, we are to inquire in what manner the subaqueous masses, whether volcanic or sedimentary, may have been formed. On this subject we have but few observations to make, for by reference to our former volumes, the reader will learn how a single stratum, whether of sand, clay, or limestone, may be thrown down at the bottom of the sea, and how shells and other organic remains

may become imbedded therein.   He will also understand how
one sheet of lava, or bed of scoriæ and volcanic sand, may be
spread out over a wide area, and how, at a subsequent period,
a second bed of sand, clay, or limestone, or a second lava-
stream may be superimposed, so that in the lapse of ages a
mountain mass may be produced.

It is enough that we should behold a single course of bricks
or stones laid by the mason upon another, in order to comprehend how a massive edifice, such as the Coliseum at Rome,
was erected ; and we can have no difficulty in conceiving that
a sea, three hundred or four hundred fathoms deep, might be
filled up by sediment and lava, provided we admit an indefinite
lapse of ages for the accumulation of the materials.

The sedimentary and volcanic masses of the newer Pliocene
era, which, in the Val di Noto, attain the thickness of two
thousand feet, are subdivided into a vast number of strata and
lava-streams, each of which were originally formed on the subaqueous surface, just as the tuffs and lavas whereof sections
are laid open in the Val del Bove, were each in their turn external additions to the Etnean cone.

It is also clear, that before any part of the mass of submarine
origin began to rise above the waters, the uppermost stratum
of the whole must have been deposited ; so that if the date of
the origin of these masses be comparatively recent, still more
so is the period of their rise above the level of the sea.

Subaqueous formations how raised.—In what manner, then,
and by what agency, did this rise of the subaqueous formations take place ? We have seen that since the commencement
of the present century, an immense tract of country in Cutch,
more than fifty miles long and sixteen broad, was permanently
upraised to the height of ten feet above its former position,
and the earthquake which accompanied this wonderful variation of level, is reported to have terminated by a volcanic eruption at Bhooi. We have also seen *, that when the Monte
Nuovo was thrown up, in the year 1538, a large fissure ap-

* Vol. i, chap. xix.

proached the small town of Tripergola, emitting a vivid light, and throwing out ignited sand and scoriæ. At length this opening reached a shallow part of the sea close to the shore, and then widened into a large chasm, out of which were discharged blocks of lava, pumice, and ashes. But no current of melted matter flowed from the orifice, although it is perfectly evident that lava existed below in a fluid state, since so many portions of it were cast up in the form of scoriæ into the air. We have shown that the coast near Puzzuoli rose, at that time, to the height of more than twenty feet above its former level, and that it has remained permanently upheaved to this day *.

On a review of the whole phenomena, it appears most probable that the elevated country was forced upwards by lava which did not escape, but which, after causing violent earthquakes, during several preceding months, produced at length a fissure from whence it discharged gaseous fluids, together with sand and scoriæ. The intruded mass then cooled down at a certain distance below the uplifted surface, and constituted a solid and permanent foundation.

If an habitual vent had previously existed near Puzzuoli, such as we may suppose to remain always open in the principal ducts of Vesuvius or Etna, the lava might, perhaps, have flowed over upon the surface, instead of heaving upwards the superficial strata. In that case, there might have been the same conversion of sea into land, the only difference being, that the lava would have been uppermost, instead of the tufaceous strata containing shells, now seen in the plain of La Starza, and on the site of the Temple of Serapis.

Subterranean lava the upheaving cause.—The only feasible theory, indeed, that has yet been proposed, respecting the causes of the permanent rise of the bed of the sea, is that which refers the phenomenon to the generation of subterranean lava. We have stated, in the first volume, that the regions now habitually convulsed by earthquakes, include within them the site of all the active volcanos. We know that the expansive force of

* Vol. i. chap. xxv.

volcanic heat is sufficiently great to overcome the resistance of columns of lava, several miles or leagues in height, forcing them up from great depths, and causing the fluid matter to flow out upon the surface. To imagine, therefore, that this same power, which is so frequently exerted in different parts of the globe, should occasionally propel a column of lava to a considerable height, yet be unable to force it through the superincumbent rocks, is quite natural.

Whenever the superimposed masses happen to be of a yielding and elastic nature, they will bend, and instead of breaking, so as to afford an escape to the melted matter through a fissure, they will allow it to accumulate in large quantities beneath the surface, sometimes in amorphous masses, and sometimes in horizontal sheets. So long as such sheets of matter retain their fluidity, and communicate with the column of lava which is still urged upwards, they must exert an enormous hydrostatic pressure on the overlying mass, tending to elevate it, and an equal force on the subjacent beds pressing them down, and probably rendering them more compact. If we consider how great is the volume of lava that sometimes flows out on the surface from volcanic vents, we must expect that it will produce great changes of level so often as its escape is impeded.

Let us only reflect on the magnitude of Iceland, an island two hundred and sixty miles long by two hundred in breadth, and which rises, at some points, to the height of six thousand feet above the level of the sea. Nearly the entire mass is represented to be of volcanic origin; but even if we suppose some parts to consist of aqueous deposits, still that portion may be more than compensated by the great volume of lava which must have been poured out upon the bottom of the surrounding sea during the growth of the entire island; for we know that submarine eruptions have been considerable near the coast during the historical era. Now if the whole of this lava had been prevented from reaching the surface, by the weight and tenacity of certain overlying rocks, it might have given rise to the gradual elevation of a tract of land nearly as large as Iceland. We say *nearly*, because the lava which cooled down beneath the surface, and under considerable pressure, would be more compact than the same when poured out in the open air, or in a sea of moderate depth, or shot up into the atmosphere by the explosive force of elastic vapours, and thus converted into sand and scoriæ.

According to this theory, we must suppose the action of the upheaving power to be intermittent, and, like ordinary volcanic eruptions, to be reiterated again and again in the same region, at unequal intervals of time and with unequal degrees of force.

If we follow this train of induction, which appears so easy and natural, to what important conclusions are we led ! The reader will bear in mind that the tertiary strata have attained in the central parts of Sicily, as at Castrogiovanni, for example, an elevation of about three thousand feet above the level of the sea, and a height of from fifty to two thousand feet in different parts of the Val di Noto. In this country, therefore, we must suppose a solid support of igneous rock to have been successively introduced into part of the earth's crust immediately subjacent, equal in volume to the upraised tract, and this generation of subterranean rock must have taken place during the latter part of the newer Pliocene period. The dimensions of the Etnean cone shrink into insignificance, in comparison to the volume of this subterranean lava; and, however staggering the inference might at first appear, that the oldest foundations of Etna were laid subsequently to the period when the Mediterranean became inhabited by the living species of testacea and zoophytes, yet we may be reconciled to such conclusions, when we find incontestable proofs of still greater revolutions beneath the surface within the same modern period.

Probable structure of the recent subterranean rocks of fusion. —Let us now inquire what form these unerupted newer Pliocene lavas of Sicily have assumed ? For reasons already explained, we may infer that they cannot have been converted into tuffs and peperinos, nor can we imagine that, under enormous pressure, they could have become porous, since we observe, that the lava which has cooled down under a moderate degree of pressure, in the dikes of Etna and Vesuvius, has a compact and porphyritic texture, and is very rarely porous or cellular. No signs of volcanic sand, scoriæ, breccia, or conglomerate are to be looked for, nor any of stratification, for all these imply formation in the atmosphere, or by the agency of water. The only proofs that we can expect to find of the *successive* origin of different parts of the fused mass, will be confined to the occasional passage of veins through portions previously consolidated. This consolidation would take place with extreme slowness, when nearer the source of volcanic heat and under enormous pressure, so that we must anticipate a perfectly crystalline and compact texture in all these subterranean products.

Now geologists have discovered, as we before stated, great abundance of crystalline and unstratified rocks in various parts of the globe, and these masses are particularly laid open to our view in those mountainous districts where the crust of the earth has undergone the greatest derangement. These rocks vary considerably in composition, and have received many names, such as granite, syenite, porphyry, and others. That they must have been formed by igneous fusion, and at many distinct eras, is now admitted; and their highly crystalline texture is such as might result from cooling down slowly from an intensely-heated state. They answer, therefore, admirably to the conditions required by the above hypothesis, and we therefore deem it probable that similar rocks have originated in the nether regions below the island of Sicily, and have attained a thickness of from one thousand to three thousand feet, since the newer Pliocene strata were deposited.

It is, moreover, very probable, that these fused masses have come into contact with subaqueous deposits far below the surface, in which case they may, in the course of ages, have greatly altered their structure, just as dikes of lava render more crystalline the stratified masses which they traverse, and obliterate all traces of their organic remains.

Suppose some of these changes to have been superinduced upon subaqueous deposits underlying the tertiary formations of Sicily, it is important to reflect that in that case no geological proofs would remain of the era when the alterations had taken place; and if, at some future period, the whole island should be uplifted, and these rocks of fusion, together with the altered strata, should be brought up to the surface, it would not be apparent that they had assumed their crystalline texture in the newer Pliocene period. For aught that would then appear, they might have acquired their peculiar mineral texture at epochs long anterior, and might be supposed to have been formed before the planet was inhabited by living beings; instead of having originated at an era long subsequent to the introduction of the *existing species*.

CHANGES OF THE SURFACE DURING AND SINCE THE EMER-GENCE OF THE NEWER PLIOCENE STRATA.

Valleys.—Geologists who are accustomed to attribute a great portion of the inequalities of the earth's surface to the excavating power of running water during a long series of ages, will probably look for the signs of remarkable freshness in the aspect of countries so recently elevated as the parts of Sicily already described. There is, however, nothing in the external configuration of that country which would strike the eye of the most practised observer, as peculiar and distinct in character from many other districts in Europe which are of much higher antiquity. The general outline of the hills and valleys would accord perfectly well with what may often be observed in regard to other regions of equal altitude above the level of the sea.

It is true that, towards the central parts of the island where the argillaceous deposits are of great thickness, as around Castrogiovanni, Caltanisetta, and Piazza, the torrents are observed annually to deepen the ravines in which they flow, and the traveller occasionally finds that the narrow mule-path, instead of winding round the head of a ravine, terminates abruptly in a deep trench which has been hollowed out, during the preceding winter, through soft clay. But throughout a great part of Italy, where the marls and sands of the Subapennine hills are elevated to considerable heights, the same rapid degradation is often perceived.

In the limestone districts of the Val di Noto, the strata are for the most part nearly horizontal, and on each side of the valley form a succession of ledges or small terraces, instead of descending in a gradual slope towards the river-plain in the manner of the argillaceous formations. When there is a bend in the valley, the exact appearance of an amphitheatre with a range of marble seats is produced. A good example of this configuration occurs near the town of Melilli, in the Val di Noto, as seen in the annexed view (No. 23). In the south of

No. 23.

Valley called Gozzo degli Martiri, below Melilli.

the island, as near Spaccaforno, Scicli and Modica, precipitous rocks of white limestone, ascending to the height of five hundred feet, have been carved out into the same form.

A careful examination of the mode of decomposition of the rock would be requisite, in order fully to explain this phenomenon. There is probably a tendency to a vertical fracture in this as in many other limestones, which, when exposed to the action of frost, scale off in small fragments at right angles to the plane of stratification. It might have been expected that, in this case, a talus composed of a breccia of the limestone would be found on each ledge, so that the slope would become gradual, but perhaps the fragments, instead of accumulating, may decompose and be washed away by the heavy rains.

The line of some of the valleys near Lentini has evidently been determined mainly by the direction of the elevatory force, as there is an anticlinal dip in the strata on either side of the valley. The same is, probably, the case in regard to the great valley of the Anapo, which terminates at Syracuse.

Sea-cliffs—proofs of successive elevation.—No decisive evidence could be looked for in the form of the valleys to determine the question, whether the subterranean movements which upheaved the newer Pliocene strata in Sicily were very numerous or few in number. But we find the signs of two periods of elevation in a long range of inland cliff on the east side of the Val di Noto, both to the north of Syracuse, beyond Melilli, and to the south beyond the town of Noto. The great limestone formation before mentioned, terminates suddenly towards the sea in a lofty precipice, *a, b,* which varies in height from five

No. 24.

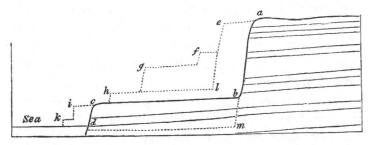

hundred to seven hundred feet, and may remind the English geologist of some of the most perpendicular escarpments of our chalk and oolite. Between the base of the precipice *a, b,* and the sea, is an inferior platform, *c, b,* consisting of similar white limestone. All the strata dip towards the sea, but are

usually inclined at a very slight angle; they are seen to extend
uninterruptedly from the base of the escarpment into the platform, showing distinctly that the lofty cliff was not produced
by a fault or vertical shift of the beds, but by the removal of
a considerable mass of rock. Hence we must conclude that
the sea, which is now undermining the cliffs of the Sicilian
coast, reached at some former period the base of the precipice
$a, b,$ at which time the surface of the terrace $c, b,$ must have
constituted the bottom of the Mediterranean. Here, then, we
have proofs of at least two elevations, but there may have been
fifty others, for the encroachment of the sea tends to obliterate
all signs of a *succession* of cliffs.

Suppose, for example, that a series of escarpments $e, f, g, h,$
once existed, and that during a long interval, free from subterranean movements, the sea had time to advance along the line
$c, b,$ all those ancient cliffs must then have been swept away one
after the other, and reduced to the single precipice $a, b.$ There
may have been an antecedent period when the sea advanced
along the line $h, l,$ substituting the single cliff $e, l,$ for the series
$e, f, g.$

We may also imagine that the present cliffs may be the
result of the union of several lines of smaller cliffs and terraces,
which may once have been produced by a succession of elevatory movements. For example, the waves may have carried
away the cliffs $k, i,$ in advancing to $c, d.$ In the same manner
they may ultimately remove the mass $c, b, m, d,$ and then the
platform $c, b,$ will disappear, and the precipice $a, m,$ will be
substituted for $a, b.$

We have stated, in the first volume, that the waves washed
the base of the inland cliff near Puzzuoli, in the Bay of Baiæ,
within the historical era, and that the retiring of the sea was
caused, in the sixteenth century, by an upheaving of the land
to an elevation of twenty feet above its original level. At that
period, a terrace twenty feet high in some parts, was laid dry
between the sea and the cliff, but the Mediterranean is hasten-
ing to resume its former position, when the terrace will be

destroyed, and every trace of the *successive* rise of the land will be obliterated.

We have been led into these observations, in order to show that the principal features in the physical geography of Sicily are by no means inconsistent with the hypothesis of the successive elevation of the country by the intermittent action of ordinary earthquakes *. On the other hand, we consider the magnitude of the valleys, and their correspondence in form with those of other parts of the globe, to lend countenance to the theory of the slow and gradual rise of subaqueous strata.

We have remarked in the first volume †, that the excavation of valleys must always proceed with the greatest rapidity when the levels of a country are undergoing alteration from time to time by earthquakes, and that it is principally when a country is rising or sinking by successive movements, that the power of aqueous causes, such as tides, currents, rivers, and land-floods, is exerted with the fullest energy.

In order to explain the present appearance of the surface, we must first go back to the time when the Sicilian formations were mere shoals at the bottom of the sea, in which the currents may have scooped out channels here and there. We must next suppose these shoals to have become small islands of which the cliffs were thrown down from time to time, as were those of Gian Greco, in Calabria, during the earthquake of 1783. The waves and currents would then continue their denuding action during the emergence of these islands, until at length, when the intervening channels were laid dry, and rivers began to flow, the deepening and widening of the val-

* Since writing the above I have read the excellent memoir of M. Boblaye, on the alterations produced by the sea on calcareous rocks on the shores of Greece. By examining the line of littoral caverns worn by the waves in cliffs composed of the harder limestones, together with the modes of decomposition of the rock, acted upon by the spray and sea air, as well as lithodomous perforations, and other markings, he has proved that there are four or five distinct ranges of ancient sea cliffs, one above the other, at various elevations in the Morea, which attest as many *successive* elevations of the country. Journal de Géologie, No. 10. Feb. 1831.

† Chap. xxiv.

leys by rivers and land-floods would proceed in the same manner as in modern times in Calabria, according to our former description *.

Before a tract could be upraised to the height of several thousand feet above the level of the sea, the joint operation of running water and subterranean movements must greatly modify the physical geography ; but when the action of the volcanic forces has been suspended, when a period of tranquillity succeeds, and the levels of the land remain fixed and stationary, the erosive power of water must soon be reduced to a state of comparative equilibrium.   For this reason, a country that has been raised at a very remote period to a considerable height above the level of the sea, may present nearly the same external configuration as one that has been more recently uplifted to the same height.

In other words, the time required for the raising of a mass of land to the height of several hundred yards must usually be so enormous (assuming as we do that the operation is effected by ordinary volcanic forces), that the aqueous and igneous agents will have time before the elevation is completed to modify the surface, and imprint thereon the ordinary forms of hill and valley, by which our continents are diversified.   But after the cessation of earthquakes these causes of change will remain dormant, or nearly so.   The greater part, therefore, of the earth's surface will at each period be at rest, simply retaining the features already imparted to it, while smaller tracts will assume, as they rise successively from the deep, a configuration perfectly analogous to that by which the more ancient lands were previously distinguished.

Migration of animals and plants.—The changes which, according to the views already explained, have been brought about in the earth's crust by the agency of volcanic heat, cannot fail to strike the imagination, when we consider how recent in the calendar of nature is the epoch to which we refer them. But if we turn our thoughts to the organic world, we shall feel,

* Chap. xxiv.

perhaps, no less surprise at the great vicissitude which it has undergone during the same period.

We have seen that a large portion of Sicily has been converted from sea to land since the Mediterranean was peopled with the living species of testacea and zoophytes. The newly emerged surface, therefore, must, during this modern zoological epoch, have been inhabited for the first time with the terrestrial plants and animals which now abound in Sicily. It is fair to infer, that the existing terrestrial species are, for the most part, of as high antiquity as the marine, and if this be the case, a large proportion of the plants and animals, now found in the tertiary districts in Sicily, must have inhabited the earth before the newer Pliocene strata were raised above the waters. The plants of the Flora of Sicily are common, almost without exception, to Italy or Africa, or some of the countries surrounding the Mediterranean *, so that we may suppose the greater part of them to have migrated from pre-existing lands, just as the plants and animals of the Phlegræan fields have colonized Monte Nuovo, since that mountain was thrown up in the sixteenth century.

We are brought, therefore, to admit the curious result, that the flora and fauna of the Val di Noto, and some other mountainous regions of Sicily, are of higher antiquity than the country itself, having not only flourished before the lands were raised from the deep, but even before they were deposited beneath the waters. Such conclusions throw a new light on the adaptation of the attributes and migratory habits of animals and plants, to the changes which are unceasingly in progress in the inanimate world. It is clear that the duration of species is so great, that they are destined to outlive many important revolutions in the physical geography of the earth, and hence those innumerable contrivances for enabling the subjects of the animal and vegetable creation to extend their range, the inhabitants

* Professor Viviani of Genoa informed me, that, considering the great extent of Sicily, it was remarkable that its flora produced scarcely any, *if any peculiar indigenous* species, whereas there are several in Corsica, and some other Mediterranean islands.

I 2

of the land being often carried across the ocean, and the aquatic
tribes over great continental spaces *. It is obviously expedient
that the terrestrial and fluviatile species should not only be fitted
for the rivers, valleys, plains, and mountains which exist at the
era of their creation, but for others that are destined to be formed
before the species shall become extinct; and, in like manner, the
marine species are not only made for the deep or shallow
regions of the ocean at the time when they are called into being,
but for tracts that may be submerged or variously altered in
depth during the time that is allotted for their continuance on'
the globe.

Recapitulation.—We may now briefly recapitulate some of
the most striking results which we have deduced from our investigation of a single district where the newer Pliocene strata
are largely developed.

In the first place, we have seen that a stratified mass of solid
limestone, attaining sometimes a thickness of eight hundred feet
and upwards, has been gradually deposited at the bottom of the
sea, the imbedded fossil shells and corallines being almost all of
recent species. Yet these fossils are frequently in the state of
mere casts, so that in appearance they correspond very closely
to organic remains found in limestones of very ancient date.

2dly. In some localities the limestone above-mentioned
alternates with volcanic rocks such as have been formed by
submarine eruptions, recurring again and again at distant intervals of time.

3dly. Argillaceous and sandy deposits have also been produced during the same period, and their accumulation has also
been accompanied by submarine eruptions. Masses of mixed
sedimentary and igneous origin, at least two thousand feet in
thickness, can thus be shown to have accumulated since the
sea was peopled with the greater number of the aquatic species
now living.

4thly. These masses of submarine origin have, since their
formation, been raised to the height of two thousand or three

* See vol. ii., chapters v., vi., and vii.

thousand feet above the level of the sea, and this elevation implies an extraordinary modification in the state of the earth's
crust at some unknown depth beneath the tract so upheaved.

5thly. The most probable hypothesis in regard to the nature
of this change, is the successive generation and forcible intrusion into the inferior parts of the earth's crust of lava which,
after cooling down, may have assumed the form of crystalline
unstratified rock, such as is frequently exhibited in those
mountainous parts of the globe where the greatest alterations
of level have taken place.

6thly. Great inequalities must have been caused on the
surface of the new-raised lands during the emergence of the
newer Pliocene strata, by the action of tides, currents, and
rivers, combined with the disturbing and dislocating force of
the elevatory movements.

7thly. There are no features in the forms of the valleys and
sea-cliffs thus recently produced, which indicate the sudden
rise of the strata to the whole or the greater part of their present altitude, while there are some proofs of distinct elevations
at successive periods.

8thly. We may infer that the species of terrestrial and
fluviatile animals and plants which now inhabit extensive
districts, formed during the newer Pliocene era, were in existence not only before the new strata were raised, but before
their materials were brought together at the bottom of the sea.

Tertiary formations of Campania—Comparison of the recorded changes in this region with those commemorated by geological monuments—Differences in the composition of Somma and Vesuvius—Dikes of Somma, their origin—Cause of the parallelism of their opposite sides—Why coarser grained in the centre—Minor cones of the Phlegræan Fields—Age of the volcanic and associated rocks of Campania—Organic remains—External configuration of the country, how produced—No signs of diluvial waves—Marine Newer Pliocene strata visible only in countries of earthquakes—Illustrations from Chili—Peru—Parallel roads of Coquimbo —West-Indian archipelago — Honduras — East-Indian archipelago—Red Sea.

TERTIARY FORMATIONS OF CAMPANIA.

Comparison of recorded changes with those commemorated by geological monuments.—IN the first volume we traced the various changes which the volcanic region of Naples is known to have undergone during the last 2000 years, and, imperfect as are our historical records, the aggregate effect of igneous and aqueous agency, during that period, was shown to be far from insignificant. The rise of the modern cone of Vesuvius, since the year 79, was the most memorable event during those twenty centuries ; but in addition to this remarkable pheno-menon, we enumerated the production of several new minor cones in Ischia, and of the Monte Nuovo, in the year 1538. We described the flowing of lava-currents upon the land and along the bottom of the sea, the showering down of volcanic sand, pumice, and scoriæ, in such abundance that whole cities were buried,—the filling up or shoaling of certain tracts of the sea, and the transportation of tufaceous sediment by rivers and land-floods. We also explained the evidence in proof of a permanent alteration of the relative levels of the land and sea in several localities, and of the same tract having, near Puz-zuoli, been alternately upheaved, and depressed, to the amount of more than 20 feet. In connexion with these convulsions, we pointed out that, on the shores of the Bay of Baiæ, there are recent tufaceous strata filled with fabricated articles, mingled with marine shells. It was also shown that the sea has been making gradual advances upon the coast, not only sweeping away the soft tuffs of the Bay of Baiæ, but excavating precipitous cliffs, where the hard Ischian and Vesuvian lavas have flowed down into the deep.

These events, we shall be told, although interesting, are the results of operations on a very inferior scale to those indicated by geological monuments. When we examine this same region, it will be said, we find that the ancient cone of Vesuvius, called Somma, is larger than the modern cone, and is intersected by a greater number of dikes,—the hills of unknown antiquity, such as Astroni, the Solfatara, and Monte Barbaro, formed by separate eruptions, in different parts of the Phlegræan fields, far outnumber those of similar origin, which are recorded to have been thrown up within the historical era. In place of modern tuffs of slight thickness, and single flows of lava, we find, amongst the older formations, hills from 500 to more than 2000 feet in height, composed of an immense series of tufaceous strata, alternating with distinct lava-currents. We have evidence that in the lapse of past ages, districts, not merely a few miles square, were upraised to the height of 20 or 30 feet above their former level, but extensive and mountainous countries were uplifted to an elevation of more than 1000 feet, and at some points more than 2000 feet above the level of the sea.

These and similar objections are made by those who compare the modern effects of igneous and aqueous causes, not with a part but with the whole results of the same agency in antecedent ages. Thus viewed in the aggregate, the leading geological features of each district must always appear to be on a colossal scale, just as a large edifice of striking architectural beauty seems an effort of superhuman power, until we reflect on the innumerable minute parts of which it is composed. A mountain mass, so long as the imagination is occu-

pied in contemplating the gigantic whole, must appear the
work of extraordinary causes, but when the separate portions
of which it is made up are carefully studied, they are seen to
have been formed successively, and the dimensions of each part,
considered singly, are soon recognized to be comparatively insignificant, and it appears no longer extravagant to liken them
to the recorded effects of ordinary causes.

Difference in the composition of Somma and Vesuvius.

As no traditional accounts have been handed down to us of
the eruptions of the ancient Vesuvius, from the times of the
earliest Greek colonists, the volcano must have been dormant
for many centuries, perhaps for thousands of years, previous to
the great eruption in the reign of Titus.   But we shall afterwards show that there are sufficient grounds for presuming this
mountain, and the other igneous products of Campania, to have
been produced during the Newer Pliocene period.

We stated in the first volume *, that the ancient and modern
cones of Vesuvius were each a counterpart of the other in
structure ; we may now remark that the principal point of
difference consists in the greater abundance in the older cone
of fragments of stratified rocks ejected during eruptions.   We
may easily conceive that the first explosions would act with the
greatest violence, rending and shattering whatever solid masses
obstructed the escape of lava and the accompanying gases, so
that great heaps of ejected pieces of sedimentary rock would
naturally occur in the tufaceous breccias formed by the earliest
eruptions.   But when a passage had once been opened and an
habitual vent established, the materials thrown out would consist of liquid lava, which would take the form of sand and
scoriæ, or of angular fragments of such solid lavas as may have
choked up the vent.

Among the angular fragments of solid rock which abound
in the tufaceous breccias of Somma, none are more common
than a saccharoid dolomite, supposed to have been derived

* Chap. xx.

from an ordinary limestone altered by heat and volcanic vapours.

Carbonate of lime enters into the composition of so many of
the simple minerals found in Somma, that M. Mitscherlich,
with much probability, ascribes their great variety to the action
of the volcanic heat on subjacent masses of limestone.

Dikes of Somma.—The dikes seen in the great escarpment
which Somma presents towards the modern cone of Vesuvius
are very numerous. They are for the most part vertical, and
traverse at right angles the beds of lava, scoriæ, volcanic
breccia, and sand, of which the ancient cone is composed. They
project in relief several inches, or sometimes feet, from the face
of the cliff, like the dikes of Etna already described (see wood-
cut No. 19), being, like them, extremely compact, and less
destructible than the intersected tuffs and porous lavas. In
height they vary from a few yards to 500 feet, and in breadth
from one to twelve feet. Many of them cut all the inclined beds
in the escarpment of Somma from top to bottom, others stop
short before they ascend above half way, and a few terminate
at both ends, either in a point or abruptly. In mineral composition they scarcely differ from the lavas of Somma, the rock
consisting of a base of leucite and augite, through which large
crystals of augite and some of leucite are scattered *. Examples are not rare of one dike cutting through another, and in
one instance a shift or fault is seen at the point of intersection.
We observed before †, when speaking of the dikes of the
modern cone of Vesuvius, that they must have been produced
by the filling up of open fissures by liquid lava. In some examples, however, the rents seem to have been filled laterally.

* Consult the valuable memoir of M. L. A. Necker, Mém. de la Soc. de Phys.
et d'Hist. Nat. de Génève, tome ii. part i., Nov. 1822.

† Vol. i. chap. xx.

No. 25.

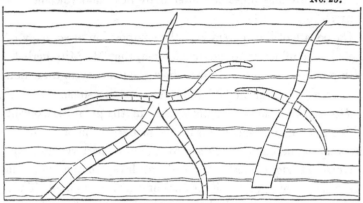

Dikes or veins at the Punto del Nasone on Somma.

The reader will remember our description of the manner in which the plain of Jerocarne, in Calabria, was fissured by the earthquake of 1783 *, so that the Academicians compared it to the cracks in a broken pane of glass. If we suppose the side walls of the ancient crater of Vesuvius to have been cracked in like manner, and the lava to have entered the rents and become consolidated, we can explain the singular form of the veins figured in the accompanying wood-cut †.

Parallelism of their opposite sides.—Nothing is more remarkable than the parallelism of the opposite sides of the dikes, which usually correspond with as much regularity as the two opposite faces of a wall of masonry. This character appears at first the more inexplicable, when we consider how jagged and uneven are the rents caused by earthquakes in masses of heterogeneous composition like those composing the cone of Somma; but M. Necker has offered an ingenious and, we think, satisfactory explanation of the phenomenon. He refers us to Sir W. Hamilton's account of an eruption of Vesuvius in the year 1779, who records the following facts. ' The lavas, when they either boiled over the crater, or broke out from the conical parts of the volcano, constantly formed channels as regular as if they had been cut by art, down the steep part of

* See vol. i. chap. xxiv., wood-cut No. 22.
† From a drawing of M. Necker, ibid.

the mountain, and, whilst in a state of perfect fusion, continued their course in those channels, which were sometimes full to the brim, and at other times more or less so according to the quantity of matter in motion.

' These channels, upon examination after an eruption, I have found to be in general from two to five or six feet wide, and seven or eight feet deep. They were often hid from the sight by a quantity of scoriæ that had formed a crust over them, and the lava, having been conveyed in a covered way for some yards, came out fresh again into an open channel. After an eruption I have walked in some of those subterraneous or covered galleries, which were exceedingly curious, the sides, top, and bottom, *being worn perfectly smooth and even* in most parts, by the violence of the currents of the red-hot lavas, which they had conveyed for many weeks successively.'

In another place, in the same memoir, he describes the liquid and red-hot matter as being received ' into a regular channel, raised upon a sort of wall of scoriæ and cinders, almost perpendicularly, of about the height of eight or ten feet, resembling much an ancient aqueduct *.'

Now, if the lava in these instances had not run out from the covered channel, in consequence of the declivity whereon it was placed—if, instead of the space being left empty, the lava had been retained within until it cooled and consolidated, it would then have constituted a small dike with parallel sides. But the walls of a vertical fissure through which lava has ascended in its way to a volcanic vent, must have been exposed to the same erosion as the four sides of the channels before adverted to. The prolonged and uniform friction of the heavy fluid as it flows upwards cannot fail to wear and smooth down the surfaces on which it rubs, and the intense heat must melt all such masses as project and obstruct the passage of the incandescent fluid.

We do not mean to assert that the sides of fissures caused by earthquakes are never smooth and parallel, but they are usually uneven, and are often seen to have been so where volcanic

* Phil. Trans., vol. lxx. 1780.

or *trap* dikes are as regular in shape as those of Somma. The solution, therefore, of this problem, in reference to the modern dikes, is most interesting, as being of very general application in geology.

Varieties in their texture.—Having explained the origin of the parallelism of the sides of a dike, we have next to consider the difference of its texture at the edges and in the middle. Towards the centre, observes M. Necker, the rock is coarser grained, the component elements being in a far more crystalline state, while at the edge the lava is sometimes vitreous and always finer grained. A thin parting band, approaching in its character to pitchstone, occasionally intervenes on the contact of the vertical dike and intersected beds. M. Necker mentions one of these at the place called Primo Monte, in the Atrio del Cavallo; I saw three or four others in different parts of the great escarpment. These phenomena are in perfect harmony with the results of the experiments of Sir James Hall and Mr. Gregory Watt, which have shown that a glassy texture is the effect of sudden cooling, and that, on the contrary, a crystalline grain is produced where fused minerals are allowed to consolidate slowly and tranquilly under high pressure.

It is evident that the central portion of the lava in a fissure would, during consolidation, part with its heat more slowly than the sides, although the contrast of circumstances would not be so great as when we compare the lava at the bottom and at the surface of a current flowing in the open air. In this case the uppermost part, where it has been in contact with the atmosphere, and where refrigeration has been most rapid, is always found to consist of scoriform, vitreous, and porous lava, while at a greater depth the mass assumes a more lithoidal structure, and then becomes more and more stony as we descend, until at length we are able to recognize with a magnifying glass the simple minerals of which the rock is composed. On penetrating still deeper, we can detect the constituent parts by the naked eye, and in the Vesuvian currents distinct crystals of augite and leucite become apparent.

The same phenomenon, observes M. Necker, may readily be exhibited on a smaller scale, if we detach a piece of liquid lava from a moving current. The fragment cools instantly, and we find the surface covered with a vitreous coat, while the interior, although extremely fine grained, has a more stony appearance.

It must, however, be observed, that although the lateral portions of the dikes are finer grained than the central, yet the vitreous parting layer before alluded to is extremely rare. This may, perhaps, be accounted for, as the above-mentioned author suggests, by the great heat which the walls of a fissure may acquire before the fluid mass begins to consolidate, in which case the lava, even at the sides, would cool very slowly. Some fissures, also, may be filled from above; and in this case the refrigeration at the sides would be more rapid than when the melted matter flowed upwards from the volcanic foci, in an intensely-heated state.

The rock composing the dikes of Somma is far more compact than that of ordinary lava, for the column of melted matter in a fissure greatly exceeds an ordinary stream of lava in weight, and the great pressure checks the expansion of those gases which give rise to vesicles in lava.

There is a tendency in almost all the Vesuvian dikes to divide into horizontal prisms, which are at right angles to the cooling surfaces *, a phenomenon in accordance with the formation of vertical columns in horizontal beds of lava.

Minor cones of the Phlegræan Fields.—In the volcanic district of Naples there are a great number of conical hills with craters on their summits, which have evidently been produced by one or more explosions, like that which threw up the Monte Nuovo in 1538. They are composed of trachytic tuff, which is loose and incoherent, both in the hills and, to a certain depth, in the plains around their base, but which is indurated below. It is suggested by Mr. Scrope, that this difference may be owing to the circumstance of the volcanic vents having burst out in a shallow sea, as was the case with Monte Nuovo, where there is a similar foundation of hard tuff, under a covering of

* See wood-cut No. 25.

loose lapilli. The subaqueous part may have become solid by
an aggregative process like that which takes place in the setting
of mortar, while the rest of the ejections, having accumulated
on dry land when the cone was raised above the water, may
have remained in a loose state *.

Age of the volcanic and associated rocks of Campania.—
If we enquire into the evidence derivable from organic remains,
respecting the age of the volcanic rocks of Campania, we find
reason to conclude that such parts as do not belong to the
recent, are referrible to the newer Pliocene period.

In the solid tuff quarried out of the hills immediately behind Naples, are found recent shells of the genera Ostrea, Cardium, Buccinum, and Patella, all referrible to species now
living in the Mediterranean †. In Ischia I collected marine
shells in beds of clay and tuff, not far from the summit of
Epomeo, or San Nichola, about 2000 feet above the level of
the sea, as also at another locality, about 100 feet below, on the
southern declivity of the mountain, and others not far above
the town of Moropano. At Casamicciol, and several places
near the sea-shore, shells have long been observed in stratified
tuff and clay. From these various points I obtained, during a
short excursion in Ischia, 28 species of shells, all of which,
with one exception, were identified by M. Deshayes with recent
species ‡.

As the highest parts of Epomeo are composed of regularly-
stratified greenish tuff, and some beds near the summit contain the fossils above-mentioned, it is clear that that mountain
was not only raised to its present height above the level of the
sea, but was also *formed* since the Mediterranean was inhabited
by the existing species of testacea.

In the Ischian tuffs we find pumice, lapilli, angular fragments
of trachytic lava, and other products of igneous ejections,
interstratified with some deposits of clay, free from any intermixture of volcanic matter. These clays might have re-

---

* Geol. Trans., vol. ii. part iii. p. 351. Second Series.
† Scrope, ibid.   ‡ See the list of these shells, Appendix II.

sulted from the decomposition of felspathic lava which abounds
in Ischia, the materials having been transported by rivers and
marine currents, and spread over the bottom of the sea where
testacea were living.   We may observe generally of these submarine tuffs, lavas, and clays, of Campania, that they strictly
resemble those around the base of Etna, and in parts of the Val
di Noto before described.

External configuration of the country how caused.—When
once we have satisfied ourselves by inspection of the marine
shells imbedded in tuffs at high elevations, that a mass of land
like the island of Ischia has been raised from beneath the waters
of the sea to its present height, we are prepared to find signs
of the denuding action of the waves impressed upon the outward
form of the island, especially if we conceive the upheaving force
to have acted by successive movements.   Let us suppose the
low contiguous island of Procida to be raised by degrees until
it attains the height of Ischia, we should in that case expect
the steep cliffs which now face Misenum to be carried upwards
and to become precipices near the summit of the central mountain.   Such, perhaps, may have been the origin of those precipices which appear on the north and south sides of the ridge
which forms the summit of Epomeo in Ischia.   The northern
escarpment is about 1000 feet in height, rising from the hollow
called the Cavo delle Neve above the village of Panella.   The
abrupt manner in which the horizontal tuffs are there cut off,
in the face of the cliff, is such as the action of the sea, working
on soft materials, might easily have produced, undermining
and removing a great portion of the mass.   A heap of shingle
which lies at the base of a steep declivity on the flanks of
Epomeo, between the Cavo delle Neve and Panella, may once,
perhaps, have been a sea-beach, for it certainly could not have
been brought to the spot by any existing torrents.

There is no difficulty in conceiving that if a large tract of the
bed of the sea near Ischia should now be gradually upheaved
during the continuance of volcanic agency, this newly-raised land
might present a counterpart to the Phlegræan fields before de-

scribed. Masses of alternating lava and tuff, the products of submarine eruptions, might on their emergence become hills and islands; the level intervening plains might afterwards appear, covered partly by the ashes drifted and deposited by water, and partly by those which would fall after the laying dry of the tract. The last features imparted to the physical geography would be derived from such eruptions in the open air as those of Monte Nuovo and the minor cones of Ischia.

No signs of diluvial waves.—Such a conversion of a large tract of sea into land might possibly take place while the surface of the contiguous country underwent but slight modification. No great wave was caused by the permanent rise of the coast near Puzzuoli in the year 1538, because the upheaving operation appears to have been effected by a long succession of minor shocks *. A series of such movements, therefore, might produce an island like Ischia without throwing a diluvial rush of waters upon low parts of the neighbouring continent. The advocates of paroxysmal elevations may, perhaps, contend that the rise of Ischia must have been anterior to the birth of all the cones of loose scoriæ scattered over the Phlegræan Fields, for, according to them, the sudden rise of marine strata causes inundations which devastate adjoining continents. But the absence of any signs of such floods in the volcanic region of Campania does not appear to us to warrant the conclusion, either that Ischia was raised previously to the production of the volcanic cones, or that it may not have been rising during the whole period of their formation.

We learn from the study of the mutations now in progress, that one part of the earth's surface may, for an indefinite period, be the scene of continued change, while another, in the immediate vicinity, remains stationary. We need go no farther than our own country to illustrate this principle; for, reasoning from what has taken place in the last ten centuries, we must anticipate that in the course of the next 4000 or 5000 years, a long strip of land, skirting the line of our eastern

---

* See vol. i. p. 457, first edition; p. 527, second edition.

coast, will be devoured by the ocean, while part of the interior, immediately adjacent, will remain at rest and entirely undisturbed. The analogy holds true in regions where the volcanic fires are at work, for part of the Philosopher's Tower on Etna has stood for the last 2000 years, at the height of more than 9000 feet above the sea, between the foot of the highest cone and the edge of the precipice which overhangs the Val del Bove, whilst large tracts of the surrounding district have been the scenes of tremendous convulsions. The great cone above has more than once been blown into the air, and again renewed; the earth has sunk down in the neighbouring Cisterna *; the cones of 1811 and 1819 have started up, on the ledge of rock below, pouring out of their craters two mighty streams of lava ; the watery deluge of 1755 has rushed down from the steep desert region, into the Val del Bove, rolling along vast heaps of rocky fragments towards the sea ; fissures, several miles in length, have opened on the flanks of Etna ; cities and villages have been shattered by partial earthquakes, or buried under lava and ashes ;—yet the tower has stood as if placed on the most perilous point in Europe, to commemorate the stability of one part of the earth's surface, while others in immediate proximity have been subject to most wonderful and terrific vicissitudes.

Marine Newer Pliocene strata only visible in countries of earthquakes.—In concluding what we have to say of the marine and volcanic formations of the newer Pliocene period, we may notice the highly interesting fact, that the marine strata of this era have hitherto been found at great elevations in those countries only where violent earthquakes have occurred during the historical ages. We do not deny that some *partial* deposits containing recent marine shells have been discovered at a considerable height in several maritime countries in Europe and elsewhere, far from the existing theatres of volcanic action ; but stratified deposits of great extent and thickness, and replete with recent species, have only been observed to enter largely into the

* See above, p. 96.

structure of the interior, as in Sicily, Calabria, and the Morea, where subterranean movements are now violent.   On the other hand, it is a still more striking fact, that there is no example of any extensive maritime district, now habitually agitated by great earthquakes, which has not, when carefully investigated, yielded traces of marine strata, either of the Recent or newer Pliocene eras, at considerable elevations.

Chili.—Conception Bay.—In illustration of the above remarks we may mention, that on the western coast of South America marine deposits occur, containing precisely the same shells as are now living in the Pacific.   In Chili, for example, as we before stated *, micaceous sand, containing the fossil remains of such species as now inhabit the Bay of Conception, are found at the height of from 1000 to 1500 feet above the level of the ocean.   It is impossible to say how much of this rise may have taken place during the *Recent* period.   We have endeavoured to show that one earthquake raised this part of the Chilian coast, in 1750, to the height of at least 25 feet above its former level.   If we could suppose a continued series of such shocks, one in every century, only 6000 years would be required to uplift the coast 1500 feet.   But we have no data for inferring that so great a quantity of elevation has taken place in that space of time, and although we cannot assume that the micaceous sand may not belong to the Recent period, we think it more probable that it was deposited during the newer Pliocene period.

Peru.—We are informed by Mr. A. Cruckshanks, that in the valley of Lima, or Rimao, where the subterranean movements have been so violent in recent times, there are indications not only of a considerable rise of the land, but of that rise having resulted from *successive* movements.   Distinct lines of ancient sea-cliffs have been observed at various heights, at the base of which the hard rocks of greenstone are hollowed out into precisely those forms which they now assume between high and low water mark on the shores of the Pacific.   Immediately below these water-worn lines are ancient beaches strewed with

* Vol. i, chap. xxv.

rounded blocks.   One of these cliffs appears in the hill behind
Baños del Pujio, about 700 feet above the level of the sea, and
200 above the contiguous valley.   Another occurs at Aman-
caes, at the height of perhaps 200 feet above the sea, and others
at intermediate elevations.

Parallel roads of Coquimbo.—We can hardly doubt that the
parallel roads of Coquimbo, in Chili, described by Captain
Hall, owe their origin to similar causes.   These roads, or shelves,
occur in a valley six or seven miles wide, which descends from
the Andes to the Pacific.   Their general width is from 20 to
50 yards, but they are, at some places, half a mile broad.
They are so disposed as to present exact counterparts of one
another, at the same level, on opposite sides of the valley.
There are three distinctly characterized sets, and a lower one
which is indistinct when approached, but when viewed from a
distance is evidently of the same character with the others.
Each resembles a shingle beach, being formed entirely of loose
materials, principally water-worn, rounded stones, from the
size of a nut to that of a man's head.   The stones are principally granite and gneiss, with masses of schistus, whinstone,
and quartz mixed indiscriminately, and all bearing marks of
having been worn by attrition under water *.

The theory proposed by Captain Hall to explain these appearances is the same as that which had been adopted to
account for the analogous parallel roads of Glen Roy in Scotland †.   The valley is supposed to have been a lake, the waters
of which stood, originally, at the level of the highest road, until
a flat beach was produced.   A portion of the barrier was then
broken down, which allowed the lake to discharge part of its
waters into the sea, and, consequently, to fall to the second
level; and so on successively till the whole embankment was
washed away, and the valley left as we now see it.

As I did not feel satisfied with this explanation, I applied to

* Captain Hall's South America, vol. ii. p. 9.
† See Sir T. D. Lauder, Ed. Roy. Soc. Trans., vol. ix., and Dr. Macculloch,
Geol. Trans., 1st Series, vol. iv. p. 314.

my friend Captain Hall for additional details, and he immediately sent me his original manuscript notes, requesting me to
make free use of them.   In them I find the following interesting
passages, omitted in his printed account.   ' The valley is completely open towards the sea; if the roads, therefore, are the
beaches of an ancient lake, it is difficult to imagine a catastrophe sufficiently violent to carry away the barrier which
should not at the same time obliterate all traces of the beaches.
I find it difficult also to account for the water-worn character of all the stones, for they have the appearance of
having travelled over a great distance, being well rounded and
dressed.   They are in immense quantity too, and much more
than one could expect to find on the beach of any lake, and
seem more properly to belong to the ocean.'

We entertained a strong suspicion, before reading these
notes, that the beaches were formed by the waves of the Pacific,
and not by the waters of a lake; in other words, that they
bear testimony to the successive rise of the land, not to
the repeated fall of the waters of a lake.   We have before
cited the proofs adduced by M. Boblaye, that in the Morea
there are four or five ranges of ancient sea-cliffs, one above the
other, at various elevations, where limestone precipices exhibit
lithodomous perforations and lines of ancient littoral caverns *.
If we discover lines of parallel upraised cliffs, we ought to find
parallel lines of elevated beaches on those coasts where the rocks
are of a nature to retain, for a length of time, the marks im-
printed on their surface.   We may expect such indications to
be peculiarly manifest in countries where the subterranean
force has been in activity within comparatively modern times,
and it is there that the hypothesis of paroxysmal elevations, and
the instantaneous rise of mountain-chains, should first have been
put to the test, before it was hastily embraced by a certain
school of geologists.

West Indian Archipelago.—According to the sketch given
by Maclure of the geology of the Leeward Islands †, the

---

* See above, p. 113.   † Quart. Journ. of Sci., vol. v. p. 311.

western range consists in great part of formations of the most modern period. It will be remembered, that many parts of this region have been subject to violent earthquakes; that in St. Vincent's and Guadaloupe there are active volcanos, and in some of the other islands boiling springs and solfataras. In St. Eustatia, there is a marine deposit, estimated at 1500 feet in thickness, consisting of coral limestone alternating with beds of shells, of which the species are, according to Maclure, the same as those now found in the sea. These strata dip to the south-west at an angle of about 45°, and both rest upon, and are covered by, cinders, pumice, and volcanic substances. Part of the madreporic rock has been converted into silex and calcedony, and is, in some parts, associated with crystalline gypsum. Alternations of coralline formations with prismatic lava and different volcanic substances also occur in Dominica and St. Christopher's, and the American naturalist remarks, that as every lava-current which runs into the sea in this archipelago is liable to be covered with corals and shells, and these again with lava, we may suppose an indefinite repetition of such alternations to constitute the foundation of each isle.

We do not question the accuracy of the opinion, that the fossil shells and corals of these formations are of recent species, for there are specimens of limestone in the Museum of the Jardin du Roi at Paris, from the Antilles, in which the imbedded shells are all or nearly all identical with those now living. Part of this limestone is soft, but some of the specimens are very compact and crystalline, and contain only the casts of shells. Of 30 species examined by M. Deshayes from this rock 28 were decidedly recent.

Honduras.—Shells sent from some of the recent strata of Jamaica, and many from the nearest adjoining continent of the Honduras, may be seen in the British Museum, and are identified with species now living in the West Indian seas.

East Indian Archipelago.—We have seen that the Indian ocean is one of the principal theatres of volcanic disturbance. We expect, therefore, that future researches in this quarter of the globe will bring to light some of the most striking examples of marine strata upraised to great heights during comparatively modern periods.

From the observations of Dr. Jack, it appears that in the island of Pulo Nias, off the west coast of Sumatra, masses of corals of recent species can be traced from the level of the sea far into the interior, where they form considerable hills. Large shells of the Chama gigas (*Tridacna*, Lamk.) are scattered over the face of the country, just as they occur on the present reefs. These fossils are in such a state of preservation as to be collected by the inhabitants for the purpose of being cut into rings for the arms and wrists *.

Madeira.—The island of Madeira is placed between the Azores and Canaries, in both of which groups there are active volcanos, and Madeira itself was violently shaken by earthquakes during the last century. It consists in great part of volcanic tuffs and porous lava, intersected in some places, as at the Brazen Head, by vertical dikes of compact lava†. Some of the marine fossil shells, procured by Mr. Bowdich from this island, are referrible to recent species.

These examples may suffice for the present, and lead us to anticipate with confidence, that in almost all countries where changes of level have taken place in our own times, the geologist will find monuments of a prolonged series of convulsions during the Recent and newer Pliocene periods. Exceptions may no doubt occur where a particular line of coast is sinking down, yet even here we may presume, from what we know of the irregular action of the subterranean forces, that some cases of partial elevation will have been caused by occasional oscillations of level, so that modern subaqueous formations will, here and there, have been brought up to view.

We shall conclude by enumerating a few exceptions to the rule above illustrated—instances of elevation where no great earthquakes have been recently experienced.

* Geol. Trans., Second Series, vol. i. part ii. p. 397.
† MS. of Captain B. Hall.

Grosœil, near Nice.—At a spot called Grosœil, near Nice, east of the Bay of Villefranche, in the peninsula of St. Hospice, a remarkable bed of fine sand occurs at an elevation of about 50 feet above the sea *. This sand rests on inclined secondary rocks, and is filled with the remains of marine species all identical with those now inhabiting the neighbouring sea. No less than 200 species of shells, and several crustacea and echini, have been obtained by M. Risso, in a high state of preservation, although mingled with broken shells. The winds have blown up large heaps of similar sand to considerable heights, upon ledges of the steep coast farther westward, but the position of the deposit at Grosœil cannot be referred to such agency, for among the shells may be seen the large Murex Triton, Linn., and a species of Cassis, weighing a pound and a half.

Uddevalla.—The ancient beaches of the Norwegian and Swedish coasts, described in the first volume †, in which the shells are of living species, present more marked exceptions as being farther removed from any line of recent convulsion. They afford evidence of a rise of 200 feet or more of parts of those coasts during the newer Pliocene, if not the Recent epoch.

West of England.—The proofs lately brought to light of analogous elevations on our western shores, in Caernarvonshire and Lancashire, during some modern tertiary period, were before pointed out ‡ ; but the data are as yet exceedingly incomplete.

Western Borders of the Red Sea.—Another exception may be alluded to, for which we are indebted to the researches of Mr. James Burton. On the western shores of the Arabian gulf, about half way between Suez and Kosire, in the 28th degree of North latitude, a formation of white limestone and calcareous sand is seen, reaching the height of 200 feet above the sea. It is replete with fossil shells, all of recent species, which are in a beautiful state of preservation, many of them retaining their colour. I have been favoured with a list of

* I examined this locality in company with Mr. Murchison in 1828.
† Chap. xiii.   ‡ See description of the map, vol. ii.

these shells, which will be found in Appendix II.*   The volcano of Gabel Tor, situate at the entrance of the Arabian gulf, is the nearest volcanic region known to us at present.

We should guard the reader against inferring, from the facts above detailed, that marine strata of the newer Pliocene period have been produced exclusively in countries of earthquakes. If we have drawn our illustrations exclusively from modern volcanic regions, it is simply for this reason, that these formations have been made visible to us in those districts only where the conversion of sea into land has taken place in times comparatively modern.   Other continents have, during the newer Pliocene period, suffered degradation, and rivers and currents have deposited sediment in other seas, but the new strata remain concealed wherever no subsequent alterations of level have taken place.

We believe, however, that to a certain limited extent the growth of new subaqueous deposits has been greatest where igneous and aqueous causes have co-operated.   It is there, as we have explained in former chapters, that the degradation of land is most rapid, and it is there only that materials ejected from below, by volcanic explosions, are added to the sediment transported by running water †.

* These fossils are now in the museum of Mr. Greenough, in London, and duplicates, presented by him, in the cabinets of the Geological Society.

† See vol. i. chap. xxiv.; and vol. ii. chap. xviii.

CHAPTER XI

Newer Pliocene fresh-water formations—Valley of the Elsa—Travertins of Rome
—Osseous breccias—Sicily—Caves near Palermo—Extinct animals in newer
Pliocene breccias—Fossil bones of Marsupial animals in Australian caves—
Formation of osseous breccias in the Morea—Newer Pliocene alluviums—
Difference between alluviums and regular subaqueous strata—The former of
various ages—Marine alluvium—Grooved surface of rocks—Erratic blocks of
the Alps—Theory of deluges caused by paroxysmal elevations untenable—
How ice may have contributed to transport large blocks from the Alps—European alluviums chiefly tertiary—Newer Pliocene in Sicily—Löss of the Valley
of the Rhine—Its origin—Contains recent shells.

FRESH WATER FORMATIONS

In this chapter we shall treat of the fresh-water formations,
and of the cave breccias and alluviums of the newer Pliocene
period.

In regard to the first of these, they must have been formed,
in greater or less quantity, in nearly all the existing lakes of
the world, in those, at least, of which the basins were formed
before the earth was tenanted by man. If the great lakes of
North America originated before that era, the sedimentary
strata deposited therein, in the ages immediately antecedent,
would, according to the terms of our definition, belong to the
newer Pliocene period.

Valley of the Elsa.—As an example of the strata of this age,
which have been exposed to view in consequence of the drainage of a lake, we may mention those of the valley of the Elsa,
in Tuscany, between Florence and Sienna, where we meet with
fresh-water marls and travertins full of shells, belonging to
species which now live in the lakes and rivers of Italy. Valleys
several hundred feet deep have been excavated through the
lacustrine beds, and the ancient town of Colle stands on a hill
composed of them. The subjacent formation consists of marine Subapennine beds, in which more than half the shells are

of recent species.   The fresh-water shells which I collected
near Colle are in a very perfect state, and the colours of the
Neritinæ are peculiarly brilliant.   The following six species,
all of which now inhabit Italy, were identified by M. Deshayes:
Paludina impura, Neritina fluviatilis, Succinea amphibia, Lim-
neus auricularis, L. pereger. and Planorbis carinatus.

   Travertins of Rome.—Many of the travertins and calcareous tufas which cap the hills of Rome may also belong to
the same period.   The terrestrial shells inclosed in these masses
are of the same species as those now abounding in the gardens
of Rome, and the accompanying aquatic shells are such as are
found in the streams and lakes of the Campagna.   On Mount
Aventine, the Vatican, and the Capitol, we find abundance of
vegetable matter, principally reeds encrusted with calcareous
tufa, and intermixed with volcanic sand and pumice.   The
tusk of a mammoth has been procured from this formation,
filled in the interior with solid travertin, wherein sparkling
crystals of augite are interspersed, so that the bone has all the
appearance of having been extracted from a hard crystalline
rock *.

   These Roman tufas and travertins repose partly on marine
tertiary strata, belonging, perhaps, to the older Pliocene era,
and partly on volcanic tuff of a still later date.   They must
have been formed in small lakes and marshes, which existed
before the excavation of the valleys which divide the seven hills
of Rome, and they must originally have occupied the lowest
hollows of the country, whereas now we find them placed upon
the summit of hills about 200 feet above the alluvial plain of
the Tiber.   We know that this river has flowed nearly in its
present channel ever since the building of Rome, and scarcely
any changes in the geographical features of the country have
taken place since that era.

   When the marine tertiary strata of this district were formed,
those of Monte Mario for example, the Mediterranean was
already inhabited by a large proportion of the existing species

---

* This fossil was shown me by Signor Riccioli at Rome.

of testacea. At a subsequent period, volcanic eruptions occurred, and tuffs were superimposed. The marine formation
then emerged from the deep, and supported lakes wherein the
fresh-water groups above described slowly accumulated, at a
time when the mammoth abounded in the country. The
valley of the Tiber was afterwards excavated, and the adjoining hills assumed their present shape, and then a long interval
may, perhaps, have elapsed before the first human settlers
arrived. Thus we have evidence of a chain of events all regarded as extremely recent by the geologist, but which, nevertheless, may have preceded, for an immense series of ages, a
very remote era in the history of nations.

OSSEOUS BRECCIAS.

Sicily.—The breccias recently found in several caves in
Sicily belong evidently to the period under consideration. We
have shown, in the sixth chapter, that the cavernous limestone
of the Val di Noto is of very modern date, as it contains a
great abundance of fossil shells of recent species. But if any
breccias are found in the caverns of this rock they must be of
still later origin.

We are informed by M. Hoffmann, that the bones of the
mammoth, and of an extinct species of hippopotamus, have
been discovered in the stalactite of caves near Sortino, of which
the situation is represented in the annexed diagram at *b*. The

No. 26.

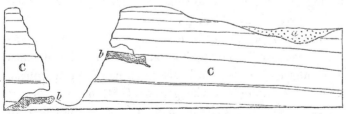

a, Alluvium,
b, *b*, Deposits in caves, } containing remains of *extinct* quadrupeds.
C, Limestone containing remains of *recent* shells.

same author also describes a breccia, containing the bones of

an extinct rhinoceros and hippopotamus, in a cave in the neighbourhood of Syracuse, where the country is composed entirely
of the Val di Noto limestone. Some of the fragments in the
breccia are perforated by lithodomi, and the whole mass is
covered by a deposit of marine clay filled with recent shells *.
These phenomena may, we think, be explained by supposing
such oscillations of level as are known to occur on maritime
coasts where earthquakes prevail, such, in fact, as have been
witnessed on the shores of the Bay of Baiæ within the last
three centuries †. For it is evident that the temporary sub-
mergence of a cave filled with osseous breccia might afford
time for the perforation of the rock by boring testacea, and for
the deposition upon it of mud, sand, and shells.

The association in these and other localities of shells of living
species with the remains of extinct mammalia is very distinct,
and corroborates the inference adverted to in a former chapter,
that the longevity of *species* in the mammalia is, upon the
whole, inferior to that of the testacea. This circumstance we
are by no means inclined to refer to the intervention of man,
and his power of extirpating the larger quadrupeds, for the
succession of mammiferous species appears to have been in like
manner comparatively rapid throughout the older tertiary
periods. Their more limited duration depends, in all probability, on physiological laws which render warm-blooded quadrupeds less capable, in general, of accommodating themselves
to a great variety of circumstances, and, consequently, of
surviving the vicissitudes to which the earth's surface is exposed in a great lapse of ages‡.

Caves near Palermo.—The caves near Palermo exhibit appearances very analogous to those above described, and much
curious information has been lately published respecting them.
According to Hoffmann, the grotto of Mardolce is distant
about two miles from Palermo, and is 20 feet high and 10

---

* Hoffmann, Archiv. für Mineralogie, p. 393. Berlin, 1831. Dr. Christie,
Proceedings of Geol. Soc., No. xxiii. p. 333.
   † Vol. i. chap. xxv.   ‡ See above, p. 48, and vol. i. chap. vi.

wide. It occurs in a secondary limestone, in the Monte Grifone, at the base of a rocky precipice about 180 feet above the sea. From the foot of this precipice an inclined plane, consisting of horizontal tertiary strata, of the newer Pliocene period, extends to the sea, a distance of about a mile.

No. 27.

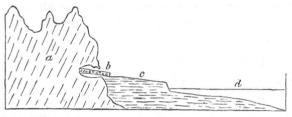

a, Monte Grifone.   c, Plain of Palermo.
b, Cave of San Ciro.   d, Bay of Palermo *.

The limestone escarpment was evidently once a sea-cliff, and the ancient beach still remains formed of pebbles of various rocks, many of which must have been brought from places far remote. Broken pieces of coral and shell, especially of oysters and pectens, are seen intermingled with the pebbles. Immediately above the level of this beach serpulæ are still found adhering to the face of the rock, and the limestone is perforated by lithodomi. Within the grotto also, at the same level, similar perforations occur, and so numerous are the holes, that the rock is compared by Hoffmann to a target pierced by musket balls. But in order to expose to view these marks of boring-shells in the interior of the cave, it was necessary first to remove a mass of breccia, which consisted of numerous fragments of rock and an immense quantity of bones imbedded in a dark brown calcareous marl. Many of the bones were rolled as if partially subjected to the action of the waves. Below this breccia, which is about 20 feet thick, was found a bed of sand filled with sea-shells of recent species, and underneath the

* This section is given by Dr. Christie, as of the Cave of San Ciro.—Ed. New Phil. Journ., No. xxiii. Its geographical position and other characters agree so precisely with that of Mardolce, described by M. Hoffmann, that it may be another name for the same cave, or one immediately adjoining.

sand again is the secondary limestone of Monte Grifone. The
state of the surface of the limestone in the cave above the level
of the marine sand is very different from that below it. *Above,*
the rock is jagged and uneven, as is usual in the roofs and sides
of limestone caverns ; *below,* the surface is smooth and polished,
as if by the attrition of the waves.

So enormous was the quantity of bones, that many ship-
loads were exported in the years 1829 and 1830, in the hope
of their retaining enough gelatine to serve for refining sugar,
for which, however, they proved useless. The bones belong
chiefly to the mammoth (*E. primigenius*), and with them are
those of an hippopotamus, smaller than the species usually
found fossil, and distinct from the recent. Several species of
deer were also found with the above*. The remains of a bear,
also, are said to have been discovered.

It is easy to explain in what manner the cavern of Mardolce
was in part filled with sea-sand, and how the surface of the
limestone became perforated by lithodomi ; but in what
manner, when the elevation of the rocks and the ancient beach
had taken place, was the superimposed osseous breccia formed ?
The extraordinary number of the imbedded animal remains
precludes, we think, at once the supposition of the whole having
been heaped up together by a single catastrophe. Let us suppose that, when the caves were at a moderate elevation above
the level of the sea, they were exposed, during a succession of
earthquakes, to be inundated again and again by waves rolling
in upon the land till they reached the base of an inland cliff,
not far from the shore. Reiterated catastrophes may thus have
occurred, like that of 1783 in Calabria, when a wave broke in
upon the coast, and after sweeping away 1400 of the inhabitants and many cattle, threw in upon the land, on its return,
the bodies of men and the carcasses of animals, mingled with
sand and pebbles. Caves so flooded might be inhabited by
some animals, and others might retreat into them during a
period of alarm. We attach no importance, however, to these

* Cuvier, Disc. Prelim., p. 345, 6th Ed.

speculations, but merely throw them out as hints for those who may re-examine these caves and be desirous of collecting additional facts.

Two other caverns are described by Dr. Christie as occurring in Mount Beliemi, about four miles west of Palermo, at a higher elevation than that of Mardolce, being more than 300 feet above the level of the sea. In one of these localities the bones are only found in a talus at the outside of the cavern; in the other, they occur both within the cave and in the talus which slopes from it to the plain below. These caves appear to be situated much above the highest point attained by the tertiary deposits in this neighbourhood, nor is there the slightest appearance in the caves themselves of the sea having been there *.

The breccias in these caves may have originated in the manner before suggested, vol. ii. chap. xiii.

Australian Breccias.— In several parts of Australia, ossifferous breccias have lately been discovered in limestone caverns, and the remains of the fossil mammalia are found to be referrible to species now living in that country, mingled with some relics of extinct animals. Many of these have been examined by Major Mitchell in the Wellington Valley, about 210 miles west from Sidney, on the river Bell, one of the principal sources of the Macquarrie, and on the Macquarrie itself.

The caverns appear to correspond closely with those which contain similar osseous breccias in Europe ; they often branch off in different directions through the rock, widening and contracting their dimensions, the roofs and floors being covered with stalactite. The bones are often broken, but do not appear water-worn. In some caves and fissures they lie imbedded in loose earth, but usually they are included in a breccia, having a red ochreous cement as hard as limestone, and like that of the Mediterranean caves.

The remains found most abundantly are those of the kangaroo. Amongst others, those of the Wombat, Dasyurus,

* Dr. T. Christie, on certain Newer Deposits in Sicily, &c.—Jameson, Ed. New Phil. Journ., No. xxiii. p. 1.

Kaola, and Phalangista, have been recognized. The greater part of them belong to existing, but several to extinct, species. One of the bones is of much greater size than the rest, and is supposed, by Mr. Clift, to belong to an hippopotamus *.

In a collection of these bones sent to Paris, Mr. Pentland thought he could recognize a species of Halmaturus of larger size than the largest living kangaroo †.

These facts are full of interest, for they prove that the peculiar type of organization which now characterizes the marsupial tribes has prevailed from a remote period in Australia, and that in that continent, as in Europe, North and South America, and India, many species of mammalia have become extinct. It also appears, although the evidence is less complete than we could have wished, that land quadrupeds, far exceeding in magnitude the wild species now inhabiting New Holland, have, at some former period, existed in that country.

Breccias now forming in the Morea.—Respecting the various ways in which fissures and caverns may become gradually filled up with osseous breccias, we may refer the reader to what we have said in a former volume ‡. It appears, however, from a recent communication of M. Boblaye, that the Morea is, of all the countries hitherto investigated, that which throws the greatest light on the mode in which the Mediterranean breccias may have originated.

In that peninsula a great many of the rivers and torrents terminate in land-locked hollows, where they are engulphed in chasms which traverse limestone. They sometimes reappear at great distances, but generally they discharge their waters below the level of the sea. 'Numerous bone caverns,' says M. Boblaye, ' may thus be filling up in our own times, and the gulphs (katavothrons) of the plain of Tripolitza have swallowed up of late years thousands of human bones, mingled

---

* Mr. Clift, Ed. New Phil. Journ., No. xx p. 394.—Major Mitchell, Proceedings of Geol. Soc., 1831, p. 321.

† Journ. de Géologie, tome iii. p. 291. The bone of an *elephant* mentioned by Mr. Pentland was the same large bone alluded to by Mr. Clift.

‡ Vol. ii. chap. xiii.

with the same ochreous clay which envelops the osseous remains
of higher antiquity *.'

NEWER PLIOCENE ALLUVIUMS.

Some writers have attempted to introduce into their classification of geological periods an *alluvial epoch*, as if the
transportation of loose matter from one part of the surface of
the land to another had been the work of one particular period.

In our opinion, they might have endeavoured, with equal
propriety, to institute a volcanic period, or a period of marine
or fresh-water deposits. We believe, on the contrary, that
alluvial formations have originated in every age, but more
particularly during those periods when land has been raised
above its former level, or depressed below it. We defined
alluvium to be such transported matter as has been thrown
down, either by rivers, floods, or other causes, upon land liable
to inundations, or which is not *permanently* submerged beneath
the waters of lakes or seas†. As examples of the *other causes*
adverted to in the above definition, we might instance a wave
of the sea raised by an earthquake, or a water-spout, or a
glacier.

We have said *permanently submerged* in order to distinguish between *alluviums* and regular subaqueous deposits.
The latter are accumulated in lakes or great submarine receptacles, the former in the channels of rivers and currents,
where the materials may be regarded as being still *in transitu,*
or on their way to a place of rest. There may be cases where
it is impossible to draw a line of demarcation between these
two classes of formations, but these exceptions are rare, and
the division is, upon the whole, convenient and natural, the
circumstances being very different under which each group
originates.

Marine alluvium.—The term ' marine alluvium ' is, perhaps,
admissible if confined to banks of shingle thrown up like the

* Journ. de Géologie, tome iii. No. x. p. 165.

† Vol. ii. chap. xiv.

Chesil bank, or to materials cast up by a wave of the sea upon the land, or those which a submarine current has left in its track. The kind last mentioned must necessarily, when the bed of the ocean has been laid dry, resemble terrestrial alluviums, with this difference, that if any fragments of organic bodies have escaped destruction they will belong to marine species.

During the gradual rise of a large area, first from beneath the waters, and then to a great height above them, several kinds of superficial gravel must be formed and transported from one place to another. When the first islets begin to appear, and the breakers are foaming upon the new-raised reefs, many rocky fragments are torn off and rolled along the bottom of the sea.

Let the reader recall to mind the action of the tides and currents off the coast of Shetland, described in the first volume*, where blocks of granite, gneiss, porphyry, and serpentine, of enormous dimensions, are continually detached from wasting cliffs during storms, and carried in a few hours to a distance of many hundred yards from the parent rocks. Suppose the floor of the ocean not far from the coast to be composed of those secondary strata of which several islands of this group consist. Such a tract, after being strewed over with detached blocks and pebbles of ancient rocks, might be converted into land, and· the geologist might then, perhaps, search in vain for the islands whence the fragments were originally derived. For the islands may have wholly disappeared, having been gradually consumed by the waves of the ocean, or submerged by subterranean movements.

Let us farther suppose this new land to be uplifted during successive convulsions to the height of 1000 feet. The marine alluvium before alluded to would be carried upwards on the summits of the hills and on the surface of elevated platforms. It might still constitute the general covering of the country, being wanting only in such valleys and ravines as may have

* Chapter xv.

been caused by earthquakes or excavated by the power of running water during the rise of the land. The alluvium in those more modern valleys would consist partly of pebbles washed out of the older gravel before mentioned, but chiefly of fragments derived from the wreck of those rocks which were removed during the erosion of the valleys.

Many of the most widely distributed of the British alluviums may we think be referred to the action of the sea previous to the elevation of the land ; and for this reason we never expect to be able to trace all the pebbles to their parent rocks. If it be objected that the high antiquity thus ascribed to many of our superficial deposits seems inconsistent with their actual state of preservation, we may observe, that they are often composed of indestructible materials, such as flint and quartz, and in many cases they have been protected for ages from the corroding action of the atmosphere by an envelope of loam or clay, from which they have been partially and slowly washed out by rain.

It must not, however, be understood that we refer the greater part of the alluviums scattered over our continents to the waves and currents of the sea, but merely some of those which have been justly regarded as most singular and anomalous, both in position and in the discordance of their contents with any known rocks in the adjacent countries.

Grooved surface of rocks.—We sometimes find the surface of large tracts hollowed out extensively in parallel grooves, such as have been described by Sir James Hall on the summits of the Corstorphine Hills, where I have myself examined them, in company with Dr. Buckland. These grooves may have been caused by the friction of blocks rolled along the floor of the ocean before the country emerged from the deep. The same appearances may be seen on a smaller scale, in the beds of many mountain-streams in Scotland, and I observed them strikingly displayed on Etna, in the defile called the Portella di Calanna, where a hard blue lava of modern date has been furrowed in this manner by the rolling of blocks down a steep declivity.

L 2

We have endeavoured, in a former volume, to point out the great power exerted by running water on the land in excavating valleys, at those periods when violent earthquakes derange, from time to time, the regular drainage of a country *. We also explained the manner in which temporary lakes are formed, and how the accumulated waters may suddenly escape, when the barriers are rent open by subsequent convulsions.

Erratic blocks.—Blocks of extraordinary magnitude have been observed at the foot of the Alps, and at a considerable height in some of the valleys of the Jura, exactly opposite the principal openings by which great rivers descend from the Alps. These fragments have been called 'erratic,' and many imaginary causes have been invented to account for their transportation. Some have talked of chasms opening in the ground immediately below, and of huge fragments having been cast out of them from the bowels of the earth. Others have referred to the deluge,—a convenient agent in which they find a simple solution of every difficult problem exhibited by alluvial phenomena. More recently, the instantaneous rise of mountain-chains has been introduced as a cause which may have given rise to diluvial waves, capable of devastating whole continents, and drifting huge blocks from one part of the earth's surface to another.

M. Elie de Beaumont has indulged in the speculation, that the sudden ' appearance of the Cordillera of the Andes' may have caused ' the historical deluge † !' Now, if we were sufficiently acquainted with the Andes to have grounds for assuming that they were not upheaved, like the Alps, at several successive periods ;—if we could assume that they have started up at once, so as to attain their actual height in an instant of time ;—if, in short, we could embrace the theory of ' paroxysmal elevations,' still we should consider the hypothesis of a connexion between the rise of the Andes and the historical

* Vol. i. chap. xxiv.

† L'Age relatif des Montagnes, sec. x.—Revue Française, No. xv., Mai, 1830, p. 55.

deluge, as most extravagant. . It cannot be disputed that, if part of the unfathomable ocean were suddenly converted into a shoal, a great body of water would be displaced, and a diluvial wave might then inundate some previously-existing continent. A line of shoals, therefore, or reefs, consisting of shattered and dislocated rocks, and surrounded on all sides by a great depth of sea, ought first to have been pointed out by the paroxysmalist as one of the protruded masses which may have caused a recent deluge. The subsequent upthrow of these same reefs to an additional height of ten, fifteen, or twenty thousand feet, converting them suddenly into a mountain ridge like the Andes, would displace a great volume of atmospheric air, not of water, and if the velocity of the movement were sufficiently great, might occasion a tremendous hurricane.

If it be said that a convulsion sufficiently violent to raise the Andes would probably extend far beyond the immediate range of the mountain chain, we reply that, according to that theory, it was not the Andes, but some other unknown tract, part perhaps of the present bed of the Pacific, which occasioned the flood. And if we indulge in conjectures as to what may have happened in contiguous regions at the time when the Cordillera arose, we ask whether those regions may not have sunk down, so as to cause a subsidence instead of an uplifting of the oceanic waters?

But leaving the farther discussion of these speculative views, let us return to the origin of the larger erratic blocks of Alpine origin. It has been often suggested, that ice may have contributed its aid towards the transfer of these enormous blocks, and, as the transporting power of ice is now so conspicuously displayed in the Alps, the idea is entitled to the fullest consideration.

Those naturalists who have seen the glaciers of Savoy, and who have beheld the prodigious magnitude of some fragments conveyed by them from the higher regions of Mont Blanc to the valleys below, to a distance of many leagues, will be prepared to appreciate the effects which a series of earthquakes might produce in this region, if the peaks or ' needles,' as they are called, of Mont Blanc were shaken as rudely as many parts of the Andes have been in our own times. The glaciers of Chamouni would immediately be covered under a prodigious load of rocky masses thrown down upon them. Let us, then, imagine one of the deep narrow gorges in the course of the Arve, between Chamouni and Cluse, to be stopped up by the sliding down of a hill-side (as the Rossberg fell in 1806 *), and a lake would fill the valley of Chamouni, and the lower parts of the glaciers would all be laid under water. The streams which flow out of arches, at the termination of each glacier, prove that at the bottom of those icy masses there are vaulted cavities through which the waters flow. Into these hollows the water of the lake would enter, and might thus float up the ice in detached icebergs, for the glaciers are much fissured, and the rents would be greatly increased during a period of earthquakes. Icebergs thus formed might, we conceive, resemble those seen by Captain Scoresby far from land in the Polar seas, which supported fragments of rock and soil, conjectured to be above fifty thousand tons in weight †. Let a subsequent convulsion, then, break suddenly the barrier of the lake, and the flood would instantly carry down the icebergs, together with their burden, to the low country at the base of the Alps.

We have stated in the first volume that blocks conveyed on floating icebergs must be deposited in different parts of the bottom of the ocean, in whatever latitudes those icebergs are dissolved ‡.

European alluviums in great part tertiary.—If those writers who speak of an ' alluvial epoch ' intend merely to say that a great part of the European alluviums are *tertiary*, we fully coincide in that opinion, for the map of Europe, given in our second volume, will show that almost every part of the existing continent of Europe has emerged from beneath the waters

* See above, vol. ii. 1st Ed., p. 229; 2d Ed. p. 235.
† See above, vol. i. p. 299, 1st Ed.; p. 342, 2d Ed.   ‡ Vol. i. ibid.

during some one or other of the tertiary periods; and it is probable, that even those districts which were land before the commencement of the tertiary epoch, may have shared in the subterranean convulsions by which the levels of adjoining countries have since been altered. During such subterranean movements new alluviums would be formed in great abundance, and those of more ancient date so modified as to retain scarcely any of their original distinguishing characters.

LOCALITIES OF NEWER PLIOCENE ALLUVIUMS.

Sicily.—Assuming, then, that almost all the European alluviums are tertiary, we have next to inquire which of them belong to the newer Pliocene period. It is clear that when a district, like the Val di Noto, is composed of rocks of this age, all the alluvium upon the surface must necessarily belong either to the newer Pliocene or to the Recent epoch. If, therefore, the elevation of the mountains of the Val di Noto was chiefly accomplished antecedently to the recent epoch, we must at once pronounce alluviums, in the position indicated at *a*, diagram No. 26 (p. 139), to belong to the newer Pliocene era. I am informed, that gravel so situated occurs at Grammichele in Sicily, containing the bones of the mammoth.

Loess of the Valley of the Rhine.—There is a remarkable alluvium filled with land-shells of recent species, which overspreads a great part of the valley of the Rhine, between Basle and Cologne, which, as it contains no remains of man or his works, we may refer to the newer Pliocene era. This deposit is provincially termed ' Loess,' or, in Alsace, ' Lehm,' and has been described by many geologists, whose observations we have lately had opportunities of verifying *.

According to M. Leonhard the loess consists chiefly of argillaceous matter combined with a sixth part of carbonate of lime and a sixth of quartzose and micaceous sand. It may be described as a pulverulent loam, of a dirty yellowish-grey colour,

* Among these we may mention MM. Leonhard, Bronn, Boué, Voltz, Steininger, Merian, Rozet, and Hibbert.

often containing calcareous sandy concretions or nodules, rarely
exceeding the size of a man's head. Its entire thickness, in
certain localities, amounts to several hundred feet; yet no signs
of stratification appear in the mass, except here and there at
the bottom, where there is a slight intermixture of materials
derived from subjacent rocks. No marine remains are anywhere imbedded in it, but land-shells of *existing species* are
extremely common, and the remains of the mammoth, horse,
and some other quadrupeds, are said to have been found in it.
The general absence of fresh-water shells is very remarkable.
I collected a few specimens in the section near the Manheim
gate of Heidelberg, and they are mentioned as having been
found at a few other spots, by several of the writers above
cited.

The loess sometimes rises to the height of 300 feet above the
alluvial plain of the Rhine, and to the height of 600 feet above
the sea; but it is confined to the valley of the Rhine and its
tributary valleys, preserving everywhere the same mineral characters, except where the lowest portion is mixed up, as before-
mentioned, with matter derived from the underlying rocks.
The loess reposes on every rock, from the granite to the gravel
of the plains of the Rhine, and must have been thrown down
from some vast body of water, densely charged with sediment,
after the country had assumed its present configuration. I
am informed by M. Studer, that it does not extend into
Switzerland, so that we may suppose the flood to have descended from near the borders of that country, perhaps from
the neighbourhood of Basle, into the valley of the Rhine, where
one of the first great obstacles to its passage would be the
Kaiserstuhl, a small group of volcanic hills which stand almost
in the middle of the plains of the Rhine, south of Strasburg,
between the chains of the Black Forest and the Vosges. These
hills are covered nearly to their summits with loess. But the
narrow gorge of Bingen and Andernach would cause the
greatest obstruction, even if we suppose that defile to have been
open when the flood descended, which was probably the case,

since we find the loess lower down the valley, on the flanks of
the Siebengebirge.

We have stated that stratification is almost entirely wanting,
but the movement of the muddy waters appears in some places
to have torn up the subjacent soil, and then to have thrown
down again the foreign matter, thus mingled with the loess, in
layers and strata. An alternation of gravel and loess has
resulted from this cause in the lower part of the section before
alluded to at Heidelberg.

I observed a similar blending of the loess, and the variegated
sandstone and red marl underlying it at Zeuten and Odenau,
in a valley on the right bank of the Rhine, at a short distance
from the Bergstrasse, between Wiesloch and Bruchsal, a locality pointed out to me by Professor Bronn. Near Andernach
there is a similar intermixture and alternation of the lower beds
of loess, with volcanic ejections such as are strewed over that
country, a phenomenon from which some observers have too
hastily inferred that the volcanic eruptions and the deposition
of the loess were contemporaneous.

The Rhine throughout a great part of its course between
the lake of Constance to the falls of Schaffhausen traverses a
tertiary deposit, called in Switzerland *molasse*, which consists in
some places of stratified yellow loam. At Stein, near Œningen, this loam is 150 feet thick, and resembles exceedingly the
löss before described, except in being regularly stratified. If
we could suppose the waters of a great lake like that of
Constance to have been suddenly let free by an earthquake,
and in their descent into the valley of the Rhine to have
intersected such strata, we might imagine the waters to have
become densely charged with loam, with which they may have
parted as soon as their velocity was diminished by spreading
over a wider space.

The catastrophe which brought down the loess must, for a
time, have desolated the country, but, in the end, it has
enriched the soil, constituting the most fertile parts of Alsace

and Lorraine, which were previously composed of barren sand and gravel.

The perfect state of preservation of the land-shells in the loess may have arisen from their having been floated in the turbid water in which there were no hard particles to injure them by friction. The occurrence of fresh-water shells is so rare as by no means to warrant the theory adopted by some, that the löss was formed in a lake instead of having been thrown down from a transient flood of muddy water. A few individual shells of aquatic species, the inhabitants, perhaps, of rivers or small ponds, may easily have been washed away and intermingled with the rest during the inundation. The names of fifteen species of recent shells, which I collected from the löss, are given in Appendix II.*

* M. Bronn of Heidelberg possesses a more extensive collection.

CHAPTER XII

Geological monuments of the *older* Pliocene period—Subapennine formations—
Opinions of Brocchi—Different groups termed by him Subapennine are not
all of the same age—Mineral composition of the Subapennine formations—
Marls—Yellow sand and gravel—Subapennine beds how formed—Illustration derived from the Upper Val d'Arno—Organic remains of Subapennine
hills—Older Pliocene strata at the base of the Maritime Alps—Genoa—Savona
—Albenga—Nice—Conglomerate of Valley of Magnan—Its origin—Tertiary
strata at the eastern extremity of the Pyrenees.

OLDER PLIOCENE FORMATIONS.

WE must now carry back our retrospect one step farther,
and treat of the monuments of the era immediately antecedent
to that last considered. We defined in the fifth chapter *, the
zoological characters by which the strata of the older Pliocene
period may be distinguished, and we shall now proceed at once
to describe some of the principal groups which answer to those
characters.

Subapennine strata.—The Apennines, it is well known, are
composed chiefly of secondary rocks, forming a chain which
branches off from the Ligurian Alps and passes down the
middle of the Italian peninsula. At the foot of these mountains, on the side both of the Adriatic and the Mediterranean,
are found a series of tertiary strata, which form, for the most
part, a line of low hills occupying the space between the older
chain and the sea. Brocchi, the first Italian geologist who
described this newer group in detail, gave it the name of the
Subapennines, and he classed all the tertiary strata of Italy,
from Piedmont to Calabria, as parts of the same system.
Certain mineral characters, he observed, were common to the
whole, for the strata consist generally of light brown or blue marl,
covered by yellow calcareous sand and gravel. There are also,

* Above, p. 54.

he added, some species of fossil shells which are found in these deposits throughout the whole of Italy.

In a catalogue, published by Lamarck, of 500 species of fossil-shells of the Paris basin, a small number only were enumerated as identical with those of Italy, and only 20 as agreeing with living species. This result, said Brocchi, is wonderful, and very different from that derived from a comparison of the fossil-shells of Italy, *more than half of which* agree with species now living in the Mediterranean, or in other seas, chiefly of hotter climates *.

He also stated, that it appeared from the observations of Parkinson, that the clay of London, like that of the Subapennine hills, was covered by sand (alluding to the Crag), and that in that upper formation of sand in England the species of shells corresponded much more closely with those now living in the ocean than did the species of the subjacent clay. Hence he inferred that an interval of time had separated the origin of the two groups. But in Italy, he goes on to say, the shells found in the marl and superincumbent sand belong entirely to the same group, and must have been deposited under the same circumstances †.

Notwithstanding the correctness of these views, Brocchi conceived that the Italian tertiary strata, as a whole, might agree with those of the basins of Paris and London, and he endeavoured to explain the discordance of their fossil contents by remarking, that the testacea of the Mediterranean differ now from those living in the ocean ‡. In attempting thus to assimilate the age of these distinct groups, he was evidently influenced by his adherence to the anciently-received theory of the gradual fall of the level of the ocean, to which, and not to the successive rise of the land, he attributed the emergence of the tertiary strata, all of which he consequently imagined to have remained under water down to a comparatively recent period.

Brocchi was perfectly justified in affirming that there were

---

* Conch. Foss. Subap., tom. i. p. 148.   † Ibid., p. 147.   ‡ Ibid., p. 166.

some species of shells common to all the strata called by him Subapennine; but we have shown that this fact is not inconsistent with the conclusion, that the several deposits may have originated at different periods, for there are species of shells common to all the tertiary eras. He seems to have been aware, however, of the insufficiency of his data, for in giving a list of species universally distributed throughout Italy, he candidly admits his inability to determine whether the shells of Piedmont were all identical with those of Tuscany, and whether those of the northern and southern extremities of Italy corresponded *.

We have already satisfactory evidence that the Subapennine beds of Brocchi belonged, at least, to three periods. To the Miocene we can refer a portion of the strata of Piedmont, those of the hill of the Superga, for example; to the older Pliocene belong the greater part of the strata of northern Italy and of Tuscany, and perhaps those of Rome; to the newer Pliocene, the tufaceous formations of Naples, the calcareous strata of Otranto, and probably the greater part of the tertiary beds of Calabria.

That there is a considerable correspondence in the arrangement and mineral composition of these different Italian groups is undeniable; but not that close resemblance which should lead us to assume an exact identity of age, even had the fossil remains been less dissimilar.

Very erroneous notions have been entertained respecting the contrast between the lithological characters of the Italian strata and certain groups of higher antiquity. Dr. Macculloch has treated of the Italian tertiary beds under the general title of ' elevated submarine alluvia,' and the overlying yellow sand and gravel may, according to him, be wholly, or in part, a terrestrial alluvium †. Had he visited Italy, we are persuaded that he would never have considered the tertiary strata of London and Paris as belonging to formations of a different order from the Subapennine groups, or as being more regu-

* Conch. Foss. Subap., tom. i. p. 143.   † Syst. of Geol., vol. i. chap. xv.

larly stratified. He seems to have been misled by Brocchi's description, who contrasts the more crystalline and solid texture of the older secondary rocks of the Apennines with the loose and incoherent nature of the Subapennine beds, which resemble, he says, the mud and sand now deposited by the sea.

We have endeavoured, in the last chapter, to restrict within definite limits the meaning of the term *alluvium ;* but if the Subapennine beds are to be designated 'marine alluvia,' the same name might, with equal propriety, be applied not only to the argillaceous and sandy groups of the London and Hampshire basins, but to a very great portion of our secondary series where the marls, clays, and sands are as imperfectly consolidated as the tertiary strata of Italy in general.

They who have been inclined to associate the idea of the more stony texture of stratified deposits with a comparatively higher antiquity, should consider how dissimilar, in this respect, are the tertiary groups of London and Paris, although admitted to be of contemporaneous date, or they should visit Sicily and behold a soft brown marl, identical in mineral character with that of the Subapennine beds, underlying a mass of solid and regularly-stratified limestone, rivalling the chalk of England in thickness. This Sicilian marl is older than the superincumbent limestone, but newer than the Subapennine marl of the north of Italy ; for in the latter the extinct shells rather predominate over the recent, in the former the recent predominate almost to the exclusion of the extinct.

We shall now consider more particularly the characters of those Subapennine beds which we refer to the older Pliocene period.

Subapennine marls.—The most important member of the Subapennine formation is a marl which varies in colour from greyish brown to blue. It is very aluminous, and usually contains much calcareous matter and scales of mica. It often exhibits no lines of division throughout a considerable thickness, but in other places it is thinly laminated. Near Parma, for example, I have counted thirty distinct laminæ in the thickness of an inch.   In some of the hills near that city the marl attains, according to Signor Guidotti, a thickness of nearly 2000 feet, and is charged throughout with shells, many of which are such as inhabit a deep sea.   They often occur in layers in such a manner as to indicate their slow and gradual accumulation.   They are not flattened but are filled with marl. Beds of lignite are sometimes interstratified, as at Medesano, four leagues from Parma ; subordinate beds of gypsum also occur in many places, as at Vigolano and Bargone, in the territory of Parma, where they are interstratified with shelly marl and sand.   At Lezignano, in the Monte Cerio, the sulphate of lime is found in lenticular crystals, in which unaltered shells are sometimes included.   Signor Guidotti, who showed me specimens of this gypsum, remarked, that the sulphuric acid must have been fully saturated with lime when the shells were enveloped, so that it could not act upon the shell.   According to Brocchi, the marl sometimes passes from a soft and pulverulent substance into a compact limestone *, but it is rarely found in this solid form.   It is also occasionally interstratified with sandstone.

The marl constitutes very frequently the surface of the country, having no covering of sand.   It is sometimes seen reposing immediately on the Apennine limestone; more rarely gravel intervenes, as in the hills of San Quirico †.   Volcanic rocks are here and there superimposed, as at Radicofani, in Tuscany, where a hill composed of marl, with some few shells interspersed, is capped by basalt.   Several of the volcanic tuffs in the same place are so interstratified with the marls as to show that the eruptions took place in the sea during the older Pliocene period.   At Acquapendente, Viterbo, and other places, hills of the same formation are capped with trachytic lava, and with tuffs which appear evidently to have been subaqueous.

Yellow Sand.—The other member of the Subapennine group, the yellow sand and conglomerate, constitutes, in most

* Conch. Foss. Subap., tom. i. p. 82.   † Ibid., p. 78.

of the places where I have seen it, a border formation near the junction of the tertiary and secondary rocks. In some cases, as near the town of Sienna, we see sand and calcareous gravel resting immediately on the Apennine limestone, without the intervention of any blue marl. Alternations are there seen of beds containing fluviatile shells, with others filled exclusively with marine species ; and I observed oysters attached to many of the pebbles of limestone. This locality appears to have been a point where a river, flowing from the Apennines, entered the sea in which the tertiary strata were formed.

Between Florence and Poggibonsi, in Tuscany, there is a great range of conglomerate of the Subapennine beds, which is seen for eleven miles continuously from Casciano to the south of Barberino. The pebbles are chiefly of whitish limestone with some sandstone. On receding from the older Apennine rocks, the conglomerate passes into yellow sand and sandstone, with shells, the whole overlying blue marl. In such cases we may suppose the deltas of rivers and torrents to have gained upon the bed of a sea where blue marl had previously been deposited.

The upper arenaceous group above described sometimes passes into a calcareous sandstone, as at San Vignone. It contains lapidified shells more frequently than the marl, owing probably to the more free percolation of mineral waters, which often dissolve and carry away the original component elements of fossil bodies and substitute others in their place. In some cases the shells imbedded in this group are silicified, as at San Vitale, near Parma, from whence I saw two species, one fresh-water and the other marine (Limnea palustris, and Cytherea concentrica, Lamk.), both *recent* and perfectly converted into flint.

On the other hand, the shells of Monte Mario, near Rome, which are probably referrible to the same formation, are changed into calcareous spar, the form being preserved notwithstanding the crystallization of the carbonate of lime.

Mode of formation of the Subapennine beds.—The tertiary strata above described have resulted from the waste of the secondary rocks which now form the Apennines, and which had become dry land before the older Pliocene beds were deposited. In the territory of Placentia we have an opportunity of observing the kind of sediment which the rivers are now bringing down from the Apennines. The tertiary marl of that district being too calcareous to be used for bricks or pottery, a substitute is obtained, by conveying into tanks the turbid waters of the rivers Braganza, Parma, Taro and Enza. In the course of a year a deposit of brown clay, much resembling some of the Subapennine marl, is procured, several feet in thickness, divided into thin laminæ of different shades of colour.

In regard to the sand and gravel, we see yellow sand thrown down by the Tiber near Rome, and by the Arno, at Florence. The northern part of the Apennines consists of a grey micaceous sandstone with an argillaceous base, alternating with shale, from the degradation of which brown clay and sand would result. If a river flow through such strata, and some one of its tributaries drains the ordinary limestone of the Apennines, the clay will become marly by the intermixture of calcareous matter. The sand is frequently yellow from being stained by oxide of iron, but this colour is by no means constant.

The similarity in composition of the tertiary strata in the basins of the Po, Arno, and Tiber, is merely such as might be expected to arise from their having been all derived from the disintegration of the same continuous chain of secondary rocks. But it does not follow that the latter rocks were all upheaved and exposed to degradation at the same time. The correspondence of the tertiary groups consists in their being all alike composed of marl, clay, and sand; but we might say the same of the London and Hampshire basins, although the English and Italian groups, thus compared, belong nearly to the two opposite extremes of the tertiary series.

The similarity in mineral character of the lacustrine deposit of the Upper Val d'Arno, and the marine Subapennine hills of northern Italy, ought, we think, to serve as a caution to the geologist, not to infer too hastily a contemporaneous origin from identity of mineral composition. The deposit of the Upper Val d'Arno occurs nearly at the bottom of a deep narrow valley, which is surrounded by precipitous rocks of secondary sandstone and shale (the macigno of the Italians and greywacke of the Germans). Hills of yellow sand, of considerable thickness, appear around the margin of the small basin, while, towards the central parts, where there has been considerable denudation, and where the Arno flows, blue clay is seen underlying the yellow sand. The shells are of freshwater origin, but we shall speak more particularly of them when we discuss the probable age of this formation in the sixteenth chapter. We desire, at present, to call the reader's attention to the fact, that we have here, in an isolated basin, such a formation as would result from the waste of the contiguous secondary rocks of the Apennines, fragments of which rocks are found in the sand and conglomerate. We should expect that if the freshwater beds were removed, and the barrier of the lake-basin closed up again, similar sediment would be again deposited, for the aqueous agents would operate in the same manner, at whatever period they might be in activity. Now, the only difference, in mineral composition, between the lacustrine deposit above alluded to, and the ordinary marine strata of the Subapennine beds, consists in the absence of calcareous matter from the clay, the torrents flowing into the lake having passed over no limestone rocks.

The lithological character of the Subapennine beds varies in different parts of the peninsula both in colour and degree of solidity. The presence, also, or absence of lignite and gypsum, and the association or non-association of volcanic rocks, are causes of great local discrepancy. The superposition of the sand and conglomerate to the marl, on the other hand, is a general point of agreement, although there are exceptions to the rule, as at San Quirico before mentioned. The cause of this arrangement may be, as we before hinted, that the arenaceous groups were first formed on the coast where rivers entered, and when these pushed their deltas farther out, they threw down the sand upon part of the bed of the sea already occupied by finer and more transportable mud.

Organic Remains.—I have been informed, by experienced collectors of the Subapennine fossils, that they invariably procure the greatest number in those winters when the rains are most abundant, an annual crop, as it were, being washed out of the soil to replace those which the action of moisture, frost, and the rays of the sun, soon reduce to dust upon the surface.

The shells in general are soft when first taken from the marl, but they become hard when dried. The superficial enamel is often well preserved, and many shells retain their pearly lustre, and even part of their external colour, and the ligament which unites the valves. No shells are more usually perfect than the microscopic, which abound near Sienna, where more than a thousand full-grown individuals are sometimes poured out of the interior of a single univalve of moderate dimensions. In some large tracts of yellow sand it is impossible to detect a single fossil, while in other places they occur in profusion.

The Subapennine testacea are referrible to species and families of which the habits are extremely diversified, some living in deep, others in shallow water, some in rivers or at their mouths. I have seen a specimen of a *fresh-water* univalve (Limnea palustris), taken from the blue marl near Parma, full of small *marine* shells. It may have been floated down by the same causes which carried wood and leaves into the ancient sea.

Blocks of Apennine limestone are found in this formation drilled by lithodomous shells. The remains not only of testacea and corals, but of fishes and crabs, are met with, as also those of cetacea, and even of terrestrial quadrupeds.

A considerable list of mammiferous species has been given by Brocchi and some other writers; and, although several mistakes have been made, and the bones of cetacea have sometimes been confounded with those of land animals, it is still indubitable that the latter were carried down into the sea when the Subapennine sand and marl were accumulated. The same

causes which drifted skeletons into lakes, such as that of the
upper Val d'Arno, may have carried down others into firths or
bays of the sea.   The femur of an elephant has been disinterred
with oysters attached to it, showing that it remained for some
time exposed after it was drifted into the sea.

Strata at the base of the Maritime Alps.—If we pass from
the Italian peninsula, and, following the borders of the Mediterranean, examine the tertiary strata at the foot of the
Maritime Alps, we find formations agreeing in zoological
characters with the Subapennine beds, and presenting many
points of analogy in their mineral composition.   The Alps, it is
well known, terminate abruptly in the sea, between Genoa and
Nice, and the steep declivities of that bold coast are continuous
below the waters, so that a depth of many hundred fathoms is
often found within stone's-throw of the beach.   Exceptions
occur only where streams and torrents enter the sea, and at
these points there is always a low level tract, intervening
between the mouth of the stream and the precipitous escarpment of the mountains.

In travelling from France to Genoa, by the new coast-road,
we are principally conveyed along a ledge excavated out of the
side of a steep slope or precipice, in the same manner as on
the roads which traverse the great interior passes of the Alps,
such as the Simplon and Mont Cenis, the difference being
that, in this case, the traveller has always the sea below him,
instead of a river.   But we are obliged occasionally to descend
by a zig-zag course into those low plains before alluded to,
which, when viewed from above, have the appearance of bays
deserted by the sea.   They are surrounded on three sides by
rocky eminences, and the fourth is open to the sea.

These leading features in the physical geography of the
country are intimately connected with its geological structure.
The rocks composing the Alpine declivities consist partly of
primary formations, but more generally of secondary, which
have undergone immense disturbance; but when we examine
the low tracts before-mentioned, we find the surface covered

with great beds of gravel and sand, such as are now annually
brought down by torrents and streams in the winter, and which
are spread in such quantity over the wide and shifting river-
channels as to render the roads for a season impassable.   The
first idea which naturally suggests itself, on viewing these plains,
is to imagine them to be deltas or spaces converted into land by
the accumulated sand and gravel brought down from the Alps
by rivers.   But, on closer inspection, we find that the apparent
lowness of the plains, which at first glance might be supposed
to be only just raised above the level of the sea, is a deception
produced by contrast.   The Alps rise suddenly to the height
of several thousand feet with a bold and precipitous outline,
while the country below is composed of horizontal strata, which
have either a flat or gently-undulating surface.   These strata
consist of gravel, sand, and marl, filled with marine shells.
They are considerably elevated, attaining sometimes the height
of 200 feet, or even more, above the level of the sea; there
must, therefore, have been a rise of the coast since they were
deposited, and they are not mere deltas or spaces reclaimed
from the sea by rivers.   Why, then, are the strata found only
at the points where rivers enter?

We must imagine that, after the coast had nearly acquired
its present configuration, the streams which flowed down into
the Mediterranean produced shoals opposite their mouths by
the continual drifting in of gravel, sand, and mud.   The Alps
were afterwards raised to a sufficient height to cause these shoals
to become land, while no perceptible alteration was produced
on intervening parts of the coast, where the sea was of great
depth near the shore.

The disturbing force appears to have acted very irregularly,
and to have produced the least elevation towards the eastern
extremity of the Maritime Alps, and a greater amount as we
proceed westward.   Thus we find the marine tertiary strata
attaining the height of about 100 feet at Genoa, 200 and 300
feet farther westward, at Albenga, and 800 or 900 feet in the
neighbourhood of Nice,

Genoa.—At Genoa the tertiary strata consist of blue marls like those of the northern Subapennines, and contain the same shells. On the immediate site of the town they rise to the height of only 20 feet above the sea, but they reach about 80 feet in some parts of the suburbs. At the base of a mountain not far from the suburbs there is an ancient

Monte d'Origina.   No. 28.

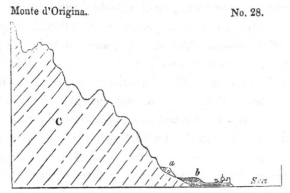

Position of Tertiary strata at Genoa.

a, Ancient sea-beach.   b, Blue marl with shells.
C, Inclined secondary strata of sandstone, shale, &c.

beach, strewed with rounded blocks of Alpine rocks, some of which are drilled by the *Modiola lithophaga,* Lamk., the whole cemented into a conglomerate*, which marks the ancient sea-beach at the height of 100 feet above the present sea.

Savona.—At Savona, proceeding westwards, we find deposits of blue marl like those of Genoa, and occupying a corresponding geological position at the base of the mountains near the sea. The shells, collected from these marls by Mr. Murchison and myself, in 1828, were examined by Signor Bonelli, of Turin, and found to agree with Subapennine fossils.

Albenga.—At Albenga these formations occupy a more extensive tract, forming the plains around that town and the low hills of the neighbourhood, which reach in some spots an elevation of 300 feet. The encircling mountains recalled to my mind those which bound the plain and bay of Palermo, and

* I have to acknowledge the assistance of Professor Viviani and Dr. Sasso who called my attention to these phenomena when I visited Genoa in Jan. 1829.

other bays of the Mediterranean, which are surrounded by bold rocky coasts.

The general resemblance of the Albenga strata to the Subapennine beds is very striking, the lowest division consisting of blue marl, which is covered by sand and yellow clay, and the highest by a mass of stratified shingle, sometimes consolidated into a conglomerate. Dr. Sasso has collected about 200 species of shells from these beds, and it appears, by his catalogue, that they agree, for the most part, with the northern Subapennine fossils, more than half of them belonging to recent species *.

Nice.—At Nice the tertiary strata are upraised to a much greater height, but they may still be said to lie at the base of the Alps which tower above them. Here, also, they consist principally of blue marl and yellow sand, which appear to have been deposited in submarine valleys previously existing in the inclined secondary strata. In one district, a few miles to the west of Nice, the tertiary beds are almost exclusively composed of conglomerate, from the point of their junction with the secondary strata to the sea.

The river Magnan flows in a deep valley which terminates at its upper extremity in a narrow ravine. Nearly vertical

Monte Calvo.   No. 29.

Section from Monte Calvo to the sea by the valley of Magnan, near Nice.

A, Dolomite and sandstone. (Green-sand formation?)

a, b, d, Beds of gravel and sand.

c, Fine marl and sand of St. Madeleine.

* Giornale Ligustico, Genoa, 1827.

precipices are laid open on each side, varying from 200 to 600
feet in height, and composed of inclined beds of shingle, sometimes separated by layers of sand, and more rarely by blue
micaceous marl.   The pebbles in these stratified shingles agree
in composition with those now brought down from the Alps by
the Var and other rivers on this coast.

The dip of the strata is remarkably uniform, being always
southwards, or towards the Mediterranean, at an angle of
about 25°.   In summer, when the bed of the river is dried up,
the geologist has a good opportunity of examining a section of
the strata, as the channel crosses for many miles the line of
bearing of the beds, which may be traced to the base of Monte
Calvo, a distance of about nine miles in a straight line from
the Mediterranean *.   It is usually impossible to determine
the exact age of such accumulations of sand and gravel, in consequence of the total absence of organic remains.   Their non-
existence may depend chiefly on the disturbed state of the
waters, where great beds of shingle are formed, which are
known to prevent testacea and fishes from living in Alpine torrents, partly on the destruction of shells by the same friction
which rounded the pebbles, and partly on the permeability of
the matrix to water, which may carry away the elements of the
decomposing fossil body, and substitute no others in their place
which might retain a cast of their form.

But it fortunately happens, in this instance, that in some
few seams of loamy marl, intervening between the pebble-beds,
and near the middle of the section, shells have been preserved in
a very perfect state of preservation, and these may furnish a
zoological date to the whole mass.   The principal of these
interstratified masses of loam occurs near the church of St.
Madeleine (at c, diagram No. 29), where the active researches
of M. Risso have brought to light a great number of shells
which agree perfectly with the species found in much greater
abundance at a spot called La Trinità, and some other locali-

* I examined this section in company with Mr. Murchison in 1828.

ties nearer to Nice. From these fossils it clearly appears that
the formation belongs to the older Pliocene era.

Such alternations of gravel and the usual thin layers of fine
sediment may easily be explained, if we reflect that the rivers
now flowing from the Maritime Alps are nearly dried up in
summer, and have only strength to drift along fine mud to the
sea; whereas, in winter, or on the melting of the snow, they
roll along large quantities of pebbles. The thicker masses of
loam, such as that of St. Madeleine, may have been produced
during a longer interval, when the river shifted for a time the
direction of its principal channel of discharge, so that nothing
but fine mud was for a series of years conveyed to that point in
the bed of the sea opposite the delta.

Uniform and continuous as the strata appear, on a general
view, in the ravine of the Magnan, we discover, if we attempt
to trace any one of them for some distance, that they thin out
and are wedge-shaped. We believe that they were thrown
down originally upon a steep slanting bank or talus, which
advanced gradually from the base of Monte Calvo to the sea.
The distance between these points is, as we have before mentioned, about nine miles, so that the accumulation of superimposed strata would be a great many miles in thickness, if they
were placed horizontally upon one another. The strata nearest
to Monte Calvo, which may be expressed by a, are certainly
older than those at b, and the group b was formed before c.
The aggregate thickness, in any one place, cannot be proved
to amount to 1000 feet, although it may, perhaps, be much
greater. But it may never exceed three or four thousand
feet; whereas, if we did not suppose that the beds were originally deposited in an inclined position, we should be forced to
imagine that a sea, many miles in depth, had been filled up by
horizontal strata of pebbles thrown down one upon another.

At no great distance on this coast the Var is annually seen
to sweep down into the sea a large quantity of gravel, which
may be spread out by the waves and currents over a considerable space. The sea at the mouth of this river is now shallow,

but it may originally have been 3000 feet deep, as it is now
close to the shore at Nice. Here, therefore, a formation resembling that of the Magnan above described may be in progress.

The time required for the accumulation of such a mass of
conglomerate as we have just considered must be immense: on
what ground such formations have been frequently referred to
diluvial waves and to periods of great disturbance, we could
never understand, for the causes now in diurnal action at the
foot of the Maritime Alps and other analogous situations seem
to us quite sufficient to explain their origin.

Tertiary strata at the eastern extremity of the Pyrenees.—
We shall conclude this chapter with one more example derived
from a region not far distant. On the borders of the Mediterranean at the eastern extremity of the Pyrenees, in the South
of France, a considerable thickness of tertiary strata are seen
in the valleys of the rivers Tech, Tet, and Gly. They bear
much resemblance to those already described, consisting partly
of a great thickness of conglomerate, and partly of clay and
sand, with subordinate beds of lignite. They abut against the
primary formation of the Pyrenees, which here consists of
mica-schist. Between Ceret and Boulon these tertiary strata
are seen inclined at an angle of between 20° and 30°. The
shells which I procured from several localities were recognized
by M. Deshayes as agreeing with Subapennine fossils.

Spain — Morea. — It appears from the recent observations
of Colonel Silvertop, that marine strata of the older Pliocene
period occur in patches at Malaga, and in Granada, in Spain.
They have also been discovered by MM. Boblaye and Virlet
in the Morea, and the names of many of the shells brought
from thence are given in the Appendix No. I.

Crag of Norfolk and Suffolk—Shown by its fossil contents to belong to the older Pliocene period—Heterogeneous in its composition—Superincumbent lacustrine deposits—Relative position of the crag—Forms of stratification—Strata composed of groups of oblique layers—Cause of this arrangement—Dislocations in the crag produced by subterranean movements—Protruded masses of chalk—Passage of marine crag into alluvium—Recent shells in a deposit at Sheppey, Ramsgate, and Brighton.

CRAG OF NORFOLK AND SUFFOLK.

THE older Pliocene strata, described in the last chapter, are all situated in countries bordering the Mediterranean, but we shall now consider a group in our own island, which belongs to the same era. We have already alluded to this deposit under the provincial name of crag *, and pointed out its superposition to the London clay, a tertiary formation of much higher antiquity †. The crag is chiefly developed in the eastern parts of Norfolk and Suffolk, from whence it extends into Essex.

Its relative age.—A collection of the shells of the ' crag' beds, which I formed in 1829, together with a much larger number sent me by my friend, Mr. Mantell of Lewes, were carefully examined by M. Deshayes, and compared to the tertiary species in his cabinet. This comparison gave the following result: out of 111 species, 66 were extinct or unknown, and 45 recent, the last, with one exception (*Voluta Lamberti*, Sow.), being now inhabitants of the German ocean. Such being the proportion of recent and extinct species, we may conclude, according to the rules before laid down ‡, that the crag belongs to the older Pliocene period.

Mineral composition.—So heterogeneous is this deposit in mineral character, that we can scarcely convey any correct notions of its appearance, without describing the beds separately in the different localities where they occur. In general, they

---

* Chap. ii. p. 19.   † See above, Diagram No. 4, p. 21.   ‡ Page 54.

consist of sand, gravel, and blue or brown marl—the shells imbedded in the sand and marl being, for the most part, broken and sometimes finely comminuted. In a few spots we find the deposit in the form of a soft stratified rock, composed almost entirely of corals, sponges, and echini *, an assemblage of species which probably lived in a tranquil sea of some depth. In other parts of our coast it consists of alternations of sand and shingle, destitute of organic remains, and more than 200 feet in thickness, as in the Suffolk cliffs, between Dunwich and Yarmouth. In others, we meet with an enormous mass, more than 300 feet in thickness, of sand, loam, and clay, containing bones of terrestrial quadrupeds and drift wood, sometimes stratified regularly, at others consisting of a confused heap of rubbish, in which fragments of the chalk and its flints are imbedded in a chalky marl.

In this aggregate are also found many fragments of older rocks, the septaria of the London clay, together with ammonites, vertebræ of ichthyosauri, and other fossils from parts of the oolitic series. It has been questioned whether all the above-mentioned beds can be considered as belonging to the same era. The subject may admit of doubt, but after examining, in 1829, the whole line of coast of Essex, Suffolk, and Norfolk, I found it impossible to draw any line of separation between the different groups. Each seemed in its turn to pass into another, and those masses which approach in character to alluvium, and contain the remains of terrestrial quadrupeds, are occasionally intermixed with the strata of the crag.

There are, however, lacustrine deposits overlying the crag, which probably belong to a distinct zoological period. These are found in small cavities, which must have existed on the surface of the crag after its elevation, and which formed small lakes or ponds wherein recent fresh-water testacea were included in loamy strata. (See wood-cut, No. 30, c.)

Relative position.—The crag is seen to rest on the chalk and on the London clay, but usually on the former. The strata

---

* R. Taylor, Geol. of East Norfolk.

are in great part horizontal, or slightly undulating; but at
some points they are much disturbed, especially where several
masses of chalk appear to have been protruded from below.

The annexed section may give a general idea of the manner
in which the crag may be supposed to rest on the chalk as we
pass from the Norfolk cliffs, at Trimmingham, into the interior,
where the country rises gradually.

No. 30.

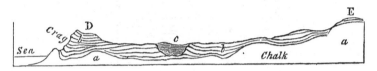

a, Chalk.   b, Crag.   c, Lacustrine deposit.
D, Trimmingham beacon.   E, Interior and higher parts of Norfolk *.

The outline of the surface of the subjacent chalk, in this
section, is imaginary, but is such as might explain the relations
of those protruded masses, three of which appear in the cliffs
near Trimmingham, and which some geologists have too hastily
assumed to be unconnected with the great mass of chalk below.
We shall treat of these presently, when we describe the disturbances which the crag appears to have suffered since its
original deposition.

In the interior, at E, there is a thick covering of sand and
gravel upon the chalk, having the characters of an alluvium,
partly, perhaps, marine, and partly terrestrial, and which seems
to pass gradually in this district into the regular marine strata
of the crag.

Forms of stratification.—In almost every formation the individual strata are rarely persistent for a great distance, the
superior and inferior planes being seldom precisely parallel to
each other; and if the materials are very coarse, the beds often
thin out if we trace them for a few hundred yards. There are
also many cases where all the layers are oblique to the general

* This section is compiled principally from one by Mr. Murchison, the others in
this chapter are from drawings by the Author.

direction of the strata, and the crag affords most interesting illustrations of this phenomenon.

In the sea-cliff near Walton, in Suffolk, opposite the Martello Tower, called R, the section represented in the annexed diagram is seen. The vertical height is about 20 feet, and

No. 31.

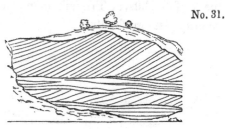

Section of shelly crag near Walton, Suffolk.

the beds consist alternately of sets of inclined and horizontal layers of sand and comminuted shells. The sand is siliceous and of a ferruginous colour, but the layers are sometimes made up of small plates of bivalve shells, arranged with their flat sides parallel to the plane of each layer, like mica in micaceous sandstones.

The number of laminæ in the thickness of an inch, both in the siliceous and shelly sand, varies from seven to ten in number, so that it is impossible to express them all in the diagram. The height of the uppermost stratum is, in this instance, remarkable, as it extends to twelve feet. The inclination of the laminæ is about 30°; but in the cliffs of Bawdesey, to the eastward, they are sometimes inclined at an angle of 45°, and even more.

No. 32.

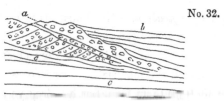

Section at the lighthouse near Happisborough.   Height sixteen feet.

a, Pebbles of chalk flint, and of rolled pieces of white chalk.
b, Loam overlying *a.*   *c, c,* Blue and brown clay.

This diagonal arrangement of the layers, sometimes called 'false stratification,' is not confined to deposits of fine sand and comminuted shells, for we find beds of shingle disposed in the same manner as is seen in the annexed section (No. 32).

The direction of the dip of the inclined layers, throughout the Suffolk coast, is so uniformly to the south, that I only saw two or three instances of a contrary nature, where the inclination was northerly. One of the best examples of this variation is exhibited in a cliff between Mismer and Dunwich, wood-cut No. 33. In this case, there are about six layers in the thickness of an inch, and the part of the cliff represented is about six feet high.

No. 33.

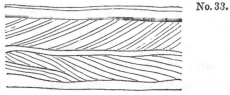

Section of part of Little Cat cliff, composed of quartzose sand, showing the inclination of the layers in opposite directions.

Another example may be seen near Walton, where the layers, which are of extreme tenuity, consist of ferruginous sand, brown loam, and comminuted shells. It is not uncommon to find in this manner sets of perfectly horizontal strata resting upon and covered by groups of wavy and transverse layers.

No. 34.

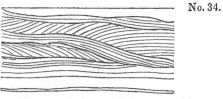

Lamination of shelly sand and loam, near the Signal-house, Walton. Vertical height four feet.

The appearances exhibited in the diagrams are not peculiar to the crag, and I have seen sand and pebble-beds of all ages, including the old red sandstone, greywacke, and clay-slate, exhibit the same arrangement,

When we inquire into the causes of such a disposition of the
No. 35.

materials of each bed or group of layers, we may, in the first place, remark,
that however numerous may be the successive layers *a, b, c*, the layer *a* must
have been deposited before *b, b* before *c*, and so of the rest.

We must suppose that each thin seam was thrown down on
a slope, and that it conformed itself to the side of the steep
bank, just as we see the materials of a talus arrange themselves
at the foot of a cliff when they have been cast down successively
from above.   If the transverse layers are cut off by a nearly
horizontal line, as in many of the above sections, it may arise
from the denuding action of a wave which has carried away the
upper portion of a submarine bank and truncated the layers of
which it was composed.   But I do not conceive this hypothesis
to be necessary; for if a bank have a steep side, it may grow by
the successive apposition of thin strata thrown down upon its
slanting side, and the removal of matter from the top may proceed simultaneously with its lateral extension. The same current
may borrow from the top what it gives to the sides, a mode of
formation which I had lately an opportunity of observing on the
rippled surface of the hills of blown sand near Calais. The undulating ridges and intervening furrows on the dunes of blown
sand resembled exactly in form those caused by the waves on
a sea-beach, and were always at right angles to the direction
of the wind which had produced them.   Each ridge had one

No. 36.

side slightly inclined and the other steep, the lee side being
always steep, as *b c, d e*, the windward side a gentle slope, as
u b, c d. When a gust of wind blew with sufficient force to
drive along a cloud of sand, all the ridges were seen to be in
motion at once, each encroaching on the furrow before it, and,
in the course of a few minutes, filling the place which the fur-

rows had occupied. Many grains of sand were drifted along the slopes *a b*, and *c d*, which, when they fell over the scarps *b c*, and *d e*, were under shelter from the wind, so that they remained stationary, resting, according to their shape and momentum, on different parts of the descent. In this manner each ridge was distinctly seen to move slowly on as often as the force of the wind augmented. We think that we shall not strain analogy too far if we suppose the same laws to govern the subaqueous and subaërial phenomena; and if so, we may imagine a submarine bank to be nothing more than one of the ridges of ripple on a larger scale, which may increase in the manner before suggested, by successive additions to the steep scarps.

The set of tides and currents, in opposite directions, may account for sudden variations in the direction of the dip of the layers, as represented in the wood-cut, No. 33, while the general prevalence of a southerly inclination in the Crag of Suffolk may indicate that the matter was brought by a current from the north.

We may refer to a drawing given in the first volume*, to show the analogy of the arrangement of the submarine strata, just considered, to that exhibited by deposits formed in the channels of rivers where a considerable transportation of sediment is in progress.

Derangement in the Crag strata.—In the above examples we have explained the want of parallelism or horizontality in the subordinate layers of different strata, by reference to the mode of their original deposition; but there are signs of disturbance which can only be accounted for by subsequent movements. The same blue and brown clay, or loam, which is often perfectly horizontal, and as regularly bedded as any of our older formations, is, in other places, curved and even folded back upon itself, in the manner represented in the annexed diagrams.

* Chap. xiv., Diag. No. 6.

No. 37.   No. 38.

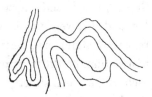

Bent strata of loam in the cliffs between   *Folding of the strata between East and*
Cromer and Runton.   *West Runton.*

In the last of these cuts a central nucleus of sand is surrounded by argillaceous and sandy layers.   This phenomenon
is very frequent, and there are instances where the materials
thus enveloped consist of broken flints mingled with pieces of
chalk, forming a white mass encircled by dark laminated clay.
The diameter of these included masses, as seen in sections laid
open in the sea-cliffs, varies from five to fifteen feet.

East of Sherringham, a heap of partially-rounded flints,
about five feet in diameter, is nearly enveloped by finely-lami-
nated strata of sand and loam, and some of the loam is entangled
in the midst of the flints.

No. 39.

Section in the Cliffs east of Sherringham.

a, Sand and loam in thin layers.

In this and similar instances, we may imagine the yielding
strata, *a,* to have subsided into a cavity, and the flints belonging to a superincumbent bed to have pressed down with their
weight, so as to cause the strata to fold round them.

That some masses of stratified sand and loam have actually
sunk down into cavities, or have fallen like landslips into ravines,
seems indicated by other appearances.   Thus, near Sherringham, the argillaceous beds, *a,* represented in the annexed
diagram (No. 40), are cut off abruptly, and succeeded by the
vertical and contorted series, *b, c.*   The face of the cliff here

represented, is 24 feet in height.   Some of the layers in *b*, *b*, are composed of pebbles, and these alternate with thin beds

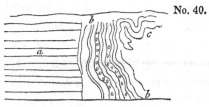

No. 40.

Section east of Sherringham, Norfolk.

a, Sand, loam, and blue clay. *b, b,* Sand and gravel. *c,* Twisted beds of loam.

of loose sand.   The whole set must once have been horizontal, and must have moved in a mass, or the relative position of the several parts would not have been preserved.   Similar appearances may, perhaps, be produced when chasms open during earthquakes and portions of yielding strata fall in from above and are engulphed.

Protruded masses of chalk.—But whatever opinion we may entertain on this point, we cannot doubt that subterranean

No. 41.

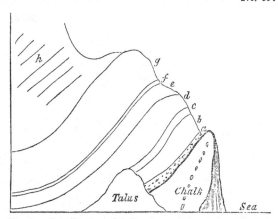

Side view of a promontory of chalk and crag, Trimmingham, Norfolk.

a, Gravel and ferruginous sand, rounded and angular pieces of chalk flint, with
     some quartz pebbles, 3 feet.
b, Laminated blue clay, 8 feet.   *c,* Yellow sand, 1 foot 6 inches.
d, Dark blue clay with fragments of marine shells, 6 feet.
e, Yellow loam and flint gravel, 3 feet.   *f,* Light blue clay, 1 foot.
g, Sand and loam, 12 feet.
h, Yellow and white sand, loam, and gravel, about 100 feet.

movements have given rise to some of the local derangements in this formation, particularly where masses of solid chalk pierce, as it were, the crag. Thus, between Mundesley and Trimmingham we see the appearances exhibited in the accompanying view (No. 41). The chalk, of which the strata are highly inclined, or vertical, projects in a promontory, because it offers more resistance to the action of the waves than the tertiary beds which, on both sides, constitute the whole of the cliff. The height of the soft crag strata immediately above the chalk is, in this place, about 130 feet. Those which are in contact (see the wood-cut) are inclined at an angle of 45°, and appear more disturbed than in other parts of the cliffs, as if they had been displaced by the movement by which the chalk was protruded.

Very similar appearances are exhibited by the northernmost of the three protuberances of chalk, of which a front view is given in the annexed diagram. It occupies a space of about 100 yards along the shore, and projects about 60 yards in advance of the general line of cliff. One of its edges, at *c*, rests upon

No. 42.

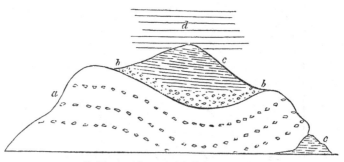

Northern protuberance of chalk, Trimmingham.

a. Chalk with flints.   *c.* Laminated blue clay.
b. Gravel, of broken and half-rounded flints.   *d.* Sand and yellow loam.

the blue clay beds of the crag, in such a manner as to imply that the mass had been undermined when the crag was deposited, unless we suppose, as some have done, that this chalk is a great detached mass enveloped by crag. For,

as one of the 'Needles', or insulated rocks of chalk, which
projected 120 feet above high water-mark, at the western
extremity of the Isle of Wight, fell into the sea in 1772 *,
so a pinnacle of chalk may have been precipitated into the
tertiary sea, at a point where some strata of the crag had
previously accumulated. The beds of flint and chalk in the
above diagram appear nearly horizontal, but they are in fact
highly inclined inwards towards the cliff. The rapid waste of
the Norfolk coast might soon enable us to understand the true
position of this mass, if observations and drawings are made
from time to time of the appearances which present themselves.

Perhaps it may be necessary to suppose, that subterranean
movements were in progress during the deposition of the crag,
and the extraordinary dislocations of the beds, in some places,
which in others are perfectly regular and horizontal, may be
most easily accounted for by introducing an alternate rise and
depression of the bed of the sea, such as we know to be usually
attendant on a series of subterranean convulsions. Several of
the contortions may also have been produced by lateral movements.

Passage of marine crag into alluvium.—By supposing the
adjoining lands to have participated in this movement, we may
explain the origin of those masses of an alluvial character
which contain the detritus of many rocks, the bones of land
animals and of drift timber, which were evidently swept down
into the sea. The land-floods which accompany earthquakes
are, as we have seen, capable of transporting such materials to
great distances †, and, as part of these alluviums must be left
somewhere upon the land, we may expect to find, on exploring
the interior, a gradual passage from the terrestrial alluvium
to that which was carried down into the sea, and which alternates with marine beds.

The fossil quadrupeds imbedded in the crag appear to be
the same as those of a great part of the alluviums of the interior

* Dodsley's Annual Register, vol. xv. p. 140.   † Vol. i. chap. 25.

of England, which may, therefore, have been formed when the testacea of the older Pliocene period were in existence.

Upon the whole, we may imagine the crag strata to bear a great resemblance to the formations which may now be in progress in the sea between the British and Dutch coasts,—a sea for the most part shallow, yet having here and there a depth of 50 or 60 fathoms, and where strong tides and currents prevail; where shells, also, and zoophytes abound, and where matter drifted from wasting cliffs must be thrown down in certain receptacles in the form of sand, shingle, and mud.

In conclusion we may observe that the history of the crag requires further elucidation, and the author is by no means satisfied with the sketch above given; but as the country is so accessible and the formation so interesting both in its structure and zoological characters, it is hoped that these remarks may excite curiosity and lead to fuller investigation.

Sheppey. — *Ramsgate.* — *Brighton.*—Deposits have lately been observed by Mr. Crow * resting on the London clay, in the Isle of Sheppey, at the height of 140 feet above the sea, and by Captain Kater at Pegwell Bay, near Ramsgate, at the height of a few yards, and by Mr. Mantell, in the cliffs near Brighton, all containing recent marine shells. But as there are only five or six species yet discovered in these localities, we cannot decide, till we obtain further information, whether these strata belong to the crag or to a more recent formation.

* Of Christ Church College, Cambridge.

Volcanic rocks of the older Pliocene period—Italy—Volcanic region of Olot in Catalonia—Its extent and geological structure—Map—Number of cones—Scoriæ—Lava currents—Ravines in the latter cut by water—Ancient alluvium underlying lava—Jets of air called 'Bufadors'—Age of the Catalonian volcanos uncertain—Earthquake which destroyed Olot in 1421—Sardinian volcanos—District of the Eifel and Lower Rhine—Map—Geological structure of the country—Peculiar characteristics of the Eifel volcanos—Lake craters—Trass—Crater of the Roderberg—Age of the Eifel volcanic rocks uncertain—Brown coal formation.

VOLCANIC ROCKS OF THE OLDER PLIOCENE PERIOD.

Italy.— It is part of our proposed plan to consider the igneous as well as the aqueous formations of each period, but we are far from being able as yet to assign to each of the numerous groups of volcanic origin scattered over Europe a precise place in the chronological series. We have already stated that the volcanic rocks of Tuscany belong, in great part at least, to the older Pliocene period,—those for example of Radicofani, Viterbo, and Aquapendente, which have been chiefly erupted beneath the sea. The same observation would probably hold true in regard to the igneous rocks of the Campagna di Roma.

, But several other districts, of which the dates are still uncertain, may be mentioned in this chapter as being possibly referrible to the period now under consideration. It will at least be useful to explain to the student the points which require elucidation before the exact age of the groups about to be described can be accurately determined.

Volcanos of Olot, in Catalonia.—I shall first direct the reader's attention to a district of extinct volcanos in the north of Spain, which is little known, and which I visited in the summer of 1830.

The whole extent of country occupied by volcanic products in Catalonia is not more than fifteen geographical miles from

north to south, and about six from east to west. The vents of
eruption range entirely within a narrow band running north
and south, and the branches which we have represented as
extending eastward in the map are formed simply of two lava-
streams, those of Castell Follit and Cellent.

No. 43.

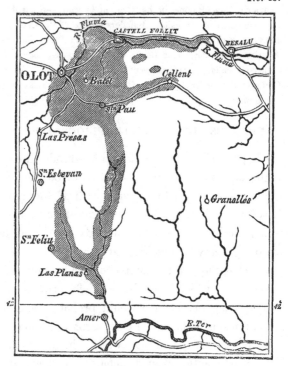

Volcanic district of Catalonia.

Dr. Maclure, the American geologist, was the first who made
known the existence of these volcanos * ; and, according to his
description, the volcanic region extended over twenty square
leagues, from Amer to Massanet. I searched in vain in the environs of Massanet, in the Pyrenees, for traces of a lava-current;
and I can say with confidence that the adjoined map gives a
correct view of the true area of the volcanic action.

* Maclure, Journ. de Phys., vol. lxvi. p. 219, 1808; cited by Daubeny, Description of Volcanos, p. 24.

Geological structure of the district.—The eruptions have burst entirely through secondary rocks, composed in great part of grey and greenish sandstone and conglomerate, with some thick beds of nummulitic limestone. The conglomerate contains pebbles of quartz, limestone, and Lydian stone. The limestone is not only replete with nummulites, but occasionally includes oysters, pectens, and other shells. This system of rocks is very extensively spread throughout Catalonia, one of its members being a red sandstone, to which the celebrated salt-rock of Cardona is subordinate. It is conjectured that the whole belongs to the age of our green-sand and chalk.

Near Amer, in the Valley of the Ter, on the southern borders of the region delineated in the map, primary rocks are seen consisting of gneiss, mica-schist, and clay-slate. They run in a line nearly parallel to the Pyrenees, and throw off the secondary strata from their flanks, causing them to dip to the north and north-west. This dip, which is towards the Pyrenees, is connected with a distinct axis of elevation, and prevails through the whole area described in the map, the inclination of the beds being sometimes at an angle of between 40 and 50 degrees.

It is evident that the physical geography of the country has undergone no material change since the commencement of the era of the volcanic eruptions, except such as has resulted from the introduction of new hills of scoriæ and currents of lava upon the surface. If the lavas could be remelted and poured out again from their respective craters, they would descend the same valleys in which they are now seen, and reoccupy the spaces which they at present fill. The only difference in the external configuration of the fresh lavas would consist in this, that they would nowhere be intersected by ravines, or exhibit marks of erosion by running water.

Volcanic cones and lavas.—There are about fourteen distinct cones with craters in this part of Spain, besides several points whence lavas may have issued ; all of them arranged along a narrow line running north and south, as will be seen in the

map.   The greatest number of perfect cones are in the immediate neighbourhood of Olot, some of which are represented in
the frontispiece, and the level plain on which that town
stands has clearly been produced by the flowing down of many
lava-streams from those hills into the bottom of a valley,
probably once of considerable depth like those of the surrounding country.

In the frontispiece an attempt is made to represent by colours
the different geological formations of which the country is
composed.   The blue line of mountains in the distance are the
Pyrenees, which are to the north of the spectator, and consist
of primary and ancient secondary rocks.   In front of these are
the secondary formations described in this chapter, coloured
grey.   Different shades of this colour are introduced, to express
various distances.   The flank of the hill, in the foreground,
called Costa de Pujou, is composed partly of secondary rocks
and partly of volcanic, the red colour expressing lava and
scoriæ.

The Fluvia, which passes near the town of Olot, has only
cut to the depth of forty feet through the lavas of the plain
before mentioned.   The bed of the river is hard basalt, and at
the bridge of Santa Madalena, are seen two distinct lava-
currents, one above the other, separated by a horizontal bed
of scoriæ eight feet thick.

In one place, to the south of Olot, the even surface of the
plain is broken by a mound of lava, called the ' Bosque de
Tosca,' the upper part of which is scoriaceous, and covered
with enormous heaps of fragments of basalt more or less porous.
Between the numerous hummocks thus formed, are deep cavities, having the appearance of small craters.   The whole precisely
resembles some of the modern currents of Etna, or that of
Côme, near Clermont, the last of which, like the Bosque de
Tosca, supports only a scanty vegetation.

Most of the Catalonian volcanos are as entire as those in the
neighbourhood of Naples, or on the flanks of Etna.   One of
these, figured in the frontispiece, called Montsacopa, is of a

1. St Michd.   2. Mone Olivet.   3. Monsacopa.   4. Puig Sacorvrra.   5. Garrinada.

View of the Volcanoes around Olot, in Catalonia.

Primary and
secondary rocks.
of Pyrenees.

Secondary
formations.

Volcanic Rocks.

very regular form, and has a circular depression or crater at the summit. It is chiefly made up of red scoriæ, undistinguish-able from that of the minor cones of Etna. The neighbouring hills of Olivet and Garrinada, also figured in the frontispiece, are of similar composition and shape. The largest crater of the whole district occurs farther to the east of Olot, and is called Santa Margarita. It is 455 feet deep, and about a mile in circumference. Like Astroni, near Naples, it is richly covered with wood, wherein game of various kinds abound.

Although the volcanos of Catalonia have broken out through sandstone, shale, and limestone, as have those of the Eifel, in Germany, to be described in the sequel, there is a remarkable difference in the nature of the ejections composing the cones in these two regions. In the Eifel, the quantity of pieces of sand-stone and shale thrown out from the vents, is often so immense as far to exceed in volume the scoriæ, pumice, and lava ; but I sought in vain in the cones near Olot for a single fragment of any extraneous rock, and Don Francisco Bolos informs me that he has never been able to detect any. Volcanic sand and ashes are not confined to the cones, but have been sometimes scattered by the wind over the country, and drifted into narrow valleys, as is seen between Olot and Cellent, where the annexed section is exposed. The light cindery volcanic matter rests in thin regular layers, just as it alighted on the slope formed by the solid conglomerate. No flood could have passed through

No. 44.

a, Secondary conglomerate.   b, Thin seams of volcanic sand and scoriæ.

the valley since the scoriæ fell, or these would have been for the most part removed.

The currents of lava in Catalonia, like those of Auvergne, the Vivarais, Iceland, and all mountainous countries, are of considerable depth in narrow defiles, but spread out into com-

paratively thin sheets in places where the valleys widen. If a
river has flowed on nearly level ground, as in the great plain
near Olot, the water has only excavated a channel of slight
depth ; but where the declivity is great, the stream has cut a
deep section, sometimes by penetrating directly through the
central part of a lava-current, but more frequently by passing
between the lava and the secondary rock which bounds the
valley. Thus, in the accompanying section, at the bridge of
Cellent, six miles east of Olot, we see the lava on one side of

No. 45.

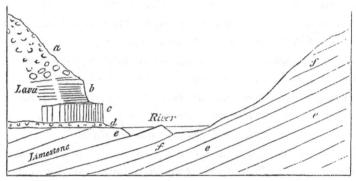

Section above the bridge of Cellent.

a, Scoriaceous lava.	*d,* Scoriæ, vegetable soil, and alluvium.
b, Schistose basalt.	*e,* Nummulitic limestone.
c, Columnar basalt.	*f,* Micaceous grey sandstone.

the small stream, while the inclined stratified rocks constitute
the channel and opposite bank. The upper part of the lava at
that place is scoriaceous ; farther down it becomes less porous,
and assumes a spheroidal structure ; still lower it divides in
horizontal plates, each about two inches in thickness, and is
more compact. Lastly, at the bottom is a mass of prismatic
basalt about five feet thick. The vertical columns often rest
immediately on the subjacent secondary rocks ; but there is
sometimes an intervention of such sand and scoriæ as cover
a country during volcanic eruptions, and which when unpro-
tected, as here, by superincumbent lava, is washed away from
the surface of the land. Sometimes the bed *d* contains a few

pebbles and angular fragments of rock ; in other places fine earth, which may have constituted an ancient vegetable soil.

In several localities, beds of sand and ashes are interposed between the lava and subjacent stratified rock, as may be seen if we follow the course of the lava-current which descends from Las Planas towards Amer, and stops two miles short of that town. The river there has often cut through the lava, and through eighteen feet of underlying limestone. Occasionally an alluvium, several feet thick, is interposed between the igneous and marine formation ; and it is interesting to remark, that in this, as in other beds of pebbles occupying a similar position, there are no rounded fragments of lava, whereas, in the modern gravel beds of rivers in this country, volcanic pebbles are abundant.

The deepest excavation made by a river through lava, which I observed in this part of Spain, is that seen in the bottom of a valley near San Feliu de Palleróls, opposite the Castell de Stolles. The lava there has filled up the bottom of a valley, and a narrow ravine has been cut through it to the depth of 100 feet. In the lower part the lava has a columnar structure. A great number of ages were probably required for the erosion of so deep a ravine ; but we have no reason to infer that this current is of higher antiquity than those of the plain near Olot. The fall of the ground, and consequent velocity of the stream, being in this case greater, a more considerable volume of rock may have been removed in an equal quantity of time.

We shall describe one more section to elucidate the phenomena of this district. A lava-stream, flowing from a ridge of hills to the east of Olot, descends a considerable slope until it reaches the valley of the river Fluvia. Here, for the first time, it comes in contact with running water, which has removed a portion, and laid open its internal structure in a precipice about 130 feet in height, at the edge of which stands the town of Castell Follit.

By the junction of the rivers Fluvia and Teronel the mass of lava has been cut away on two sides ; and the insular mass

B (No. 46) has been left, which was probably never so high as the cliff A, as it may have constituted the lower part of the sloping side of the original current.

No. 46.

Section at Castell Follit.

A, Church and town of Castell Follit, overlooking precipices of basalt.

B, Small island, on each side of which branches of the river Teronel flow to meet the Fluvia.

c, Precipice of basaltic lava, chiefly columnar.

d, Ancient alluvium underlying the lava-current.

e, Inclined strata of secondary sandstone.

From an examination of the vertical cliffs, it appears that the upper part of the lava on which the town is built is scoriaceous, passing downwards into a spheroidal basalt; some of the huge spheroids being no less than six feet in diameter. Below this is a more compact basalt with crystals of olivine. There are in all about four distinct ranges of prismatic basalt, separated by thinner beds not columnar, and some of which are schistose. The whole mass rests on alluvium, ten or twelve feet in thickness, composed of pebbles of limestone and quartz, but without any intermixture of igneous rocks; in which circumstance alone it appears to differ from the modern gravel of the Fluvia.

Bufadors.—The volcanic rocks near Olot have often a cavernous structure like some of the lavas of Etna; and in many parts of the hill of Batet, in the environs of the town, the sound returned by the earth, when struck, is like that of an archway. At the base of the same hill are the mouths of several subterranean caverns, about twelve in number, which are called in the country 'bufadors,' from which a current of cold air issues during summer; but which in winter is said to be scarcely perceptible. I visited one of these bufadors in the beginning of August, 1830, when the heat of the season was unusually intense, and found a cold wind blowing from it, which may easily be explained, for as the external air when rarefied by heat ascends, the cold air from the interior of the mountain rushes in to supply its place.

Age of the Catalonian volcanos uncertain.—It now only remains to offer some remarks on the probable age of these Spanish volcanos. Attempts have been made to prove, that in this country, as well as in Auvergne and the Eifel, the earliest inhabitants were eye-witnesses to the volcanic action. In the year 1421 it is said, when Olot was destroyed by an earthquake, an eruption broke out near Amer, and consumed the town. The researches of Don Francisco Bolos have, I think, shown, in the most satisfactory manner, that there is no good historical foundation for the latter part of this story; and any geologist who has visited Amer must be convinced that there never was any eruption on that spot. It is true that, in the year above-mentioned, the whole of Olot, with the exception of a single house, was cast down by an earthquake; one of those shocks which at distant intervals, during the last five centuries, have shaken the Pyrenees, and particularly the country between Perpignan and Olot, where the movements, at the period alluded to, were most violent.

Some houses are said to have sunk into the earth; and this account has been corroborated by the fact that, within the memory of persons now living, the buried arches of a Benedictine monastery were found at a depth of six feet beneath the surface; and still later, some houses were dug out in the street called Aigua. Don Bolos informed me, that he was present when the latter excavation was made, and when the roof of a buried house, nearly entire, was found six feet beneath the surface, the interior being in a great part empty, so that it was

necessary to fill it up with earth and stones, in order to form a
sure foundation for the new edifice.

The annihilation of the ancient Olot may, perhaps, be ascribed, not to the extraordinary violence of the movement on
that spot, but to the cavernous nature of the subjacent rocks;
for Catalonia is beyond the line of those modern European
earthquakes which destroy towns throughout extensive areas.

As we have no historical records, then, to guide us in regard
to the extinct volcanos, we must appeal to geological monuments. We have little doubt that some fossil land-shells, and
bones of quadrupeds, will hereafter reward the industry of collectors. If such remains are found imbedded in volcanic ejections, the period of the eruptions may be inferred; but at
present we have no evidence beyond that afforded by superposition, in regard to which the annexed diagram will present
to the reader, in a synoptical form, the results obtained from
numerous sections.

No. 47.

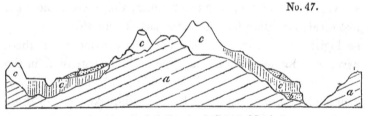

Superposition of rocks in the volcanic district of Catalonia.

a, Sandstone and nummulitic limestone.   *c,* Cones of scoriæ and lava.
b, Older alluvium with volcanic pebbles.   *d,* Newer alluvium.

The more modern alluvium *d* is partial, and has been formed
by the action of rivers and floods upon the lava; whereas the
older gravel, *b,* was strewed over the country before the volcanic eruptions. In neither have any organic remains been
discovered, so that we can merely affirm, as yet, that the volcanos broke out after the elevation of some of the newest rocks
of the secondary series, and before the formation of an alluvium, *d,* of unknown date. The integrity of the cones merely
shows that the country has not been agitated by violent earth-

quakes, nor subjected to the action of any great transient flood since their origin.

East of Olot, on the Catalonian coast, marine tertiary strata occur, which, near Barcelona, attain the height of about 500 feet. It appears probable, from a small number of shells which I collected, that these strata may correspond with the Sub-apennine beds, so that if the volcanic district had extended thus far, we might be able to determine the age of the igneous products, by observing their relation to these older Pliocene formations *.

Sardinian volcanos.—The line of extinct volcanos in Sardinia, described by Captain Smyth †, is also of uncertain date, as, notwithstanding the freshness of some of the cones and lavas, they may be of high antiquity. They rest, however, on a tertiary formation, supposed by some to correspond to the Sub-apennine strata, but of which the fossil remains have not been fully described.

VOLCANIC ROCKS OF THE EIFEL.

The volcanos of the Lower Rhine and the Eifel are of no less uncertain date than those of Catalonia; but we are desirous of pointing out some of their peculiar characters, and shall, therefore, treat of them in this chapter, trusting that future investigations will determine their chronological relations more accurately.

For the geographical details of this volcanic region, we refer the reader to the annexed map, for which I am indebted to Mr. Leonard Horner, whose residence in the country has enabled him to verify the maps of MM. Nöeggerath and Von Oyenhausen, from which that now given has been principally compiled.

---

* For some account of the Olot volcanos see ' Noticia de Los Estinguidos Volcanes de la Villa de Olot,' by Francisco Bolos. Barcelona. No date,—but the observations, I am told, preceded those of Dr. Maclure.

† Present state of Sardinia, &c., pp. 69, 70.

ENGLISH MILES.

1   2   3   4   5   No. 48.

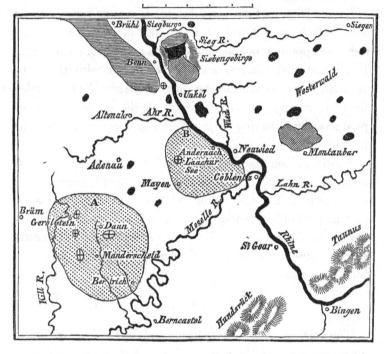

Volcanic } A, Upper Eifel.
District } B, Lower Eifel.

Points of eruption, with craters and scoriæ.

Trachyte.

Basalt.

Brown coal.

N.B. The country in that part of the map which is left blank is almost entirely composed of graywacke.

There has been a long succession of eruptions in this country, and some of them must have occurred when its physical geography was in a very different state, while others have happened when the whole district had nearly assumed its present configuration.

The fundamental rock of the Eifel is an ancient secondary sandstone and shale, to which the obscure and vague appellation of 'graywacke' has been given. The formation has precisely the characters of a great part of those grey and red sandstones and shales which are called 'old red sandstone' in

England and Scotland, where they constitute the inferior member of the carboniferous series.   In the Eifel they occupy the same geological position, and in some parts alternate with a limestone, containing trilobites and other fossils of our mountain and transition limestones.   The strata are inclined at all angles from the horizontal to the vertical, and must have undergone reiterated convulsions before the country was moulded into its present form.

Lake-Craters.—The volcanos have broken out sometimes at the bottom of deep valleys, sometimes on the summit of hills, and frequently on intervening platforms.   The traveller often falls upon them unexpectedly in a district otherwise extremely barren of geological interest.   Thus, for example, he might arrive at the village of Gemunden, immediately south of Daun, without suspecting that he was in the immediate vicinity of some of the most remarkable vents of eruption.   Leaving a stream which flows at the bottom of a deep valley in a sandstone country, he climbs the steep acclivity of a hill where he observes the edges of strata of sandstone and shale dipping inwards towards the mountain.   When he has ascended to a considerable height he sees fragments of scoriæ sparingly scattered over the surface, till at length on reaching the summit he finds himself suddenly on the edge of a *tarn,* or deep circular lake-basin.

No. 49.

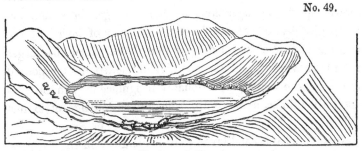

The Gemunden Maar.

This, which is called the Gemunden Maar, is the first of three lakes which are in immediate contact, the same ridge forming the barrier of two neighbouring cavities (see diag. No. 50).   On

viewing the first of these we recognize the ordinary form of a
crater, for which we have been prepared by the occurrence of

No. 50.

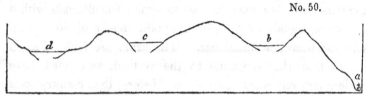

a, Village of Gemunden.   c, Weinfelder Maar.
b, Gemunden Maar.   d, Schalkenmehren Maar.

scoriæ scattered over the surface of the soil. But on examining the walls of the crater, we find precipices of sandstone and
shale which exhibit no signs of the action of heat, and we look
in vain for those beds of lava and scoriæ, dipping in opposite
directions on every side, which we have been accustomed to consider as characteristic of volcanic craters. As we proceed, however, to the opposite side of the lake, and afterwards visit the
craters c and d, we find a considerable quantity of scoriæ and
some lava, and see the whole surface of the soil sparkling with
volcanic sand and ejected fragments of half-fused shale, which
preserves its laminated texture in the interior, while it has a
vitrified or scoriform coating.

We cannot, therefore, doubt that these great hollows have
been formed by gaseous explosions; in other words, that parts
of the summits of hills composed of sandstone and shale were
blown up during a copious discharge of gas or steam, accompanied by the escape of a small quantity of lava. It is a
peculiar feature of the Eifel volcanos that aëriform discharges
have been violent, and the quantity of melted matter poured
out from the vents proportionably insignificant. In this respect they differ, as a group, from any assemblage of extinct
volcanos which I have seen in France, Italy, or Spain.

In some of the Eifel lavas, as in Auvergne and the Vivarais,
fragments of granite, gneiss, and clay-slate are found inclosed;
pieces of these rocks having probably been torn off by the
melted matter and gases as they rose from below.

A few miles to the south of the lakes above-mentioned

occurs the Pulvermaar of Gillenfeld, an oval lake of very
regular form, and surrounded by an unbroken ridge of fragmentary materials, consisting of ejected shale and sandstone,
and preserving an uniform height of about one hundred and
fifty feet above the water. The side slope in the interior is
at an angle of about 45°; on the exterior, of 35°. Volcanic
substances are intermixed very sparingly with the ejections
which in this place entirely conceal from view the stratified
rocks of the country *.

The Meerfelder Maar is a cavity of far greater size and
depth, hollowed out of similar strata; the sides presenting some
abrupt sections of inclined secondary rocks, which in other
places are buried under vast heaps of pulverised shale. I
could discover no scoriæ amongst the ejected materials, but
balls of olivine, and other volcanic substances are mentioned as
having been found †. This cavity, which we must suppose to
have discharged an immense volume of gas, is nearly a mile in
diameter, and is said to be more than one hundred fathoms
deep. In the neighbourhood is a mountain called the Moseberg, which consists of red sandstone and shale in its lower
parts, but supports on its summit a triple volcanic cone, while
a distinct current of lava is seen descending the flanks of the
mountain. The edge of the crater of the largest cone reminded
me much of the form and characters of that of Vesuvius.

If we pass from the Upper to the Lower Eifel we find the
celebrated lake-crater of Laach, which has a greater resemblance than any of those before-mentioned to the Lago di
Bolsena, and others in Italy—being surrounded by a ridge of
gently sloping hills, composed of loose tuffs, scoriæ, and blocks
of a variety of lavas.

Trass and its origin.—It appears that in the Lower Eifel
eruptions of trachytic lava preceded the emission of currents
of basalt, and that immense quantities of pumice were thrown
out wherever trachyte issued. In this district, also, we find

* Scrope, Edin. Journ. of Sci., June 1826, p. 145.
† Hibbert, Extinct Volcanos of the Rhine, p. 24.

the tufaceous alluvium of the Rhine volcanos called *trass*, which has covered large areas, and choked up some valleys now partially re-excavated. This trass is, like the loess, unstratified. The base is composed almost entirely of pumice, in which are included fragments of basalt and other lavas, pieces of burnt shale, slate, and sandstone, and numerous trunks and branches of trees.

If an eruption, attended by a copious evolution of gases, should now happen in one of the lake basins, we might suppose the water to remain for weeks in a state of violent ebullition, until it became of the consistency of mud, just as the sea became charged with red mud round the new island of Sciacca, in the Mediterranean, in the year 1831. If a breach should then be made in the side of the cone, the flood would sweep away great heaps of ejected fragments of shale and sandstone, which would be borne down into the adjoining valleys. Forests would be torn up by such a flood, which would explain the occurrence of the numerous trunks of trees dispersed irregularly through the trass.

Crater of the Roderberg.—One of the most interesting volcanos on the left bank of the Rhine is called the Roderberg. It forms a circular crater nearly a quarter of a mile in diameter, and one hundred feet deep, now covered with fields of corn. The highly inclined graywacke strata rise even to the rim of one side of the crater, but they are overspread by quartzose gravel, and this again is covered by volcanic scoriæ and tufaceous sand. The opposite wall of the crater is a scoriaceous rock, like that at the summit of Vesuvius. It is quite evident that the eruption in this case burst through the graywacke and alluvium which immediately overlies it; and I observed some of the quartz pebbles mixed with scoriæ on the flanks of the mountain, so placed as if they had been cast up into the air, and had fallen again with the volcanic ashes.

On the opposite, or right bank of the Rhine, are the Siebengebirge, a group of mountains wherein analogous phenomena are exhibited. There also trachytic lavas have flowed out and covered the graywacke ; and basaltic currents of a somewhat later date have followed.

There is, however, such a connexion between these rocks that a suite might be procured from the Siebengebirge, showing an-insensible gradation from highly crystalline tra-chyte into compact basalt, with the accompanying passage of the hornblende in the former, into augite in the latter.

Age of the volcanic rocks of the Eifel uncertain.—Besides the ancient inclined graywacke, we have in the immediate vicinity of the valley of the Rhine, a nearly horizontal tertiary formation, called brown coal, from the association with it of beds of lignite worked for coal.   The great mass of the igneous rocks are seen to be newer than this formation ; and thus we obtain a relative date of much local importance for the volcanos of the whole region.   This brown coal consists of beds of sand and sandstone, with nodules of clay-ironstone, and siliceous conglomerate.   Beds of lignite of various thickness are inter-stratified with the clays and sands, and often irregularly dif-fused through them.   This deposit was classed with the plastic clay at a time when every group of tertiary strata was referred to the age of some one of the subdivisions of the Paris basin, but as no shells, either marine, fresh-water, or land have yet been found imbedded, it is not easy to decide the age of the formation.   Near Marienforst, in the vicinity of Bonn, large blocks are found on the surface of a white opaque quartz rock, containing numerous casts of fresh-water shells which appear to belong to *Planorbis rotundatus* and *Limnea longiscatus,* two well-known Eocene species[*] ; but this rock is not in situ, and may possibly have been a local deposit in some small lake, fed by a spring holding silica in solution.   Yet, as there are beds of the brown coal at Marienforst, and this for-mation contains in other places subordinate beds of silex, it seems to me most probable that the quartzose blocks alluded to were derived from some member of that tertiary group.

[*] M. Deshayes, to whom I showed the specimens, said he felt as confident of the above identifications as *mere casts* would warrant.

The other organic remains of the brown coal are principally fishes; they are found in a bituminous shale, called paper-coal, from being divisible into extremely thin leaves. The individuals are extremely numerous, but they appear to belong to about five species, which M. Agassiz informs me are all extinct, and hitherto peculiar to the brown coal. They belong to the fresh-water genera Leuciscus, Aspius, and Perca. The remains of frogs also, of an extinct species, have been discovered in the paper coal, and a perfect series may be seen in the museum at Bonn, from the most imperfect state of the tadpole to that of the full-grown animal. With these a salamander, scarcely distinguishable from the recent species, has been found.

All the distinguishable remains of plants in the lignite and associated beds are said to belong to dicotyledonous trees and shrubs, bearing a close resemblance to those now existing in the country. The same is declared to be the case with the remains found in the trachytic tuffs and in the trass; but the absolute identification of species on which some geologists have insisted must be received with great caution.

As trachytic tuff has been observed at several places interstratified with the clay beds of the brown coal formation, and containing the same impressions of plants, there can be no doubt that the oldest eruptions began when the fresh-water deposits were still in progress, and when the geographical features of the country must have been extremely different from those which it has now assumed.

We have stated that the volcanic ejections of the Roderberg repose upon a bed of gravel. This gravel forms part of an ancient alluvium which is quite distinct in character from that now found in the plains of the valley of the Rhine. It consists chiefly of quartz pebbles, and is found at considerable elevations both on the graywacke and brown coal beds. It forms indeed a general capping to the latter, varying from ten to thirty-five feet in thickness, and was probably an alluvium formed at that period when the ancient lake, in which the brown coal strata were deposited, was drained; for the disappearance of that great body of fresh water may naturally be supposed to have taken place when the country was undergoing great changes in its physical geography.

Beds and large veins of quartz are found in the Hundsruck, Taunus, and Eifel, the nearest mountain-chains which border this part of the Rhine, and their degradation may have supplied the quartz found in this gravel called *Kiesel gerolle* by the Germans.

It has been supposed by some writers that the *latest* volcanic eruptions of the Eifel and Rhine coincided in epoch with the deposition of the Loess before described (chap. xi.). Such an association, if established, would give a comparatively recent date to the most modern igneous eruptions; but I looked in vain for any clear indications of such a connexion, and all the sections which I saw appeared to indicate the posteriority of the Loess. The integrity of the volcanic cones is, for reasons before explained, a character to which we attach no value.

We have, therefore, in this region, graywacke covered by brown coal, and some volcanic formations so blended with the latter as to prove the igneous eruptions to have been contemporaneous. Yet when we endeavour to assign a chronological position to any one part of the series by reference to organic remains, we discover that the evidence is vague and inconclusive. I have as yet been unable to obtain satisfactory proof that any one species of fossil animal or plant has been found in the brown coal, or superimposed formations which was common to a tertiary group of known date in any other part of Europe; whereas the reader will bear in mind that the relative age of different tertiary formations, of which we have before spoken, was usually determined by reference to a comparison of several hundred, often more than a thousand, species of testacea*:

---

* A memoir has lately been communicated to the Geological Society of London, by Mr. Horner, on the geology of this district. For fuller details consult Nöeggerath's Rheinland Westphalen, and the works of Von Dechen, Oyenhausen, Von Buch, Steininger, Van der Wyck, Scrope, Daubeny, Leonhard, and Hibbert.

Miocene period—Marine formations—Faluns of Touraine—Comparison of the Faluns of the Loire and the English Crag—Basin of the Gironde and Landes —Fresh-water limestone of Saucats—Position of the limestone of Blaye— Eocene strata in the Bordeaux basin—Inland cliff near Dax—Strata of Piedmont—Superga—Valley of the Bormida—Molasse of Switzerland—Basin of Vienna—Styria—Hungary—Volhynia and Podolia—Montpellier.

MIOCENE FORMATIONS—MARINE.

HAVING treated in the preceding chapters of the older and newer Pliocene formations, we shall next consider those members of the tertiary series which we have termed Miocene. The distinguishing characters of this group, as derived from its imbedded fossil testacea, have been explained in the fifth chapter (p. 54). In regard to the relative *position* of the strata, they underlie the older Pliocene, and overlie the Eocene formations, when any of these happen to be present.

The area covered by the marine, fresh-water, and volcanic rocks of the Miocene period, in different parts of Europe, can already be proved to be very considerable, for they occur in Touraine, in the basin of the Loire, and still more extensively in the south of France, between the Pyrenees and the Gironde. They have also been observed in Piedmont, near Turin, and in the neighbouring valley of the Bormida, where the Apennines branch off from the Alps. They are largely developed in the neighbourhood of Vienna and in Styria; they abound in parts of Hungary; and they overspread extensive tracts in Volhynia and Podolia.

Shells characteristic of the Miocene strata are found in all these countries, figures of some of which are given in Plate 2 in this volume. They characterize the period, because they are either wanting or extremely rare in the Eocene or Pliocene formations.

We shall now proceed to notice briefly some of the countries before enumerated as containing monuments of the era under consideration.

Touraine.—We have already alluded to the proofs of super-position adduced by M. Desnoyers, to show that the shelly strata provincially called 'the Faluns of the Loire' were pos-terior to the most recent fresh-water formation of the basin of the Seine. Their position, therefore, shows that they are of newer origin than the Eocene strata,—more recent, at least, than the uppermost beds of the Paris basin. But an exami-nation of their fossil contents proves also that they are refer-rible to that type which distinguishes the Miocene period. When three hundred of the Touraine shells were compared with more than eleven hundred of the Parisian species, there were scarcely more than twenty which could be identified ; and, on the other hand, the fossil shells of the Touraine beds agree far less with the testacea now inhabiting our seas than does the group occurring in the older Pliocene strata of northern Italy.

The Miocene strata of the Loire have been observed to repose on a great variety of older rocks between Sologne and the sea, in which line they are seen to rest successively upon gneiss, clay-slate, coal-measures, Jura limestone, greenstone, chalk, and lastly upon the upper fresh-water deposits of the basin of the Seine. They consist principally of quartzose gravel, sand, and broken shells. The beds are generally inco-herent, but sometimes agglutinated together by a calcareous or earthy cement, so as to serve as a building-stone. Like the shelly portion of the crag of Norfolk and Suffolk, the *faluns* and associated strata are of slight thickness, not exceeding seventy feet. They often bear a close resemblance to the crag in appearance, the shells being stained of the same ferru-ginous colour, and being in the same state of decay ; serving in Touraine, just as in Norfolk and Suffolk, to fertilize the arable land. Like the crag, also, they contain mammiferous remains, which are not only intermixed with marine shells, but sometimes encrusted with serpulæ, flustra, and balani. These

terrestrial quadrupeds belong to the genera Mastodon, Rhinoceros, Hippopotamus, &c., the assemblage, considered as a
whole, being very distinct from those of the Paris gypsum.

I examined several detached patches of the Touraine beds,
where they rest on primary strata in the environs of Nantes,
particularly one locality at Les Cleons, about eight miles south-
east of that town, and was struck with the evidence afforded by
them of the emergence of large intervening tracts of granitic
schist since the Miocene era, which we might otherwise have
supposed to have been raised at a very remote epoch. It is
probable that these patches of tertiary deposits were originally
local, having been thrown down wherever the set of the tides
and currents permitted an accumulation to take place.

The faluns and contemporary strata of the basin of the Loire
may be considered generally as having been formed in a shallow
sea, into which a river, flowing perhaps from some of the lands
now drained by the Loire, introduced from time to time fluviatile shells, wood, and the bones of quadrupeds, which may
have been washed down during floods. Some of these bones
have precisely the same black colour as those found in the
peaty shell-marl of Scotland; and we might imagine them to
have been dyed black in *Miocene peat* which was swept down
into the sea during the waste of cliffs, did we not find the
remains of cetacea in the same strata, bones, for example, of
the lamantine, morse, sea-calf, and dolphin, having precisely
the same colour.

*Comparison of the Faluns of the Loire and the English
Crag.*—The resemblance which M. Desnoyers has pointed out
as existing between the English *crag* and the French *faluns*
is one which ought by no means to induce us to ascribe a contemporaneous origin to these two groups, but merely a similarity of geographical circumstances at the respective periods
when each was deposited. In every age, where there is land
and sea, there must be shores, shallow estuaries, and rivers;
and near the sea-coasts banks of marine shells and corals may
accumulate. It must also be expected that rivers will drift in

fresh-water shells, together with sand and pebbles, and occasionally, perhaps, sweep down the carcasses of land quadrupeds into the sea. If the sand and shells, both of the 'crag' and the 'faluns' have each acquired the same ferruginous colour, such a coincidence would merely lead us to infer that, at each period, there happened to be springs charged with iron, which flowed into some part of the sea or basin of the river, by which the sediment was carried down into the sea.

Even had the French and English strata which we are comparing shared a greater number of mineral characters in common, that identity could not have justified us in inferring the synchronous date of the two groups, where the discordance of fossil remains is so marked. The argument which infers a contemporaneous origin from correspondence of mineral contents, proceeds on the supposition that the materials were either washed down from a common source, or from different sources into a common receptacle. If, according to the latter hypothesis, the crag and the faluns were thrown down in one continuous sea, the testacea could not have been so distinct in two very contiguous regions, unless we assume that the laws which regulated the geographical distribution of species were then distinct from those now prevailing. But if it be said that the two basins may have been separated from each other, as are those of the Mediterranean and Red Sea, by an isthmus, and that distinct assemblages of species may have flourished in each, as in the example above-mentioned is actually the case *, we may reply that such narrow lines of demarcation are extremely rare now, and must have been infinitely more so in remoter tertiary epochs, because there can be no doubt that the proportion of land to sea has been greatly on the increase in European latitudes during the more modern geological eras.

In the *faluns*, and in certain groups of the same age, which occur, not far to the west of Orleans, M. Desnoyers has discovered the following mammiferous quadrupeds. *Palæothe-*

* See above, chap. x.

rium magnum, Mastodon angustidens, Hippopotamus major,
and *H. minutus, Rhinoceros leptorhinus,* and *R. minutus, Tapir gigas, Anthracotherium* (small species), *Sus, Equus* (small
species), *Cervus,* and an undetermined species of the Rodentia.

The first species on this list is common to the Paris gypsum,
and is therefore an example of a land quadruped common to
the Miocene and Eocene formations, an exception perfectly in
harmony with the results obtained from the study of fossil
shells *.

Basin of the Gironde and district of the Landes.—A great
extent of country between the Pyrenees and the Gironde is
overspread by tertiary deposits which have been more particularly studied in the environs of Bordeaux and Dax, from
whence about 600 species of shells have been obtained. These
shells belong to the same type as those of Touraine.—See Appendix I.†

Most of the beds near Dax, whence these shells are procured, consist of incoherent quartzose sand, mixed for the most
part with calcareous matter, which has often bound together
the sand into concretionary nodules. A great abundance
of fluviatile shells occur in many places intermixed with the
marine; and in some localities microscopic shells are in great
profusion.

The tertiary deposits in this part of France are often very
inconstant in their mineralogical character, yet admit generally
of being arranged in four groups, which are enumerated in the
explanation of diagram No. 51.

In some places the united thickness of these groups is considerable, but in the country between the Pyrenees and the
valley of the Adour around Dax, the disturbed secondary rocks

* For further details respecting the basin of the Loire, see M. Desnoyers, Ann.
des Sci. Nat., tome xvi. pp. 171 and 402, where full references to other authors
are given.

† M. de Basterot has given a description of more than 300 shells of Bordeaux
and Dax, and figures of the greater number of them. Mém. de la Soc. d'Hist.
Nat. de Paris, tome ii.

are often covered by a thin pellicle only of tertiary strata, which rests horizontally on the chalk and does not always conceal it.

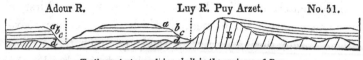

Adour R.　　　　Luy R. Puy Arzet.　　　No. 51.

Tertiary strata overlying chalk in the environs of Dax.

a, Siliceous sand without shells.　　　c, Sand and marl with shells.
b, Gravel.　　　　　　　　　　　　　d, Blue marl with shells.
E, Chalk and volcanic tuff.

In the valleys of the Adour and Luy, sections of all the members of the tertiary series are laid open, but the lowest blue marl, which is sometimes 200 feet thick, is not often penetrated. On the banks of the Luy, however, to the south of Dax, the subjacent white chalk is exposed in inclined and vertical strata. In the hill called Puy Arzet the chalk, characterized by its peculiar fossils, is accompanied by beds of volcanic tuff, which are conformable to it, and which may be considered as the product of submarine eruptions which took place in the sea wherein the chalk was formed.

About a mile west of Orthès, in the Bas Pyrenees, the blue marl is seen to extend to the borders of the tertiary formation, and rises to the height probably of six or seven hundred feet. In that locality many of the marine Miocene shells preserve their original colours. This marl is covered by a considerable thickness of ferruginous gravel, which seems to increase in volume near the borders of the tertiary basin on the side of the Pyrenees.

In an opposite direction, to the north of Dax, the shelly sands often pass into calcareous sandstone, in which there are merely the casts of shells as at Carcares, and into a shelly breccia resembling some rocks of recent origin which I have received from the coral reefs of the Bermudas.

Fresh-water limestone at Saucats.—Associated with the Miocene strata near Bordeaux, at a place called Saucats, is a

compact fresh-water limestone, of slight thickness, which is perforated on the upper surface by marine shells, for the most
part of extinct species.   It is evident that the space must have
been alternately occupied by salt and fresh water.   First, a
lagoon may have been formed, in which the water may have
become fresh ; then a barrier of sand, by which the sea was
excluded for a time, may have been breached, whereby the
salt water again obtained access.

   Eocene strata in the Bordeaux basin.—The relations of some
of the members of the tertiary series, in the basin of the Gironde,
have of late afforded matter of controversy.   A limestone, resembling the calcaire grossier of Paris, and from 100 to 200
feet in thickness, occurs at Pauliac and Blaye, and extends on
the right bank of the Gironde, between Blaye and La Roche.
It contains many species of fossils identical with those of the
Paris basin.   This fact was pointed out to me by M. Deshayes
before I visited Blaye in 1830 ; but although I recognized the
mineral characters of the rock to be very different from those
of the Miocene formations in the immediate neighbourhood
of Bordeaux, I had not time to verify its relative position.   I
inferred, however, the inferiority of the Blaye limestone to the
Miocene strata, from the order in which each series presented
itself as I receded from the chalk and passed to the central
parts of the Bordeaux basin.

   Upon leaving the white chalk with flints, in travelling from
Charente by Blaye to Bordeaux, I first found myself upon
overlying red clay and sand (as at Mirambeau) ; I then came
upon the tertiary limestone above alluded to, at Blaye; and
lastly, on departing still farther from the chalk, reached the
strata which at Bordeaux and Dax contain exclusively the
Miocene shells.

   The occurrence both of Eocene and Miocene fossils in the
same basin of the Gironde, had been cited by M. Boué as a
fact which detracted from the value of zoological characters as
a means of determining the chronological relations of tertiary

groups. But on farther inquiry, the fact, on the contrary, has furnished additional grounds of confidence in these characters.

M. Ch. Desmoulins replied, in answer to M. Boué's objections, that the assemblage of Eocene shells are never intermixed with those found in the ' moellon,' as he calls the sandy calcareous rock of the environs of Bordeaux and Dax; and M. Dufrénoy farther stated, that the hills of limestone which border the right bank of the Gironde, from Marmande as far as Blaye, present several sections wherein the Parisian (or Eocene) limestone is seen to be separated from the shelly strata called ' faluns,' or ' moellon,' by a fresh-water formation of considerable thickness. It appears, therefore, that as the marine faluns of Touraine rest on a fresh-water formation, which overlies the marine calcaire grossier of Paris, so the marine Miocene strata of Bordeaux are separated from those of Blaye by a fresh-water deposit *.

The following diagram, therefore, will express the order of position of the groups above alluded to.

No. 52.

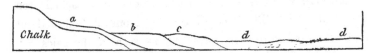

a, Red clay and sand.
b, Limestone like calcaire grossier, sometimes alternating with green marl and containing Eocene shells.
c, Fresh-water formation, same as that of the department of Lot and Garonne.
d, Tertiary strata of the Landes, with Miocene fossils.

Inland cliff near Dax.—A few miles west from Dax, and at the distance of about twelve miles from the sea, a steep bank is seen running in a direction nearly north-east and south-west, or parallel to the contiguous coast. This steep declivity, or *brae*, which is about 50 feet in height, conducts us from the higher platform of the Landes to a lower plain which extends to the sea. The outline of the ground might suggest to every geologist the opinion, that the bank in question was once

* Bulletin de la Soc. Géol. de France, tome ii. p. 440.

a sea-cliff, when the whole country stood at a lower level

No. 53.

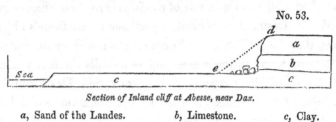

Section of Inland cliff at Abesse, near Dax.

a, Sand of the Landes.   *b*, Limestone.   *c*, Clay.

relatively to the sea. But this can no longer be regarded
as matter of conjecture. In making excavations recently for
the foundation of a building at Abesse, a quantity of loose
sand, which formed the slope *d, e*, was removed, and a perpendicular cliff exposed about 50 feet in height. The bottom of
this cliff consists of limestone, *b*, which contains shells and
corals of Miocene species, and is probably a calcareous form of
the division *c* (diagram No. 51, p. 207). Immediately below
this limestone is the clay *c* (probably *d*, diagram No. 51, p. 207),
and above it the usual tertiary sand *a* of the department
of the Landes. At the base of the precipice are seen large,
partially-rounded, masses of rock, evidently detached from
the stratum *b*. The face of the limestone is hollowed out
and weathered into such forms as are seen in the calcareous
cliffs of the adjoining coast, especially at Biaritz, near Bayonne*.
It is evident that, when the country was at a somewhat lower
level, the sea advanced along the surface of the argillaceous
stratum *c*, which, by its yielding nature, favoured the waste
and undermining of the more solid superincumbent limestone
b. Afterwards, when the country had been elevated, part of
the sand *a* fell down, or was drifted by the winds, so as to
form the talus *d, e*, which masked the inland cliff until it was
artificially laid open to view.

The situation of this cliff is interesting, as marking one of
the pauses which intervened between the successive movements
of elevation whereby the marine tertiary strata of this country

---

* This spot was pointed out to me by the proprietor of the lands of Abesse in
1830.

were upheaved to their present height, a pause which allowed time for the sea to advance and strip off the upper beds *a, b,* from the denuded clay *c.*

Hills of Mont Ferrat and the Superga.—The late Signor Bonelli of Turin was the first who remarked that the tertiary shells found in the green sand and marl of the Superga near Turin differed, as a group, from those generally characteristic of the Subapennine beds. The same naturalist had also observed, that many of the species peculiar to the Superga were identical with those occurring near Bordeaux and Dax. The strata of which the hill of the Superga is composed, are inclined at an angle of more than 70 degrees. They consist partly of fine sand and marl, and partly of a conglomerate composed of primary boulders, which forms a lower part of the series, and not, as represented by M. Brongniart by mistake *, an unconformable and overlying mass †. This same series of beds is more largely developed in the chain of Mont Ferrat, especially in the basin of the Bormida. The high road which leads from Savona to Alessandria intersects them in its northern descent, and the formation may be well studied along this line at Carcare, Cairo, and Spinto, at all which localities fossil shells occur in a bright green sand. At Piana, a conglomerate, interstratified with this green sand, contains rounded blocks of serpentine and chlorite schist, larger than those near the summit of the Superga, some of the blocks being not less than nine feet in diameter.

When we descend to Aqui, we find the green sand giving place to bluish marls, which also skirt the plains of the Tanaro at lower levels. These newer marls are associated with sand, and are nearly horizontal, and appear to belong to the older Pliocene Subapennine strata ‡. The shells which characterize the latter, abound in various parts of the country near Turin; but that region has not yet been examined with sufficient care to enable us to give exact sections to illustrate the superpo-

* Terrains du Vicentin, p. 26.
† I examined the Superga in company with Mr. Murchison in 1828.
‡ See section, wood-cut No. 4, p. 21.

sition of the Miocene and older Pliocene beds. It is, however, ascertained, that the highly-inclined green sand, which comes immediately in contact with the primary rocks, is the oldest part of the series *.

Molasse of Switzerland.—If we cross the Alps, and pass from Piedmont to Savoy, we find there, at the northern base of the great chain and throughout the lower country of Switzerland, a soft green sandstone, much resembling some of the beds of the basin of the Bormida, above described, and associated in a similar manner with marls and conglomerate. This formation is called, in Switzerland, 'molasse,' said to be derived from 'mol,' 'soft,' because the stone is easily cut in the quarry. It is of vast thickness, but shells have so rarely been found in it that they do not supply sufficient data for correctly determining its age. M. Studer, in his treatise on the 'molasse,' enumerates some fossil shells found near Lucerne, agreeing, apparently, with the testacea of the Subapennine hills. The correspondence in mineral character between the green sand of Piedmont and that of Switzerland can in nowise authorise us to infer identity of age, but merely to conclude that both have been derived from the degradation of similar ancient rocks.

Until the place of the 'molasse' in the chronological series of tertiary formations has been more rigorously determined, the application of this provincial name to the tertiary groups of other countries must prove a source of ambiguity, and we regret that the term has been so vaguely employed by M. Boué.

Styria, Vienna, Hungary, &c.—Of the various groups which have hitherto been referred to the Miocene era, none are so important in thickness and geographical extent as those which are found at the eastern extremity of the Alps, in what have been termed the basins of Vienna and Styria, and which spread thence into the plains of Hungary. The collection of shells formed by M. Constant Prevost, in the neighbourhood of

* We trust that MM. Pareto, Passini, Sismonda, and La Marmora, will devote their attention to the relative position of the several groups of tertiary strata in Piedmont, by instituting a comparison between their respective organic remains.

Vienna, and described by him in 1820[*], were alone sufficient to identify a great part of the formations of that country with the Miocene beds of the Loire, Gironde, and Piedmont. The fossil remains subsequently procured by that indefatigable observer M. Boué have served to show the still greater range of the same beds through Hungary and Transylvania.

It appears from the recently published memoirs of Professor Sedgwick and Mr. Murchison[†], that the formations in Styria may be divided into groups corresponding to those adopted by M. Partsch for the Vienna beds; the basin of Vienna exhibiting nearly the same phenomena as that of Styria. These regions have evidently formed, during the Miocene period, two deep bays of the same sea, separated from each other by a great promontory connected with the central ridge of the eastern Alps.

The English geologists, above mentioned, describe a long succession of marine strata intervening between the Alps and the plains of Hungary, which are divisible into three natural groups, each of vast thickness, and affording a great variety of rocks. All these groups are of marine origin, and lie in nearly horizontal strata, but have a slight prevailing easterly dip, so that, in traversing them from west to east, we commence with the oldest and end with the youngest beds. At their western extremity they fill an irregular trough-shaped depression, through which the waters of the Mur, the Raab, and the Drave, make their way to the lower Danube[‡]. They here consist of conglomerate, sandstone, and marls, some of the marls containing marine shells. Beds also of lignite occur, showing that wood was drifted down in large quantities into the sea. In parts of the series there are masses of rounded siliceous pebbles resembling the shingle banks which are forming on some of our coasts.

The second principal group is characterized by coralline

* Journal de Physique, Novembre, 1820.
† Geol. Trans., Second Series, vol. iii. p. 301.   ‡ Ibid., p. 382.

and concretionary limestone of a yellowish white colour : it is
finely exposed in the escarpments of Wildon, and in the hills
of Ehrenhausen, on the right bank of the Mur *. This coralline limestone is not less than 400 feet thick at Wildon, and
exceeds, therefore, some of the most considerable of our
secondary groups in England, as, for example, the 'Coral
Rag †.'

Beds of sandstone, sand, and shale, and calcareous marls, are
associated with the above-mentioned limestone.

The third group, which occurs at a still greater distance
from the mountains, is composed of sandstone and marl, and of
beds of limestone, exhibiting here and there a perfectly oolitic
structure. In this system fossil shells are numerous.

It is by no means clear that the coralline limestones of the
second group, are posterior in origin to all the beds of the first
division ; they may possibly have been formed at some distance
from land, while the head of the gulf was becoming filled up
with enormous deposits of gravel, sand, and mud, which may,
in that quarter, have rendered the waters too turbid for the
fullest development of testaceous and coralline animals.

In regard to the age of the formations above described, we
may observe that the middle group, both in the basins of Styria
and Vienna, belongs indisputably to the Miocene period, for
the species of shells are the same as those of the Loire, Gironde,
and other contemporary basins before noticed. Whether the
lowest and uppermost systems are referrible to the same, or to
distinct tertiary epochs, is the only question. We cannot doubt
that the accumulation of so vast a succession of beds required
an immense lapse of ages, and we are prepared to find some
difference in the species characterizing the different members
of the series ; nevertheless, all may belong to different subdivisions of the Miocene period. Professor Sedgwick and Mr.
Murchison have suggested that the inferior, or first group,
which comprises the strata between the Alps and the coralline

* Geol. Trans., Second Series, vol. iii. p. 355.   † Ibid., p. 390.

limestone of Wildon, may correspond in age to the Paris basin ; but the list of fossils which they have given, seems rather to favour the supposition, that the deposit is of the Miocene era. They enumerate four characteristic Miocene fossils,—Mytilus Brardii, Cerithium pictum, C. pupæforme, and C. plicatum,— and if there are some few of the associated shells common to the Paris basin, such a coincidence is no more than holds true in regard to all the European Miocene formations.

On the other hand, the third or newest system, which over-lies the coralline limestone, contains fossils which do not appear to depart so widely from the Miocene type as to authorize us to separate them. They appear to agree with the tertiary strata of a great part of Hungary and Transylvania, which will be seen, by the tables of shells in Appendix I., to be re-ferrible to the Miocene period.

Volhynia and Podolia.—We may expect to find many other districts in Europe composed of Miocene strata, and there ap-pears already to be sufficient evidence that the marine deposits of the platform of Volhynia and Podolia were of this era. The fossils of that region, which is bounded by Galicia on the west, and the Ukraine on the east, and comprises parts of the basins of the Bog and the Dniester, has been investigated by Von Buch, Eichwald, and Du Bois, and the latter has given excellent plates of more than one hundred fossil shells of the country, which M. Deshayes finds to agree decidedly with the fossils of the Miocene period *.

The formation consists of different rocks, sand and sand-stone, clay, coarse limestone, and a white oolite, the last of which is of great extent.

Montpellier.—The tertiary strata of Montpellier contain many of the Dax and Bordeaux species of shells, so that they are probably referrible to the Miocene epoch ; but in the cata-logue given by M. Marcel de Serres, many *Pliocene* species, similar to those of the Subapennine beds, are enumerated.

---

* Conch. Foss. du Plateau Wolhyni-Podo, par F. du Bois. Berlin, 1831.

This subject requires fuller investigation, and it would be highly interesting if the Montpellier beds should be found to indicate a passage from the fossils of the Miocene type to those of the older Pliocene. We are fully prepared for the discovery of such intermediate links, and we have endeavoured to provide a place for them in the classification proposed in the fifth chapter *.

* Page 57.

CHAPTER XVI.

Miocene alluviums—Auvergne—Mont Perrier—Extinct quadrupeds—Velay—
Orleanais—Alluviums contemporaneous with Faluns of Touraine—Miocene
fresh-water formations—Upper Val d'Arno—Extinct mammalia—Coal of Cadibona—Miocene volcanic rocks—Hungary—Transylvania—Styria—Auvergne
—Velay.

In the present chapter we shall offer some observations on the
alluviums and fresh-water formations of the Miocene era, and
shall afterwards point out the countries in Europe where the
volcanic rocks of the same period may be studied.

MIOCENE ALLUVIUMS.

Auvergne.—The annexed drawing will explain to the reader
the position of two ancient beds of alluvium, *c* and *e,* in Au-

No. 54.

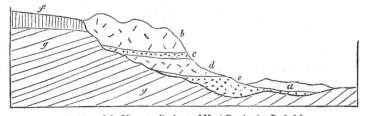

Position of the Miocene alluviums of Mont Perrier (or Boulade).

a, Newer alluvium.

b, Second trachytic breccia.

c, Second Miocene alluvium with bones.

d, First trachytic breccia.

e, First Miocene alluvium with bones.

f, Compact basalt.

g, Eocene lacustrine strata.

vergne, in which the remains of several quadrupeds characteristic of the Miocene period have been obtained. In order to
account for the situation of these beds of rounded pebbles and
sand, we must suppose that after the tertiary strata *g,* covered
by the basaltic lava *f,* had been disturbed and exposed to
aqueous denudation, a valley was excavated, wherein the alluvium *e* accumulated, and in which the remains of quadrupeds

then inhabiting the country were buried.   The trachytic
breccia *d* was then superimposed ; this breccia is an aggregate
of shapeless and angular fragments of trachyte, cemented by
volcanic tuff and pumice, resembling some of the breccias
which enter into the composition of the neighbouring extinct
volcano of Mont Dor in Auvergne, or those which are found
in Etna.   Upon this rests another alluvium *c*, which also contains the bones of Miocene species, and this is covered by
another enormous mass of tufaceous breccia.   We suppose the
breccias to have resulted from the sudden rush of large bodies
of water down the sides of an elevated volcano at its moments
of eruption, when snow perhaps was melted by lava.   Such
floods occur in Iceland, sweeping away loose blocks of lava
and ejections surrounding the crater, and then strewing the
plains with fragments of igneous rocks, enveloped in mud or
' moya.'   The abrupt escarpment presented by the above-
described beds, *b, c, d, e*, towards the valley of the Couze, must
have been caused by subsequent erosion, whereby a large portion of those masses has been carried away *.

In the alluviums *c* and *e*, MM. Croizet, Jobert, Chabriol, and
Bouillet have discovered the remains of about forty species
of extinct mammalia, the greater part of which are peculiar
as yet to this locality ; but some of them characteristic of the
Miocene period, being common to the faluns of Touraine, and
associated in other localities with marine Miocene strata.
Among these species may be enumerated Mastodon minor
and M. arvernensis, Hippopotamus major, Rhinoceros leptorhinus and Tapir arvernensis.   The Elephas primigenius, a
species common to so many tertiary periods, is also stated to
accompany the rest.   In some cases the remains are not sufficiently characteristic to indicate the exact species, but the
following genera can be determined: the boar, horse, ox,
hyæna (two species), felis (three or four species), bear (three

---

* For an account of the position and age of the volcanic breccias of Mont
Perrier and Boulade, see Lyell and Murchison on the beds of Mont Perrier, Ed.
New Phil. Journ., July, 1829, p. 15.

species), deer, a great variety, canis, otter, beaver, hare, and water-rat *.

Velay.—In Velay a somewhat similar group of mammiferous remains were found by Dr. Hibbert †, in a bed of volcanic scoriæ and tuff, inclosed between two beds of basaltic lava, at Saint-Privat d'Allier. Some of the bones were found adhering to the slaggy lava. Among the animals were Rhinoceros leptorhinus, Hyæna spelæa, and another species allied to the spotted hyæna of the Cape, together with four undetermined species of deer ‡.

At Cussac and Solilhac, one league from Puy en Velay, M. Robert discovered, in an ancient alluvium covered with lava, the remains of Elephas primigenius, Rhinoceros leptorhinus, Tapir arvernensis, horse (two species), deer (seven species), ox (two species), and an antelope.

Orleanais.—In the Orleanais, at Avaray, Chevilly, les Aides, and les Barres, fossil land quadrupeds have been found associated with fluviatile shells and reptiles, identical with those found in the marine faluns of Touraine §. These are supposed, with great probability, by M. Desnoyers, to mark the passage of streams which flowed towards the sea in which the faluns were deposited. They bear the same relation to the Miocene strata of Touraine, as part of the ancient gravel and silt of England, containing the bones of elephants and other extinct animals, probably bear to the crag.

MIOCENE FRESH-WATER FORMATIONS.

Upper Val d'Arno.—There are a great number of isolated tertiary formations, of fresh-water origin, resting on primary and secondary rocks in different parts of Europe, in the same

---

* Recherches sur les Oss. Foss. du Dépt. du Puy de Dome, 4to., 1828.—Essai Géol. et Mineral. sur les Environs d'Issoire, Dépt. du Puy de Dome, folio, 1827.

† Edin. Journ. of Sci., No. 4, New Series, p. 276.

‡ Figures of some of these remains are given by M. Bertrand de Doue, Ann. de la Soc. d'Agricult. de Puy, 1828.

§ MM. Desnoyers and Lockart, Bulletin de la Soc. Géol., tom. ii. p. 336.

manner as we now find small lakes scattered over our continents and islands wherein deposits are forming, quite detached from all contemporary marine strata. To determine
the age of such groups with reference to the great chronological series established for the marine strata, must often be a
matter of difficulty, since we cannot always enjoy an opportunity of studying a locality where the fresh-water species are
intermixed with marine shells, or where they occur in beds
alternating with marine strata.

The deposit of the Upper Val d'Arno before alluded to,
(p. 161) was evidently formed in an ancient lake; but although
the fossil testaceous and mammiferous remains preserved
therein are very numerous, it is scarcely possible, at present,
to decide with certainty the precise era to which they belong.
I collected six species of lacustrine shells, in an excellent
state of preservation, from this basin belonging to the genera
Anodon, Paludina and Neritina; but M. Deshayes was unable
to identify them with any recent or fossil species known
to him. If the beds belonged to the older Pliocene formations we might expect that several of the fossils would agree
specifically with living testacea ; and we are therefore disposed
to believe that they belong to an older epoch. If we consider
the terrestrial mammalia of the same beds, we immediately
perceive that they cannot be assimilated to the Eocene type,
as exhibited in the Paris basin, or in Auvergne and Velay :
but some of them agree with Miocene species. Mr. Pentland
has obligingly sent me the following list of the fossil mammifers
of the Upper Val d'Arno which are in the museums of
Paris.——*Feræ.*—Ursus cultridens, Viverra Valdarnensis, Canis
lupus, and another of the size of the common fox. Hyæna
radiata, H. fossilis. Felis (a new species of the size of the
panther). *Rodentia.*—Histrix, nearly allied to dorsalis, Castor.
Pachydermata.—Elephas Italicus, Mastodon angustidens, M.
Taperoides, Tapir——, Equus ——, Sus scrofa, Rhinoceros
leptorhinus, Hippopotamus major, fossilis. *Ruminantia.*—

Cervus megaceros, (?) C. Valdarnensis, C. ——, new species, Bos, bubalo affinis, B. urus and B. taurus.

Cuvier also mentions the remains of a species of lophiodon as occurring among the bones in the Upper Val d'Arno [*]. The elephant of this locality has been called by Nesti [†] *meridionalis,* and is considered by him as distinct from the Siberian fossil species *E. primigenius,* with which, however, some eminent comparative anatomists regard it as identical.   The skeletons of the hippopotamus are exceedingly abundant; no less than forty had been procured when I visited Florence in 1828. Remains of the elephant, stag, ox, and horse, are also extremely numerous.   In winter the superficial degradation of the soil is so rapid, that bones which the year before were buried are seen to project from the surface of the soil, and are described by the peasants as growing.   In this manner the tips of the horns of stags, or of the tusks of hippopotamuses often appear on the surface, and thus lead to the discovery of an entire head or skeleton.

Cadibona.—Another example of an isolated lacustrine deposit, belonging possibly to the Miocene period, is that which occurs at Cadibona, between Savona and Carcare.   Its position is described in the annexed section, which does not however

No. 55.

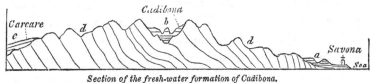

Section of the fresh-water formation of Cadibona.

a. Blue marl and yellow sand (older Pliocene).
b. Sand, shale and coal of Cadibona (Miocene?).
c. Green sand, &c. of the Bormida (Miocene).
d. Chloritic and micaceous schist, serpentine, &c.

pretend to accuracy in regard to the relative heights of the different rocks, or the distances of the places from each other.

* Oss. Foss., vol. v. p. 504.
† Lettere sopra alcune Ossa Fossili del Val d'Arno, &c.   Pisa, 1825.

The lacustrine strata are composed of gravel, grit, and mica-ceous sandstone, of such materials as were derivable from the surrounding primary rocks; and so great is the thickness of this mass, that some valleys intersect it to the depth of seven or eight hundred feet without penetrating to the subjacent for-mations.   In one part of the series, carbonaceous shales occur, and several seams of coal from two to six feet in thickness, but no impressions of plants of which the species could be deter-mined, and no shells have been discovered.   Many entire jaws and other bones of an extinct mammifer, called by Cuvier An-thracotherium, have been found in the coal-beds, the bone being itself changed into a kind of coal; but as this species does not occur elsewhere in association with organic remains of known date, it affords us no aid in our attempt to assign a place to the lignites of Cadibona *.

MIOCENE VOLCANIC ROCKS.

Hungary.—M. Beudant, in his elaborate work on Hungary, describes five distinct groups of volcanic rocks, which, although rarely of great extent, form striking features in the physical geography of that country, rising as they do abruptly from extensive plains composed of tertiary strata.   They may have constituted islands in the ancient sea, as Santorin and Milo now do in the Grecian archipelago; and M. Beudant has re-marked that the mineral products of the last-mentioned islands resemble remarkably those of the Hungarian extinct volcanos, where many of the same minerals, as opal, calcedony, resinous silex (*silex resinite*), pearlite, obsidian, and pitchstone abound.

The Hungarian lavas are chiefly felspathic, consisting of different varieties of trachyte; many are cellular and used as millstones; some so porous and even scoriform as to resemble those which have issued in the open air.   Pumice occurs in great quantity, and there are conglomerates, or rather

* The author visited Cadibona in August, 1828, in company with Mr. Murchison.

breccias, wherein fragments of trachyte are bound together by pumiceous tuff or sometimes by silex.

It is probable that these rocks were permeated by the waters of hot springs, impregnated, like the Geysers, with silica ; or, in some instances perhaps, by aqueous vapours, which, like those of Lancerote, may have precipitated hydrate of silica *.

By the influence of such springs or vapours the trunks and branches of trees washed down during floods, and buried in tuffs on the flanks of the mountains, may have become silicified. It is scarcely possible, says M. Beudant, to dig into any of the pumiceous deposits of these mountains without meeting with opalized wood, and sometimes entire silicified trunks of trees of great size and weight.

It appears from the species of shells collected principally by M. Boué, and examined by M. Deshayes, that the fossil remains imbedded in the volcanic tuffs, and in strata alternating with them in Hungary, are of the Miocene type, and no identical, as was formerly supposed, with the fossils of the Paris basin.

Transylvania.—The igneous rocks of the eastern part of Transylvania described by M. Boué, are probably of the same age. They cover a considerable area, and bear a close resemblance to the Hungarian lavas, being chiefly trachytic. Several large craters, containing shallow lakes like the Maars of the Eifel, are met with in some regions ; and a rent in the trachytic mountains of Budoshagy exhales hot sulphureous vapours, which convert the trachyte into alum-stone, a change which that rock has undergone at remote periods in several parts of Hungary.

Styria.—Many of the volcanic groups of this country bear a similar relation to the Styrian tertiary deposits, as do the Hungarian rocks to the marine strata of that country. The shells are found imbedded in the volcanic tuffs in such a manner as to show that they lived in the sea when the volcanic eruptions were in progress, as many of the Val di Noto lavas

* See above, vol. i. chap. xxii.

in Sicily, before described, were shown to be contemporaneous with newer Pliocene strata *.

Auvergne—Velay.—We believe that part of the volcanic eruptions of Auvergne took place during the Miocene period ; those, for example, which cover, or are interstratified with the alluviums mentioned in this chapter, and some of the ancient basaltic cappings of hills in Auvergne, which repose on gravel characterized by similar organic remains. A part also of the igneous rocks of Velay belong to this epoch, but to these we shall again refer when we treat more fully of the volcanic rocks of Central France ; the older part of which are referrible to the Eocene period.

* Sedgwick and Murchison, Geol. Trans. Second Series, vol. iii. p. 400.— Daubeny, Extinct Volcanos, p. 92.

CHAPTER XVII

Eocene period — Fresh-water formations — Central France — Map — Limagne d'Auvergne—Sandstone and conglomerate—Tertiary Red marl and sandstone like the secondary 'new red sandstone'—Green and white foliated marls—Indusial limestone—Gypseous marls—General arrangement and origin of the Travertin—Fresh-water formation of the Limagne—Puy en Velay—Analogy of the strata to those of Auvergne—Cantal—Resemblance of Aurillac limestone and its flints to our upper chalk—Proofs of the gradual deposition of marl—Concluding Remarks.

EOCENE FRESH-WATER FORMATIONS.

WE have now traced back the history of the European formations to that period when the seas and lakes were inhabited by a few only of the existing species of testacea, a period which we have designated *Eocene*, as indicating the *dawn* of the present state of the animate creation. But although a small number only of the living species of animals were then in being, there are ample grounds for inferring that all the great classes of the animal kingdom, such as they now exist, were then fully represented. In regard to the testacea, indeed, it is no longer a matter of inference, for 1400 species of this class have been obtained from that small number of detached Eocene deposits which have hitherto been examined in Europe.

The celebrated Paris basin, the position of which was pointed out in the former part of this volume, (see wood-cut, p. 16) first presents itself, and seems to claim our chief attention when we treat of the phenomena of this era. But in order more easily to explain to the student the peculiar nature and origin of that group, it will be desirable, first, to give a brief sketch of certain deposits of Central France, which afford many interesting points of analogy, both in organic remains and mineral composition, and where the original circumstances under which the strata were accumulated may more easily be discerned.

Q

Auvergne.—We allude to the lacustrine basins of Auvergne, Cantal, and Velay, the site of which may be seen in the

No. 56.

annexed Map *. They appear to be the monuments of ancient
lakes which may have resembled in geographical distribution
some of those now existing in Switzerland, and may like them
have occupied the depressions in a mountainous country, and
have been each fed by one or more rivers and torrents. The
country where they occur is almost entirely composed of
granite, and different varieties of granitic schist, with here
and there a few patches of secondary strata much dislocated,
and which have probably suffered great denudation. There
are also some vast piles of volcanic rock, (see the Map,) the
greater part of which are newer than the fresh-water strata,
often resting upon them, whilst a small part were evidently of
contemporaneous origin. Of these igneous rocks we shall treat
more particularly in the nineteenth chapter, and shall first turn
our attention exclusively to the lacustrine beds.

The most northern of the fresh-water groups is situated in
the valley-plain of the Allier, which lies within the department
of the Puy de Dome, being the tract which went formerly by
the name of the Limagne d'Auvergne. It is inclosed by two
parallel primitive ranges,—that of the Forèz, which divides the
waters of the Loire and Allier, on the east, and that of the Monts
Domes, which separates the latter river from the Sioule, on the
west †. The average breadth of this tract is about 20 miles,
and it is for the most part composed of nearly horizontal strata
of sand, sandstone, calcareous marl, clay, limestone, and some
subordinate groups, none of which observe a fixed and invariable order of superposition. The ancient borders of the lake,
wherein the fresh-water strata were accumulated, may generally
be traced with precision, the granite and other ancient rocks
rising up boldly from the level country. The precise junction, however, of the lacustrine and granitic beds is rarely seen,
as a small valley usually intervenes between them. The fresh-

---

* The following account of the fresh-water formations of Central France is the
result of observations made in the summer of 1828, in company with Mr. Murchison.

        † Scrope, Geology of Central France, p. 15.

water strata may sometimes be seen to retain their horizontality within a very slight distance of the border-rocks, while in some places they are inclined, and in a few instances vertical. The principal divisions into which the lacustrine series may be separated are the following : 1st, Sandstone, grit, and conglomerate. 2ndly, green and white foliated marls. 3dly, limestone or travertin, oolite, &c. 4thly, gypseous marls.

1. *Sandstone and conglomerate.*—Strata of sand and gravel, sometimes bound together into a solid rock, are found in great abundance around the confines of the lacustrine basin, containing, in different places, pebbles of all the ancient rocks of the adjoining elevated country, namely, granite, gneiss, mica-schist, clay-slate, porphyry, and others. But the arenaceous strata do not form one continuous band around the margin of the basin, being rather disposed like the independent deltas which grow at the mouths of torrents along the borders of existing lakes *.

At Chamalieres, near Clermont, we have an example of one of these littoral groups of local extent where the pebbly beds slope away from the granite as if they had formed a talus beneath the waters of the lake near the steep shore. A section, of about 50 feet in vertical height, has been laid open by a torrent, and the pebbles are seen to consist throughout of rounded and angular fragments of granite, quartz, primary slate, and red sandstone, but without any intermixture of those volcanic rocks which now abound in the neighbourhood. Partial layers of lignite and pieces of wood are found in these beds, but no shells, a fact which probably indicates that testacea could not live where the turbid waters of a stream were frequently hurrying down uprooted trees, together with sand and pebbles, or, that if they existed, they were triturated by the transported rocks.

There are other localities on the margin of the basin where quartzose grits are found, composed of white sand bound together by a siliceous cement.

* See vol. i. chap. xiv. p. 249; and 2nd. Ed. p. 286.

Occasionally, when the grits rest on granite, as at Chamalieres before mentioned, and many other places, the separate
crystals of quartz, mica, and felspar, of the disintegrated
granite, are bound together again by the silex, so that the
granite seems regenerated in a new and even more solid form,
and thus so gradual a passage may sometimes be traced between a crystalline rock and one of mechanical origin, that we
can scarcely distinguish where one ends and the other begins.

In the Puy de Jussat, and the neighbouring hill of La
Roche, are white quartzose grits, cemented into a sandstone
by calcareous matter, which is sometimes so abundant as to
form imbedded nodules. These sometimes constitute spheroidal concretions six feet in diameter, and pass into beds of
solid limestone resembling the Italian travertins, or the deposits of mineral springs.

In the hills above mentioned, we have the advantage of seeing a section continuously exposed for about 700 feet in thickness. At the bottom are foliated marls, white and green,
about 400 feet thick, and above, resting on the marls, are the
quartzose grits before mentioned with the associated travertins.
This section is observed close to the confines of the basin, so
that the lake must here have been filled up near the shore with
fine mud, before the coarse superincumbent sand was introduced. There are other cases where sand is seen below the
marl.

2. *Red marl and sandstone.*—But the most remarkable of the
arenaceous groups is a red sandstone and red marl, identical in
all their characters with the secondary *new red sandstone* and
marl of England. In the latter, the red ground is sometimes
variegated with light greenish spots, and the same may be seen
in its tertiary counterpart of fresh-water origin at Coudes, on
the Allier. The marls are sometimes of a purplish-red colour,
as at Champheix, and are accompanied by a reddish limestone,
like the well-known 'cornstone,' which is associated with the
old red sandstone of English geologists. The red sandstone
and marl of Auvergne have evidently been derived from the

degradation of gneiss and mica-schist, which are seen *in situ* on the adjoining hills, decomposing into a soil very similar to the tertiary red sand and marl. We also find pebbles of gneiss, mica-schist, and quartz, in the coarser sandstones of this group, clearly pointing to the parent rocks from which the sand and marl were derived. The red beds, although destitute of organic remains, pass upwards into strata containing Eocene fossils, and are certainly an integral part of the lacustrine formation.

3. *Green and white foliated marls.*—A great portion of what we term clay in ordinary language, consists of the same materials as sandstone, but the component parts are in a finer state of subdivision. The same primary rocks, therefore, of Auvergne, which, by the partial degradation of their harder parts, gave rise to the quartzose grits and conglomerates before mentioned, would, by the reduction of the same into powder, and by the decomposition of their felspar, mica, and hornblende, produce aluminous clay, and, if a sufficient quantity of carbonate of lime was present, calcareous marl. This fine sediment would naturally be carried out to a greater distance from the shore, as are the various finer marls now deposited in Lake Superior *. And, as in the American lake, shingle and sand are annually amassed near the northern shores, so in Auvergne the grits and conglomerates before mentioned were evidently formed near the borders.

The entire thickness of these marls is unknown, but it certainly exceeds, in some places, 700 feet. They are for the most part either light-green or white, and usually calcareous. They are thinly foliated, a character which frequently arises from the innumerable thin plates or scales of that small animal called *cypris*, a genus which comprises several species, of which some are recent, and may be seen swimming rapidly through the waters of our stagnant pools and ditches. This animal resides within two small valves like those of a bivalve shell, and it moults its integuments annually, which the conchiferous

* See vol. i. chap. xiii.

molluscs do not. This circumstance may partly explain the countless myriads of the shells of cypris which were shed in the Eocene lakes, so as to give rise to divisions in the marl as thin as paper, and that too in stratified masses several hundred feet thick. A more convincing proof of the tranquillity and clearness of the waters, and of the slow and gradual process by which the lake was filled up with fine mud, cannot be desired. We may easily suppose that, while in the deep and central parts of the basin, this fine sediment was thrown down, gravel, sand, and rocky fragments were hurried into the lake near the shore, and formed the group first described.

Not far from Clermont the green marls, containing the cypris in abundance, approach to within a few yards of the granite which forms the borders of the basin. The annexed section occurs at Champradelle, in a small ravine north of La petite Baraque, and above the bridge.

No. 57.

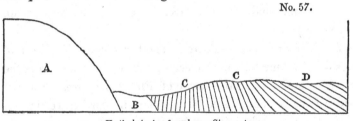

Vertical strata of marl near Clermont.

A, Granite.   C, Green marl, vertical and inclined.
B, Space of 60 feet in which no section is seen.   D. White marl.

The occurrence of these marls so near the ancient margin may be explained by considering that, at the bottom of the ancient lake in spaces intermediate between the points where rivers and torrents entered, no coarse ingredients were deposited, but finer mud only was drifted by currents. The *verticality* of some of the beds in the above section bears testimony to considerable local disturbance subsequent to the deposition of the marls, but such inclined and vertical strata are very rare.

4. *Limestone, travertin, &c.*—Both the preceding members of the lacustrine deposit, the marls and grits, pass occasionally

into limestone.   Sometimes only concretionary nodules abound in them ; but these, by an· additional quantity of calcareous matter, unite, as already noticed (p. 229), into regular beds.

On each side of the basin of the Limagne, both on the east at Gannat, and on the west at Vichy, a white oolitic limestone is quarried.   At Vichy, the oolite resembles our Bath stone in appearance and beauty, and, like it, is soft when first taken from the quarry, but soon hardens on exposure to the air.   At Gannat, the stone contains land-shells and bones of quadrupeds, resembling those of the Paris gypsum.   In several places in the neighbourhood of Gannat, at Marculot among others, this stone is divided by layers of clay.

At Chadrat, in the hill of La Serre, the limestone is pisolitic, and in this and other respects resembles the travertin of Tivoli. It presents the same combination, of a radiated and concentric structure, and the coats of the different segments of spheroids have the same undulating surface.   (See wood-cut No. 5, chap. xii. vol. i.)

Indusial limestone.—There is another remarkable form of fresh-water limestone in Auvergne, called ' indusial,' from the cases, or *indusiæ*, of the larvæ of Phryganea, great heaps of which have been encrusted, as they lay, by hard travertin, and formed into a rock.   We may often see, in our ponds, some of the living species of these insects, covered with small fresh-water shells, which they have the power of fixing to the outside of their tubular cases, in order, probably, to give them weight and strength.   It appears that, in the same manner, a large species which swarmed in the Eocene lakes of Auvergne, was accustomed to attach to its dwelling the shells of a small spiral univalve of the genus *Paludina*.   A hundred of these minute shells are sometimes seen arranged around one tube, part of the central cavity of which is still occasionally empty, the rest being filled up with thin concentric layers of travertin.   When we consider that ten or twelve tubes are packed within the compass of a cubic inch, and that some single strata of this limestone are six feet thick, and may be traced over a considerable area, we may form some idea of the countless number of insects and mollusca which contributed their integuments and shells to compose this singularly constructed rock. It is unnecessary to suppose that the Phryganeæ lived on the spots where their cases are now found ; they may have multiplied in the shallows near the margin of the lake, and their buoyant cases may have been drifted by a current far into the deep water.

The calcareous strata of the Limagne, like the other members of the lacustrine formation, are for the most part horizontal, or inclined at a very slight angle, but instances of local dislocation are sometimes seen. At the town of Vichy, for example, the strata dip at an angle of between 30 and 40 degrees ; in an ancient quarry behind the convent of Celestines, and near the hot spring at the same place, the beds of limestone are seen first inclined at an angle of 80°, and then vertical.

5. *Gypseous marls.*—More than 50 feet of thinly-laminated gypseous marls, exactly resembling those in the hill of Montmartre, at Paris, are worked for gypsum at St. Romain, on the right bank of the Allier. They rest on a series of green cypriferous marls which alternate with grits, the united thickness of this inferior group being seen, in a vertical section on the banks of the river, to exceed 250 feet.

General arrangement and origin of the fresh-water formations of Auvergne.—The relations of the different groups above described cannot be learnt by the study of any one section, and he who sets out with the expectation of finding a fixed order of succession may perhaps complain that the different parts of the basin give contradictory results. The arenaceous division, the marls and the limestone, may all be seen in some localities to alternate with each other, yet it can by no means be affirmed that there is no order of arrangement. The sands, sandstone, and conglomerate, constitute in general a littoral group; the foliated white and green marls a contemporaneous central deposit, and the limestone is for the most part subordinate to the newer portions of the above groups.

We never meet with calcareous rocks covered by a considerable thickness of quartzose sand or green marl, and the uppermost marls and sands are more calcareous than the lower.
From the resemblance of the Eocene limestones of Auvergne
to the Italian travertins, we may conclude that they were derived from the waters of mineral springs,—such springs as now
exist in Auvergne, and which rising up through the granite
precipitate travertin. They are sometimes thermal, but this
character is by no means constant.

We suppose that, when the ancient lake of the Limagne first
began to be filled with sediment, no volcanic action had produced lava and scoriæ on any part of the surface of Auvergne.
No pebbles, therefore, of lava were transported into the lake,—
no fragments of volcanic rocks imbedded in the conglomerate.
But at a later period, when a considerable thickness of sandstone and marl had accumulated, eruptions broke out, and lava
and tuff were alternately deposited, at some spots, with the lacustrine strata.   Of this we shall give proofs in the 19th chapter.
It is not improbable that cold and thermal springs, holding
different mineral ingredients in solution, increased in number
during the successive convulsions attending this development
of volcanic agency, and thus carbonate and sulphate of lime,
silex, and other minerals, were produced. Hence these minerals predominate in the uppermost strata. The subterranean
movements may then have continued until they altered the
relative levels of the country and caused the waters of the lakes
to be drained off, and the farther accumulation of regular
fresh-water strata to cease. The occurrence of these convulsions anterior to the Miocene epoch, and prolonged during
a succession of after-ages, may explain why no fresh-water formations more recent than the Eocene are now found in this
country.

We may easily conceive a similar series of events to give rise
to analogous results in any modern basin, such as that of Lake
Superior, for example, where numerous rivers and torrents are
carrying down the detritus of a chain of mountains into the

lake. The transported materials must be arranged according to their size and weight, the coarser near the shore, the finer at a greater distance from land; but in the gravelly and sandy beds of Lake Superior no pebbles of modern volcanic rocks can be included, since there are none of these at present in the district. If the igneous action should break out in that country and produce lava, scoriæ, and thermal springs, the deposition of gravel, sand, and marl, might still continue as before; but in addition, there would then be an intermixture of volcanic gravel and tuff, and rocks precipitated from the waters of mineral springs.

Although the fresh-water strata of the Limagne approach generally to a horizontal position, the proofs of local disturbance are sufficiently numerous and violent to allow us to suppose great changes of level since the Eocene period. We are unable to assign a northern barrier to the ancient lake, although we can still trace its limits to the east, west, and south, where they were formed of bold granitic eminences. But we need not be surprised at our inability to restore the physical geography of the country after so great a series of volcanic eruptions. It is by no means improbable that one part of the district may have been moved upwards bodily, while the others remained at rest, or even suffered a movement of depression.

Puy en Velay.—In the department of the Haute Loire, a fresh-water formation, very analogous to that of Auvergne, is situated in the basin of the Loire, and is exposed in the valley in which stands the town of Le Puy. Since the deposition of the lacustrine strata, there have been so many volcanic eruptions in this country, and such immense quantities of lava and scoriæ poured out upon the surface, that the aqueous rocks are almost buried and concealed. We are indebted, however, to the researches of M. Bertrand de Doue for having distinctly ascertained the succession of strata, and we have had opportunities of verifying his observations during a visit to Le Puy.

In this basin we find, as in Auvergne, two great divisions, consisting of grits and marls ; the former composed of quartzose grit, sometimes granitiform, reddish and mottled sands and conglomerates, all evidently derived from the degradation of granitic rocks, and resembling exceedingly the arenaceous group of the Limagne before described. This formation is almost confined to the borders of the basin, and was evidently a littoral deposit. The other member of the formation, the *marls*, are more or less calcareous, and are associated with limestone and gypsum, which last is worked for agricultural uses, and exactly resembles that of Paris.

The analogy in the mineral character of the Velay and Paris basins is rendered more complete by the presence in both of silex in regular beds. In the limestone I found gyrogonites, or seeds of the Chara, of the same species as those most common in the Paris basin; and M. Bertrand de Doue has discovered the bones of several mammiferous animals of the same genera as those which characterize the basins of Auvergne and Paris [*]. The shells also of this formation correspond specifically with those of Eocene formations in other parts of France.

The sand and conglomerate of the fresh-water basin of Velay is entirely free from volcanic pebbles, agreeing in this respect with the analogous group of the Limagne; but the fact is the more striking in Velay, because the masses of trachyte, clinkstone, and other igneous rock now abounding in that country, have an aspect of extremely high antiquity, and constitute a most prominent feature in the geological structure of the district. Yet the non-intermixture of volcanic products with the lacustrine sediment, is just what we should expect when we have ascertained that the imbedded organic remains of those strata are Eocene; whereas the lavas belong in part, if not entirely, to the Miocene period [†].

Cantal.—Near Aurillac, in Cantal, another series of fresh-water strata occurs, which resembles, in mineral character and organic remains, those of Auvergne and Velay already described. The leading feature of this group, as distinguished

---

[*] Descrip. Géognos. des Env. du Puy en Velay, 1823.
[†] See above, p. 219, and below, Chap. xix.

from the two former, is the immense abundance of silex associated with the calcareous marls and limestone, which last, like
the limestone of Auvergne, constitutes an upper member of the
fresh-water series.

The formation of the Cantal may be divided into two
groups, the lowest composed of gravel, sand, and clay, such as
might have been derived from the wearing down and decomposition of the granitic schists of the surrounding country; the
upper system consisting of siliceous and calcareous marls,
contains subordinately gypsum, silex, and limestone—deposits
such as the waters of springs charged with carbonate and
sulphate of lime, and with silica, may have produced.

Fresh-water limestone and flints resembling chalk.—To the
English geologist, the most interesting feature in the Cantal
is the resemblance of the fresh-water limestone, and its accompanying flint, to our upper chalk, a resemblance which,
like that of the red sandstone of Auvergne to our secondary
' new red,' is the more important, as being calculated to
put the student upon his guard against too implicit a reliance on lithological characters as tests of the relative ages of
rocks. When we approach Aurillac from the west, we pass
over great heathy plains, where the sterile mica-schist is barely
covered with vegetation. Near Ytrac, and between La Capelle
and Viscamp, we begin to see the surface strewed over with
loose broken flints, some of them black in the interior, but with
a white external coating, others stained with tints of yellow and
red, and looking precisely like the flint gravel of our chalk districts. When heaps of this gravel have thus announced our
approach to a new formation, we arrive at length at the escarpment of the lacustrine beds. At the bottom of the hill we see
strata of clay and sand resting on mica-schist; and above, in
the quarries of Belbet, Leybros, and Bruel, a white limestone,
in horizontal strata, the surface of which has been hollowed
out into irregular furrows, since filled up with broken flint,
marl, and vegetable mould. We recognize in these cavities,
filled with dark mould and flint gravel, an exact counterpart to

the appearances so frequently presented on the furrowed surface of our white chalk. Proceeding onwards from these quarries, along a road made of the white limestone, which reflects as glaring a light in the sun, as do our roads composed of chalk, we reach, at length, in the neighbourhood of Aurillac, hills of limestone and calcareous marl, in horizontal strata, separated in some places by regular layers of flint in nodules, the coating of each nodule being of an opaque white colour, like the exterior of the flinty nodules of our chalk. In these last the hard white substance has been ascertained to consist, in some instances, wholly of siliceous matter, and sometimes to contain a small admixture of carbonate of lime *, and the analysis of those of the Cantal would probably give the same results. The Aurillac flints have precisely the appearance of having separated from their matrix after the siliceous and calcareous matter had been blended together. The calcareous marl sometimes occupies small sinuous cavities in the flint, and the siliceous nodule, when detached, is often as irregular in form as those found in our chalk.

By what means, then, can the geologist at once decide that the limestone and silex of Aurillac are referrible to an epoch entirely distinct from that of the English chalk? It is not by reference to position, for we can merely say of the lacustrine beds, as we should have been able to declare of the true chalk had it been present, that they overlie the granitic rocks of this part of France. It is by reference to the organic remains that we are able to pronounce the formation to belong to the Eocene tertiary period. Instead of the marine Alcyonia of our cretaceous system, the silicified seed-vessels of the Chara, a plant which grows at the bottom of lakes, abound in the flints of Aurillac, both in those which are *in situ* and those forming the gravel. Instead of the Echinus and marine testacea of the chalk, we find in the marls and limestones the shells of the Planorbis, and other lacustrine testacea, all of

* Phillips, Geol. Trans. First Series, vol. v. p. 22.—Outlines of Geology, p. 95.

them, like the gyrogonites, agreeing specifically with species of the Eocene type.

Proofs of the gradual deposition of marl.—Some sections of the foliated marls in the valley of the Cer, near Aurillac, attest, in the most unequivocal manner, the extreme slowness with which the materials of the lacustrine series were amassed. In the hill of Barrat, for example, we find an assemblage of calcareous and siliceous marls, in which, for a depth of at least 60 feet, the layers are so thin that thirty are sometimes contained in the thickness of an inch ; and when they are separated we see preserved in each the flattened stems of Charæ, or other plants, or sometimes myriads of small *paludinæ* and other fresh-water shells. These minute foliations of the marl resemble precisely some of the recent laminated beds of the Scotch marl lakes, and when divided may be compared to the pages of a book, each containing a history of a certain period of the past. The different layers may be grouped together in beds from a foot to a foot and a half in thickness, which are distinguished by differences of composition and colour, the latter being white, green, and brown. Occasionally there is a parting layer of pure flint, or of black carbonaceous vegetable matter, one inch thick, or of white pulverulent marl. We find several hills in the neighbourhood of Aurillac composed of such materials for the height of more than 200 feet from their base, the whole sometimes covered by rocky currents of trachytic or basaltic lava *.

Concluding remarks.—So wonderfully minute are the separate parts of which some of the most massive geological monuments are made up ! When we desire to classify, it is necessary to contemplate entire groups of strata in the aggregate ; but if we wish to understand the mode of their formation, and to explain their origin, we must think only of the minute subdivisions of which each mass is composed. We must bear in mind how many thin, leaf-like seams of matter, each con-

* Lyell and Murchison, sur les Dépôts Lacust. Tertiaires du Cantal, &c. Ann. des Sci. Nat., Oct. 1829.

taining the remains of myriads of testacea and plants, frequently enter into the composition of a single stratum, and how great a succession of these strata unite to form a single group! We must remember, also, that volcanos like the Plomb du Cantal, which rises in the immediate neighbourhood of Aurillac, are equally the result of successive accumulation, consisting of reiterated flows of lava and showers of scoriæ; and we have shown, when we treated of the high antiquity of Etna, how many distinct lava-currents and heaps of ejected substances are required to make up one of the numerous conical envelopes whereof a volcano is composed.—Lastly, we must not forget that continents and mountain-chains, colossal as are their dimensions, are nothing more than an assemblage of many such igneous and aqueous groups, formed also in succession during an indefinite lapse of ages, and superimposed upon each other.

Marine formations of the Eocene period—Strata of the Paris basin how far analogous to the lacustrine deposits of Central France—Geographical connexion of the Limagne d'Auvergne and the Paris basin—Chain of lakes in the Eocene period—Classification of groups in the Paris basin—Observations of M. C. Prevost—Sketch of the different subdivisions of the Paris basin—Contemporaneous marine and fresh-water strata—Abundance of Cerithia in the Calcaire grossier—Upper marine formation indicates a subsidence—Part of the Calcaire grossier destroyed when the upper marine strata originated—All the Parisian groups belong to one great epoch—Microscopic shells—Bones of quadrupeds in gypsum—In what manner entombed—Number of species—All extinct—Strata with and without organic remains alternating—Our knowledge of the physical geography, fauna, and flora of the Eocene period considerable—Concluding remarks.

EOCENE FORMATIONS—PARIS BASIN.

THE geologist who has studied the lacustrine formations described in the last chapter cannot enter the tract usually termed ' the Paris Basin ' without immediately recognizing a great variety of rocks with which his eye has already become familiar. The green and white marls of Auvergne, Cantal, and Velay, again present themselves, together with limestones and quartzose grits, siliceous and gypseous marls, nodules and layers of flint, and saccharoid gypsum ; lastly, in addition to all this identity of mineral character, we find an assemblage of the same species of fossil animals and plants.

When we consider the geographical proximity of the two districts, we are the more prepared to ascribe this correspondence in the mineral composition of these groups to a combination of similar circumstances in the same era. From the map (No. 56, p. 226) in the last chapter, it will be seen that the united waters of the Allier and Loire, after descending from the valleys occupied by the fresh-water formations of Central France, flow on till they reach the southern extremity of what is called the Paris basin. M. Omalius d'Halloy long ago suggested the very natural idea that there existed formerly a chain of lakes, reaching from the highest part of the central mountain-group of France, and terminating in the basin of Paris, which he supposes was at that time an arm of the sea.

Notwithstanding the great changes which the physical geography of that part of France must since have undergone, we may easily conceive that many of the principal features in the configuration of the country may have remained unchanged, or but slightly modified. Hills of volcanic matter have indeed been formed since the Eocene formations were accumulated, and the levels of large tracts have been altered in relation to the sea ; lakes have been drained, and a gulf of the sea turned into dry land, but many of the reciprocal relations of the different parts of the surface may still remain the same. The waters which flowed from the granitic heights into the Eocene lakes may now descend in the same manner into valleys once the basins of those lakes. Let us, for example, suppose the great Canadian lakes, and the gulf into which their waters are discharged, to be elevated and laid dry by subterranean movements. The whole hydrographical basin of the St. Lawrence might be upraised during these convulsions, yet that river might continue, after so extraordinary a revolution, to drain the same elevated regions, and might continue to convey its waters in the same direction from the interior of the continent to the Atlantic. Instead of traversing the lakes, it would hold its course through deposits of lacustrine sand and shelly marl, such as we know to be now forming in Lakes Superior and Erie ; and these fresh-water strata would occupy the site and bear testimony to the pristine existence of the lakes. Marine strata might also be brought into view in the space where an inlet of the sea, like the estuary of the St. Lawrence, had once received the continental waters ; and in such formations we might discover shells of lacustrine and fluviatile species intermingled with marine testacea and zoophytes.

Subdivisions of strata in the Paris basin.—The area which has been called the Paris basin is about one hundred and eighty miles in its greatest length from north-east to south-west, and about ninety miles from east to west. This space may be described as a depression in the chalk (see diagram No. 2, p. 16), which has been filled up by alternating groups of marine and fresh-water strata. MM. Cuvier and Brongniart attempted in 1811 to distinguish five different formations, and to arrange them in the following order, beginning with the lowest :—

1. First fresh-water formation.... $\begin{cases}\text{Plastic clay.}\\\text{Lignite.}\\\text{First sandstone.}\end{cases}$

2. First marine formation .... Calcaire grossier.

3. Second fresh-water formation .. $\begin{cases}\text{Siliceous limestone.}\\\text{Gypsum, with bones of animals.}\\\text{Fresh-water marls.}\end{cases}$

4. Second marine formation....... $\begin{cases}\text{Gypseous marine marls.}\\\text{Upper marine sands and sandstones.}\\\text{Upper marine marls and limestones.}\end{cases}$

5. Third fresh-water formation. $\begin{cases}\text{Siliceous millstone, without shells.}\\\text{Siliceous millstone, with shells.}\\\text{Upper fresh-water marls.}\end{cases}$

These formations were supposed to have been deposited in succession upon the chalk ; and it was imagined that the waters of the ocean had been by turns admitted into and excluded from the same region. But the subsequent investigations of

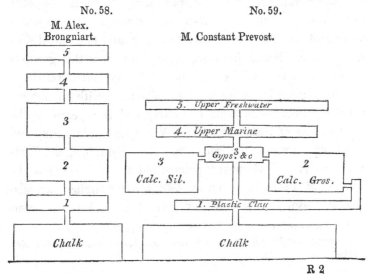

No. 58.   No. 59.

M. Alex.
Brongniart.   M. Constant Prevost.

several geologists, especially of M. Constant Prevost *, have
led to great modifications in the theoretical views entertained
respecting the order in which the several groups were formed ;
and it now appears that the formations Nos. 1, 2, and 3, of
the table of MM. Cuvier and Brongniart, instead of having
originated one after the other, are divisible into four nearly
contemporaneous groups.

Superposition of different formations in the Paris basin.—
A comparison of the two accompanying diagrams will enable
the reader to comprehend at a glance the different relations
which the several sets of strata bear to each other, according
to the original, as well as the more modern classification.
We shall now proceed to lay before the reader a brief sketch
of the several sets of strata referred to in the above systems.

Immediately upon the chalk a layer of broken chalk flints,
often cemented into a breccia by siliceous sand, is very commonly found.   These flints probably indicate the action of the
sea upon reefs of chalk when a portion of that rock had emerged
and before the regular tertiary beds were superimposed.   To
this partial layer no reference is made in the annexed sections.

Plastic clay and sand.—Upon this flinty stratum, or, if it
be wanting, upon the chalk itself, rests frequently a deposit of
clay and lignite (No. 1 of the above tables).   It is composed
of fresh-water shells and drift-wood, and was, at first, regarded
as a proof that the Paris basin had originally been filled with
fresh water.   But it has since been shown that this group is
not only of very partial extent, but is by no means restricted
to a fixed place in the series ; for it alternates with the marine
calcaire grossier (No. 2 of the tables), and is repeated in the
very middle of that limestone at Veaugirard, Bagneux, and
other places, where the same Planorbes, Paludinæ, and Limnei occur †.   M. Desnoyers pointed out to me a section in the
suburbs of Paris, laid open in 1829, where a similar intercalation was seen in a still higher part of the calcaire grossier.

* Bulletin des Sci. de la Soc. Philom.; May, 1825, p. 74.

† Prevost, Sur les Submersions Itératives, &c.   Mem. de la Soc. d'Hist. Nat.
de Paris, tome iv. p. 74.

These observations relieve us from the difficulty of seeking a cause why vegetable matter, and certain species of fresh-water shells and a particular kind of clay, was first introduced into the basin, and why the same space was subsequently usurped by the sea. A minute examination of the phenomena leads us simply to infer, that a river charged with argillaceous sediment entered a bay of the sea and drifted down, from time to time, fresh-water shells and wood.

Calcaire grossier.—The calcaire grossier above alluded to, is composed of a coarse limestone, often passing into sand, such as may perhaps have been derived from the aqueous degradation of a chalk country. It contains by far the greater number of the fossil shells which characterize the Paris basin. No less than 400 distinct species have been derived from a single locality near Grignon. They are imbedded in a calcareous sand, chiefly formed of comminuted shells, in which, nevertheless, individuals in a perfect state of preservation, both of marine, terrestrial, and fresh-water species, are mingled together, and were evidently transported from a distance. Some of the marine shells may have lived on the spot, but the Cyclostoma and Limnea must have been brought there by rivers and currents, and the quantity of triturated shells implies considerable movement in the waters.

Nothing is more remarkable in this assemblage of fossil testacea than the astonishing proportion of species referrible to the genus Cerithium *. There occur no less than 137 species of this genus in the Paris basin, and almost all of them in the calcaire grossier. Now the living testacea of this genus inhabit the sea near the mouths of rivers, where the waters are brackish, so that their abundance in the marine strata of the Paris basin is in perfect harmony with the hypothesis before advanced, that a river flowed into the gulf, and gave rise to the beds of clay and lignite before mentioned. But there are ample data for inferring that the gulf was supplied with fresh water by more than one river, for while the calcaire grossier occupies the northern

* See the tables of M. Deshayes, Appendix I., p. 26.

part of the Paris basin, another contemporaneous deposit, of fresh-water origin, appears at the southern extremity.

Calcaire siliceux.—This group (No. 3 of the foregoing tables) is a compact siliceous limestone, which resembles a precipitate from the waters of mineral springs. It is, for the most part, devoid of organic remains, but in some places it contains fresh-water and land species, and never any marine fossils. The siliceous limestone and the calcaire grossier occupy distinct parts of the basin, the one attaining its fullest development in those places where the other is of slight thickness. They also alternate with each other towards the centre of the basin, as at Sergy and Osny, and there are even points where the two rocks are so blended together, that portions of each may be seen in hand specimens. Thus in the same bed, at Triel, we have the compact fresh-water limestone, characterized by its Limnei, mingled with the coarse marine limestone through which the small multilocular shell, called milliolite, is dispersed in countless numbers. These microscopic testacea are also accompanied by Cerithia and other shells of the calcaire grossier. It is very extraordinary that, although in this instance both kinds of sediment must have been thrown down together on the same spot, each still contains its own peculiar organic remains*.

These facts lead irresistibly to the conclusion, that while to the north, where the bay was probably open to the sea, a marine limestone was formed, another deposit of fresh-water origin was introduced to the southward, or at the head of the bay. For it appears that during the Eocene period, as now, the ocean was to the north, and the continent, where the great lakes existed, to the south. From the latter region we may suppose a body of fresh water to have descended charged with carbonate of lime and silica, the water being perhaps in sufficient volume to convert the upper end of the bay into fresh water, like some of the gulfs of the Baltic.

Gypsum and marls.—The next group to be considered is

* M. Prevost has pointed out this limestone to me, both in situ at Triel, and in hand specimens in his cabinet.

the gypsum, and the white and green marls, subdivisions of No. 3 of the table of Cuvier and Brongniart. These were once supposed to be entirely subsequent in origin to the two groups already considered; but M. Prevost has pointed out that in some localities they alternate repeatedly with the calcaire siliceux, and in others with some of the upper members of the calcaire grossier. The gypsum, with its associated marls and limestone, is in greatest force towards the centre of the basin, where the two groups before mentioned are less fully developed; and M. Prevost infers, that while those two principal deposits were gradually in progress, the one towards the north, and the other towards the south, a river descending from the east may have brought down the gypseous and marly sediment.

It must be admitted, as highly probable, that a bay or narrow sea, 180 miles in length, would receive, at more points than one, the waters of the adjoining continent; at the same time we must observe, that if the gypsum and associated green and white marls of Montmartre were derived from an hydrographical basin distinct from that of the southern chain of lakes before adverted to, this basin must nevertheless have been placed under circumstances extremely similar; for the identity of the rocks of Velay and Auvergne with the fresh-water group of Montmartre, is such as can scarcely be appreciated by geologists who have not carefully examined the structure of both these countries.

Some of our readers may think that the view above given of the arrangement of four different sets of strata in the Paris basin is far more obscure and complicated than that first presented to them in the system of MM. Cuvier and Brongniart. We admit that the relations of the several sets of strata are less simple than the first observers supposed, being much more analogous to those exhibited by the lacustrine groups of Central France before described.

The simultaneous deposition of two or more groups of strata in one basin, some of them fresh-water and others marine, must always produce very complex results; but in proportion as it is more difficult in these cases to discover any fixed order of superposition in the associated mineral masses, so also is it more easy to explain the manner of their origin and to reconcile their relations to the agency of known causes. Instead of the successive irruptions and retreats of the sea, and changes in the chemical nature of the fluid and other speculations of the earlier geologists, we are now simply called upon to imagine a gulf, into one extremity of which the sea entered, and at the other a large river, while other streams may have flowed in at different points, whereby an indefinite number of alternations of marine and fresh-water beds were occasioned.

Second or Upper marine group.—The next group, called the second or Upper marine formation (No. 4 of the tables), consists in its lower division of green marls which alternate with the fresh-water beds of gypsum and marl last described. Above this division the products of the sea exclusively predominate, the beds being chiefly formed of micaceous sand, 80 feet or more in thickness, surmounted by beds of sandstone with scarcely any limestone. The summits of a great many platforms and hills in the Paris basin consist of this upper marine series, but the group is much more limited in extent than the calcaire grossier. Although we fully agree with M. C. Prevost that the alternation of the various marine and fresh-water formations before described admit of a satisfactory explanation without supposing different retreats and subsequent returns of the sea, yet we think a subsidence of the soil may best account for the position of the upper marine sands. Oscillations of level may have occurred whereby for a time the sea and a river prevailed each in their turn, until at length a more considerable sinking down of part of the basin took place, whereby a tract previously occupied by fresh water was converted into a sea of moderate depth.

In one part of the Paris basin there are decisive proofs that during the Eocene period, and before the upper marine sand was formed, parts of the calcaire grossier were exposed to the action of denuding causes. At Valmondois, for example, a deposit of the upper marine sandstone is found *, in which rolled blocks of the calcaire grossier with its peculiar fossils, and fragments of a limestone resembling the calcaire siliceux, occur. These calcareous blocks are rolled and pierced by perforating shells belonging to no less than fifteen distinct species, and they are imbedded, as well as worn shells washed out from the calcaire grossier, with the ordinary fossils of the upper marine sand.

We have seen that the same earthquake in Cutch could raise one part of the delta of the Indus and depress another, and cause the river to cut a passage through the upraised strata and carry down the materials removed from the new channel into the sea. All these changes, therefore, might happen within a short interval of time between the deposition of two sets of strata in the same delta †.

It is not improbable, then, that the same convulsions which caused one part of the Paris basin to sink down so as to let in the sea upon the area previously covered by gypsum and fresh-water marl, may have lifted up the calcaire grossier and the siliceous limestone, so that they might be acted upon by the waves, and fragments of them swept down into the contiguous sea, there to be drilled by boring testacea.

It is observed that the older marine formation at Laon is now raised 300 metres above the sea, whereas the upper marine sands never attain half that elevation. Such may possibly have been the relative altitude of the two groups when the newest of them was deposited.

Third fresh-water formation.—We have still to consider another formation, the third fresh-water group (No. 5 of the preceding tables). It consists of marls interstratified with beds of flint and layers of flinty nodules. One set of siliceous layers is destitute of organic remains, the other replete with them.

* M. Deshayes, Memoires de la Soc. d'Hist. Nat. de Paris, tom. i. p. 243.— The sandstone is called, by mistake, gres marin *inferieur*, instead of *superieur*, to which last the author has since ascertained it to belong.

† Vol. i. 2d Edit. chap. xxiii.; vol. ii. 1st Edit. chap. xvi.

Gyrogonites, or fossil seed-vessels of charæ, are found abundantly in these strata, and all the animal and vegetable remains
agree well with the hypothesis, that after the gulf or estuary
had been silted up with the sand of the upper marine formation, a great number of marshes and shallow lakes existed, like
those which frequently overspread the newest parts of a delta.
These lakes were fed by rivers or springs which contained, in
chemical solution or mechanical suspension, such kinds of sediment as we have already seen to have been deposited in the
lakes of Central France during the Eocene period.

The Parisian groups all Eocene.—Having now given a rapid
sketch of the different groups of the Paris basin, we may
observe generally that they all belong to the Eocene epoch,
although the entire series must doubtless have required an
immense lapse of ages for its accumulation. The shells of the
different fresh-water groups, constituting at once some of the
lowest and uppermost members of the series, are nearly all
referrible to the same species, and the discordance between the
marine testacea of the calcaire grossier and the upper marine
sands is very inconsiderable.

A curious observation has been made by M. Deshayes, in
reference to the changes which one species, the *Cardium porulosum,* has undergone during the long period of its existence in
the Paris basin. Different varieties of this cardium are characteristic of different strata. In the oldest sand of the Sois-
sonnais (a marine formation underlying the regular beds of the
calcaire grossier), this shell acquires but a small volume, and
has many peculiarities which disappear in the lowest beds
of the calcaire grossier. In these the shell attains its full size,
and many peculiarities of form, which are again modified in the
uppermost beds of the calcaire grossier, and these last characters are preserved throughout the whole of the ' upper
marine' series *.

Microscopic shells.—In some parts of the calcaire grossier
microscopic shells are very abundant, and of distinct species

* Coquilles characterist. des Terrains, 1831.

from those before mentioned of the older Pliocene beds of Italy. We may remind those readers who are not familiar with these minute fossil bodies, that they belong to the order *Cephalopoda*, the animals of which are most free in their movements, and most advanced in their organization, of all the mollusca.   The multilocular cephalopods have been separated, by d'Orbigny, into two subdivisions : first, those having a syphon or internal tube connecting the different chambers, such as the nautilus and ammonite; and, secondly, those without a syphon, to which the microscopic species now under consideration belong.   They are often in an excellent state of preservation, and their forms are singularly different from those of the larger testacea.   We have given a plate of some of these, from unpublished drawings by M. Deshayes, who has carefully selected the most remarkable types of form.

The *natural size* of each species figured in plate 4, is indicated by minute points, to which we call the reader's attention, as they might be easily overlooked.

Bones of quadrupeds in gypsum.—We have already considered the position of the gypsum which occurs in the form of a saccharoid rock in the hill of Montmartre at Paris, and other central parts of the basin.   At the base of that hill it is seen distinctly to alternate with soft marly beds of the calcaire grossier, in which cerithia and other marine shells occur. But the great mass of gypsum may be considered as a purely fresh-water deposit, containing land and fluviatile shells, together with fragments of palm-wood, and great numbers of skeletons of quadrupeds and birds, an assemblage of organic remains which has given great celebrity to the Paris basin.   The bones of fresh-water fish, also, and of crocodiles, and many land and fluviatile reptiles occur in this rock.   The skeletons of mammalia are usually isolated, often entire, the most delicate extremities being preserved as if the carcasses clothed with their flesh and skin had been floated down soon after death, and while they were still swoln by the gases generated by their first decomposition.   The few accompanying shells are of those light kinds which frequently float on the surface of rivers together with wood.

M. Prevost has, therefore, suggested that a river may have swept away the bodies of animals, and the plants which lived on its borders, or in the lakes which it traversed, and may have carried them dówn into the centre of the gulf into which flowed the waters impregnated with sulphate of lime. We know that the Fiume Salso in Sicily enters the sea so charged with various salts that the thirsty cattle refuse to drink of it. A stream of sulphureous water, as white as milk, descends into the sea from the volcanic mountain of Idienne, on the east of Java ; and a great body of hot water, charged with sulphuric acid, rushed down from the same on one occasion, and inundated a large tract of country, destroying, by its noxious properties, all the vegetation *. In like manner the Pusanibio, or ' Vinegar river' of Colombia, which rises at the foot of Puracé, an extinct volcano 7500 feet above the level of the sea, is strongly impregnated with sulphuric and muriatic acid, and with oxide of iron. We may easily suppose the waters of such streams to have properties noxious to marine animals, and in this manner we may explain the entire absence of marine remains in the ossifferous gypsum †.

There are no pebbles or coarse sand in the gypsum, a circumstance which agrees well with the hypothesis that these beds were precipitated from water holding sulphate of lime in solution, and floating the remains of different animals. The bones of land quadrupeds however are not confined entirely to the fresh-water formation to which the gypsum belongs, for the remains of a Palæotherium, together with some fresh-water shells have been found in a marine stratum belonging to the calcaire grossier at Beauchamp.

In the gypsum the remains of about fifty species of quadrupeds have been found all extinct and nearly four-fifths

---

* Leyde Magaz. voor Wetensch Konst en Lett., partie v. cahier i. p. 71. Cited by Rozet, Journ. de Geologie, tom. i, p. 43.

† M. C. Prevost, Submersions Itératives, &c. Note 23.

of them belonging to a division of the order Pachydermata, which is now only represented by four living species, namely by three tapirs and the daman of the Cape. A few carnivorous animals are associated, among which are a species of fox and gennet. Of the Rodentia, a dormouse and a squirrel; of the Insectivora, a bat; and of the Marsupialia, (an order now confined to America, Australia, and some contiguous islands,) an opossum, have been discovered.

Of birds about ten species have been ascertained, the skeletons of some of which are entire. None of them are referrible to existing species *. The same remark applies to the fish, according to MM. Cuvier and Agassiz, as also to the reptiles. Among the last are crocodiles and tortoises of the genera Emys and Trionix.

The tribe of land quadrupeds most abundant in this formation is such as now inhabits alluvial plains and marshes and the banks of rivers and lakes, a class most exposed to suffer by river inundations. Whether the disproportion of carnivorous animals can be ascribed to this cause, or whether they were comparatively small in number and dimensions, as in the indigenous fauna of Australia, when first known to Europeans, is a point on which it would be rash perhaps to offer an opinion in the present state of our knowledge.

We have no reason to be surprised that all the species of vertebrated animals hitherto observed are extinct, when we recollect that out of 1122 species of fossil testacea obtained from the Paris basin, 38 only can be identified with species now living. We have more than once adverted to the fact that extinct mammalia are often found associated with assemblages of *recent* shells, a fact from which we have inferred the inferior duration of species in mammalia as compared to the testacea; and it is not improbable that the higher order of animals in general may more readily become extinct than the marine molluscs. Some of the thirty-eight species of testacea above alluded to, as having survived from the Eocene period to our own times, have now a

* Cuvier, Oss. Foss. tom. iii. p. 255.

wide geographical range, as, for example, *Lucina divaricata,*
and are therefore fitted to exist under a great variety of circumstances.   On the other hand, the great proportion of the
Eocene marine testacea which have become extinct sufficiently
demonstrates that the loss of species has been due to general
laws, and that a sudden catastrophe, such as the invasion of a
whole continent by the sea—a cause which could only anni-
hilate the terrestrial and fresh-water tribes, is an hypothesis
wholly inadequate to account for the phenomenon.

Strata with and without organic remains alternating.—Between the gypsum of the Paris basin and the upper marine
sands a thin bed of oysters is found, which is spread over a
remarkably wide area.   From the manner in which they lie, it
is inferred that they did not grow on the spot, but that some
current swept them away from a bed of oysters formed in some
other part of the bay.   The strata of sand which immediately
repose on the oyster-bed are quite destitute of organic remains ;
and nothing is more common in the Paris basin and in
other formations, than alternations of shelly beds with others
entirely devoid of them.   The temporary extinction and
renewal of animal life at successive periods have been inferred
from such phenomena, which may nevertheless be explained,
as M. Prevost justly remarks, without appealing to any such
extraordinary revolutions in the state of the animate creation.
A current one day scoops out a channel in a bed of shelly sand
and mud, and the next day, by a slight alteration of its course,
ceases to prey upon the same bank.   It may then become
charged with sand unmixed with shells, derived from some dune,
or brought down by a river.   In the course of ages an indefinite number of transitions from shelly strata to those without shells may thus be caused.

Concluding remarks.—It will be seen by our observations
on Auvergne and other parts of Central France, and on the
district round Paris, that geologists have already gained a considerable insight into the state of the physical geography of
part of Europe during the Eocene period.   We can point to

some districts where lakes and rivers then existed, and to the
site of some of the lands encircling those lakes, and to the
position of a great bay of the sea, into which their surplus
waters were discharged. We can also show, as we shall endeavour to explain in the next chapter, the points where some
volcanic eruptions took place. We have acquired much information respecting the quadrupeds which inhabited the land
at that period, and concerning the reptiles, fishes, and testacea
which swarmed in the waters of lakes and rivers; and we have
a collection of the marine Eocene shells more complete than has
yet been obtained from any existing sea of equal extent in
Europe. Nor are the contemporary fossil plants altogether
unknown to us, which, like the animals, are of extinct species,
and indicate a warmer climate than that now prevailing in the
same latitudes.

When we reflect on the tranquil state of the earth implied
by some of the lacustrine and marine deposits of this age, and
consider the fullness of all the different classes of the animal
kingdom, as deduced from the study of the fossil remains, we
are naturally led to conclude, that the earth was at that period
in a perfectly settled state, and already fitted for the habitation
of man.

The heat of European latitudes during the Eocene period
does not seem to have been superior if equal to that now experienced between the tropics; some *living* species of molluscous animals both of the land, the lake, and the sea, existed
when the strata of the Paris basin were formed, and the contrast in the organization of the various tribes of Eocene
animals when compared to those now co-existing with man,
although striking, is not, perhaps, so great as between the
living Australian and European types. At the same time we
are fully aware that we cannot reason with any confidence on
the capability of our own or any other contemporary species to
exist under circumstances so different as those which might
be caused by an entirely new distribution of land and sea;
and we know that in the earlier tertiary periods the physical

geography of the northern hemisphere was very distinct.   Our inability to account for the atmospheric and other latent causes, which often give rise to the most destructive epidemics, proves the extent of our ignorance of the entire assemblage of conditions requisite for the existence of any one species on the globe.

CHAPTER XIX.

Volcanic rocks of the Eocene period—Auvergne—Igneous formations associated with lacustrine strata—Hill of Gergovia—Eruptions in Central France at successive periods—Mont Dor an extinct volcano—Velay—Plomb du Cantal—Train of minor volcanos stretching from Auvergne to the Vivarais—Monts Domes—Puy de Côme—Puy Rouge—Ravines excavated through lava—Currents of lava at different heights—Subjacent alluviums of distinct ages—The more modern lavas of Central France may belong to the Miocene period—The integrity of the cones not inconsistent with this opinion—No eruptions during the historical era—Division of volcanos into ante-diluvian and post-diluvian inadmissible—Theories respecting the effects of the Flood considered—Hypothesis of a partial flood—Of a universal deluge—Theory of Dr. Buckland as controverted by Dr. Fleming—Recapitulation.

EOCENE VOLCANIC ROCKS.

WHEN we treated in the seventeenth chapter of the lacustrine deposits of Central France, we purposely omitted to give a detailed account of the associated volcanic rocks, to which we now recall the attention of the reader. (See the Map, p. 226.)

We stated that, in the arenaceous and pebbly group of the lacustrine basins of Auvergne, Cantal, and Velay, no volcanic pebbles had ever been detected, although massive piles of igneous rocks are now found in the immediate vicinity. As this observation has been confirmed by minute research, we are warranted in inferring, as we before explained, that the volcanic eruptions had not commenced when the older subdivisions of the fresh-water groups originated.

In Cantal and Velay we believe no decisive proofs have yet been brought to light that any of the igneous out-bursts happened during the deposition of the fresh-water strata; but there can be no doubt that in Auvergne some volcanic explosions took place before the drainage of the lakes, and at a time when the Eocene species of animals and plants still flourished. We shall first advert to these proofs as relating to the history

of the period under consideration, and shall then proceed to
show that there are in the same country volcanic rocks of
much newer date, some of which appear to be referrible to the
Miocene era.

Volcanic rocks associated with Lacustrine in Auvergne.—The
first locality to which we shall call the reader's attention is Pont
du Chateau near Clermont, where a section is seen in a precipice on the right bank of the river Allier *. Beds of volcanic tuff
alternate with a fresh-water limestone, which is in some places
pure, but in others spotted with fragments of volcanic matter,
as if it were deposited while showers of sand and scoriæ were
projected from a neighbouring vent †. This limestone contains
the *Helix Ramondi* and other shells of Eocene species. It is
immaterial to our present argument whether the volcanic sand
was showered down from above, or drifted to the spot by a
river, for the latter opinion must presuppose the country to
have been covered with volcanic ejections during the Eocene
period.

Another example occurs in the Puy de Marmont, near
Veyres, where a fresh-water marl alternates with volcanic tuff
containing Eocene shells. The tuff or breccia in this locality
is precisely such as is known to result from volcanic ashes
falling into water, and subsiding together with ejected fragments of marl and other stratified rocks. These tuffs and
marls are highly inclined, and traversed by a thick vein of
basalt which, as it rises in the hill, divides into two branches.

Gergovia.—The hill of Gergovia near Clermont affords a
third example. We agree with MM. Dufrénoy and Jobert
that there is no alternation here of lava and fresh-water strata,
in the manner supposed by some other observers ‡; but the
position and contents of some of the tuffs prove them to have
been derived from volcanic eruptions which occurred during
the deposition of the Eocene formations.

---

* This place, and all the others in Auvergne, mentioned in this chapter, were
examined by the author, in company with Mr. Murchison, in 1828.

  † See Scrope's Central France, p. 21.   ‡ Scrope, ibid. p. 7.

The bottom of the hill consists of slightly inclined beds of white and greenish marls, more than three hundred feet in thickness, which are intersected by a dike of basalt, which may be studied in the ravine above the village of Merdogne. The dike here cuts through the marly strata at a considerable angle,

No. 60.

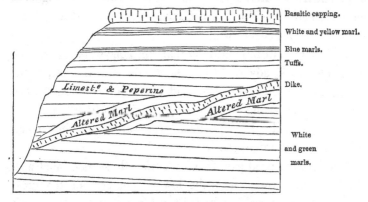

Hill of Gergovia.

producing, in general, great alteration and confusion in them for some distance from the point of contact. Above the white and green marls, a series of beds of limestone and marl, containing fresh-water shells, are seen to alternate with volcanic tuff. In the lowest part of this division, beds of pure marl alternate with compact fissile tuff resembling some of the subaqueous tuffs of Italy and Sicily called *peperinos*. Occasionally fragments of scoriæ are visible in this rock. Still higher is seen another group of some thickness, consisting exclusively of tuff, upon which lie other marly strata intermixed with volcanic matter.

There are many points in Auvergne where igneous rocks have been forced by subsequent injection through clays and marly limestones, in such a manner that the whole has become blended in one confused and brecciated mass, between which and the basalt there is sometimes no very distinct line of demarcation. In the cavities of such mixed rocks we often find calcedony and crystals of mesotype, stilbite and arragonite. To

formations of this class may belong some of the breccias immediately adjoining the dike in the hill of Gergovia; but it
cannot be contended that the volcanic sand and scoriæ interstratified with the marls and limestones in the upper part of
that hill were introduced, like the dike, subsequently by intrusion from below. They must have been thrown down like
sediment from water, and can only have resulted from igneous
action which was going on contemporaneously with the deposition of the lacustrine strata.

The reader will bear in mind that this conclusion agrees well
with the proofs, adverted to in the seventeenth chapter, of the
abundance of silex, travertin and gypsum precipitated when
the upper lacustrine strata were formed: for these rocks, as
we have pointed out, are such as the waters of mineral and
thermal springs might generate.

The igneous products above mentioned, as associated with
the lacustrine strata, form the lowest members of the great
series of volcanic rocks of Auvergne, Cantal, and Velay, which
repose for the most part on the granitic mountains (see Map,
above, p. 226). There was evidently a long succession of
eruptions, beginning with those of the Eocene period, and
ending, so far as we can yet infer from the evidence derived
from fossil remains, with those of the Miocene epoch. The
oldest part of the two principal volcanic masses of Mont Dor
and the Plomb du Cantal may perhaps belong to the Eocene
period,—the newer portion of the same mountains to the Miocene; just as Etna commenced its operations during the newer
Pliocene era, and has continued them down to the Recent
epoch, and still retains its energy undiminished. There are
some parts of the Mont Mezen, in Velay, which are perhaps
of the same antiquity as the oldest parts of Mont Dor. Besides these ancient rocks, of which the lavas are in a great
measure trachytic, there are many minor cones in Central
France, for the most part of posterior origin, which extend from
Auvergne, in a direction north-west and south-east, through
Velay, into the Vivarais, where they are seen in the basin of

the Ardêche. This volcanic line does not pass by the Plomb du Cantal; it was formed, as nearly as we can conjecture in the present imperfect state of our knowledge, during the Miocene period; but there may probably be found, among these cones and their accompanying lavas, rocks of every intermediate age between the oldest and newest volcanic formations of Central France.

We shall first give a brief description of the Mont Dor and the Plomb du Cantal, and then pass on to the train of newer cones, examining the evidence at present obtained respecting their relative ages, and the light which they throw on the successive formation of alluviums and on the excavation of valleys.

Mont Dor.—Mont Dor, the most conspicuous of the volcanic masses of Auvergne, rests immediately on the granitic rocks standing apart from the fresh-water strata *. This volcano rises suddenly to the height of several thousand feet above the surrounding platform, and retains the shape of a flattened and somewhat irregular cone, all the sides sloping more or less rapidly, until their inclination is gradually lost in the high plain around. It is composed of layers of scoriæ, pumice-stones, and their fine detritus, interstratified with beds of trachyte and basalt, which descend often in uninterrupted currents, till they reach and spread themselves around the base of the mountain †. Conglomerates also, composed of angular and rounded fragments of igneous rocks, are observed to alternate with the above; and the various masses are seen to dip off from the central axis, and to lie parallel to the sloping flanks of the great cone, in the same manner as we have described when treating of Etna.

The summit of the mountain terminates in seven or eight rocky peaks, where no regular crater can be traced, but where we may easily imagine one to have existed which may have been shattered by earthquakes, and have suffered degradation by aqueous causes. Originally, perhaps, like the highest

* See the Map, p. 226.   † Scrope's Central France, p. 98.

crater of Etna, it may have formed an insignificant feature in the great pile, and may frequently have been destroyed and renovated.

We cannot at present determine the age of the great mass of Mont Dor, because no organic remains have yet been found in the tuffs, except impressions of the leaves of trees of species not determined. Some of the lowest parts of the great mass are formed of white pumiceous tuffs, in which animal remains may perhaps be one day found. In the mean time, we conclude that Mont Dor had no existence when the grits and conglomerates of the Limagne, which contain no volcanic materials, were formed; but some of the earliest eruptions may perhaps have been contemporary with those described in the commencement of this chapter. To the latest of these eruptions, on the other hand, we refer those trachytic breccias of Mont Perrier which were shown in the sixteenth chapter, p. 217, to alternate with Miocene alluviums.

Velay.—The observations of M. Bertrand de Doue have not yet established that any of the most ancient volcanos of Velay were in action during the Eocene period, although it is very probable that some of them may have been contemporaneous with the oldest of the Auvergne lavas. There are beds of gravel in Velay, as in Auvergne, covered by lava at different heights above the channels of the existing rivers. In the highest and most ancient of these alluviums the pebbles are exclusively of granitic rocks; but in the newer, which are found at lower levels, they contain an intermixture of volcanic substances. We have already shown, in the sixteenth chapter, that, in the volcanic ejections and alluviums covered by the lavas of Velay, the bones of animals of Miocene species have been found, in which respect the phenomena accord perfectly with those of Auvergne.

Plomb du Cantal. In regard to the age of the igneous rocks of the Cantal we are still less informed, and at present can merely affirm that they overlie the Eocene lacustrine strata of that country. The Plomb du Cantal (see Map, wood-cut

No. 56) is a conical mass, which has evidently been formed, like the cone of Etna, by a long series of eruptions. It is composed of trachytic and basaltic lavas, tuffs, and conglomerates, or breccias, forming a mountain several thousand feet in height. This volcano evidently broke out precisely on the site of the lacustrine deposit before described (Chapter xvii.), which had accumulated in a depression of a tract composed of micaceous schist. In the breccias, even to the very summit of the mountain, we find ejected masses of the fresh-water beds, and sometimes fragments of flint, containing Eocene shells. Deep valleys radiate in all directions from the central heights of the mountain, especially those of the Cer and Jourdanne, which are more than twenty miles in length, and lay open the geological structure of the mountain. No alternation of lavas with undisturbed Eocene strata have been observed, nor any tuffs containing fresh-water shells; on the northern side of the Plomb du Cantal, at La Vissiere, near Murat, we have pointed out on the Map (wood-cut, p. 226) a spot where fresh-water limestone and marl are seen covered by a thickness of about 800 feet of volcanic rock. Shifts are here seen in the strata of limestone and marl *.

Although it appears that the lavas of the Cantal are more recent than the fresh-water formation of that country, it does not follow that they may not belong to the Eocene period. The lake may possibly have been drained by the earthquakes which preceded or accompanied the first eruptions, but the Eocene animals and plants may have continued to exist for a long series of ages, while the cone went on increasing in dimensions.

Train of minor Volcanos.—We shall next consider those minor volcanos before alluded to, which stretch in a long range from Auvergne to the Vivarais, and which appear for the most part to be of newer origin than the mountains above described. They have been thrown up in a great number of isolated points, and much resemble those scattered over the

* See Lyell and Murchison, Ann. des Sci. Nat., Oct. 1829.

Phlegræan fields and the flanks of Etna. They have given rise chiefly to currents of basaltic lava, whereas those of Mont Dor and the Cantal are in great part trachytic. There are perhaps about three hundred of these minor cones in Central France; but a part of them only occur in Auvergne, where some few are found at the bottom of valleys excavated through the more ancient lavas of Mont Dor, as the Puy de Tartaret, for example, whence issues a current of lava which, flowing into the bed of the river Couze, gave rise to the lake of Chambon. Here the more ancient columnar basalts of Auvergne are seen forming the upper portion of the precipices which bound the valley.

But the greater part of the minor cones of Auvergne are placed upon the granitic platform, where they form an irregular ridge about eighteen miles in length and two in breadth. They are usually truncated at the summit, where the crater is often preserved entire, the lava having issued from the base of the hill. But frequently the crater is broken down on one side, where the lava has flowed out. The hills are composed of loose scoriæ, blocks of lava, lapilli, and puzzuolana, with fragments of trachyte and granite.

The lavas may be often traced from the crater to the nearest valley, where they usurp the channel of the river, which has often excavated a deep ravine through the basalt. We have thus an opportunity of contrasting the enormous degradation which the solid and massive rock has suffered by aqueous erosion and the integrity of the cone of sand and ashes which has, in the mean time, remained uninjured on the neighbouring platform, where it was placed beyond the reach of the power of running water.

Puy de Côme.—We may mention the Puy de Côme and its lava current, near Clermont, as one of the numerous illustrations of the phenomenon here alluded to. This conical hill rises from the granitic platform at an angle of about 40 degrees to the height of more than 900 feet. Its summit presents two distinct craters, one of them with a vertical depth of

250 feet.   A stream of lava takes its rise at the western base
of the hill, instead of issuing from either crater, and descends
the granitic slope towards the present site of the town of Pont
Gibaud.   Thence it pours in a broad sheet down a steep
declivity into the valley of the Sioule, filling the ancient river-
channel for the distance of more than a mile.   The Sioule,
thus dispossessed of its bed, has worked out a fresh one between
the lava and the granite of its western bank ; and the excavation has disclosed, in one spot, a wall of columnar basalt about
fifty feet high *.

The excavation of the ravine is still in progress, every winter
some columns of basalt being undermined and carried down
the channel of the river, and in the course of a few miles rolled
to sand and pebbles.   Meanwhile the cone of Côme remains
stationary, its loose materials being protected by a dense vegetation, and the hill standing on a ridge not commanded by
any higher ground whence floods of rain-water may descend.

Puy Rouge.—At another point, farther down the course of
the Sioule, we find a second illustration of the same phenomenon in the Puy Rouge, a conical hill to the north of
the village of Pranal.   The cone is composed entirely of
red and black scoriæ, tuff, and volcanic bombs.   On its western
side there is a worn-down crater whence a powerful stream
of lava has issued and flowed into the valley of the Sioule.
The river has since excavated a ravine through the lava and
subjacent gneiss, to the depth of 400 feet.

On the upper part of the precipice forming the left side of
this ravine, we see a great mass of black and red scoriaceous
lava ; below this a thin bed of gravel, evidently an ancient
river-bed, now at an elevation of 50 feet above the channel of
the Sioule.   The gravel again rests upon gneiss, which has
been eroded to the depth of 50 feet †.   It is quite evident in
this case that, while the basalt was gradually undermined and

---

* Scrope's Central France, p. 60, and plate.

† See Lyell and Murchison on the Excavation of Valleys, Edin. New Phil
Journ., July 1829.

carried away by the force of running water, the cone whence
the lava issued escaped destruction, because it stood upon a
platform of gneiss several hundred feet above the level of the
valley in which the force of running water was exerted.

It is needless to multiply examples, or the Vivarais would
supply many others equally striking. Among many we may
instance the cone of Jaujac, and its lava current *, which is a
counterpart of that near Pranal last mentioned.

Lavas and Alluviums of different Ages.—We have seen that
on the flanks of Etna, since the commencement of the present
century, several currents of lava have flowed at the bottom of
the Val del Bove, at the foot of precipices formed of more
ancient lavas and tuffs. So we find in Auvergne that some
streams of melted matter have flowed in valleys, the sides of
which consist partly of older lavas. These are often seen
capping the hills in broad sheets, resting sometimes on granite,
sometimes on fresh-water strata.

Many of the earlier lavas of Auvergne have flowed out upon
the platform of granite before all the existing valleys had been
excavated; others again spread themselves in broad sheets over
the horizontal lacustrine deposit, when these had been covered
with gravel, probably soon after the drainage of the lakes.
Great vicissitudes in the physical geography of the country
must have taken place since the flowing of these ancient lavas;
and it is evident that the changes were gradual and successive, caused probably by the united agency of running
water and subterranean movements. We frequently observe
one mass of lava capping a hill, and a second at a lower elevation, forming a terrace on the side of a valley; or sometimes
occupying the bed of a river.

It is a most interesting fact that we almost invariably find
in these cases beds of gravel underlying the successive currents
of lava, as in Catalonia before described (pp. 189, 190). Occasionally, when the highest platform of lava is 700 or 800 feet
above the lowest, we cannot fail to be struck with the won-

* See Scrope's Central France, plate 14.

derful alterations effected in the drainage of the country since
the first current flowed; for the most elevated alluviums must

No. 61.

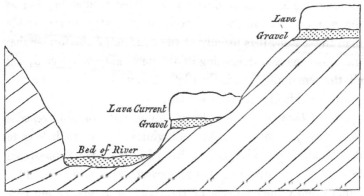

Lavas of Auvergne resting on alluviums of different ages.

originally have been accumulated in the lowest levels of the
then existing surface.   As some geologists have referred
almost all the superficial gravels to one era, and have supposed
them to be the result of one sudden catastrophe, the phenomena of Auvergne above alluded to are very important.
The flows of volcanic matter have preserved portions of the
surface in the state in which they existed at successive periods,
so that it is impossible to confound together the alluviums of
different ages.   The reader will see at once by reference to
the wood-cut (No. 61) that a considerable interval of time
occurred between the formation of the uppermost bed of gravel
and that next below it; during which interval the uppermost
lava was poured out and a valley excavated, at the bottom of
which the second bed of gravel accumulated.   In like manner
the pouring out of a second current of lava, and a farther
deepening of the valley, took place between the date of the
second gravel and that of the modern alluvium which now fills
the channel of the river *.

* For localities in Central France where lavas or sheets of basalt repose on
alluviums at different elevations above the present valleys, consult the works of
MM. Le Grand d'Aussi, Montlosier, Ramond, Scrope, Bertrand de Doue, Croizet,
Jobert, Bouillet, and others.

When rivers are dispossessed of their channels by lava, they usually flow between the mass of lava and one side of the original valley. They then eat out a passage, partly through the volcanic and partly through the older formation; but as the soft tertiary marls in Auvergne give way more readily than the basalt, it is usually at the expense of the former that the enlarging and deepening of the new valley is effected, and all the remaining lava is then left on one side, in the manner represented in the above wood-cut.

Age of the more modern lavas.—The only organic remains found as yet in the ancient alluviums appear to belong to the Miocene period; but we have heard of none discovered in the gravel underlying the newest lavas, — those which either occupy the channels of the existing rivers or are very-slightly elevated above them. We think it not improbable that even these may be of Miocene date, although the conjecture will appear extremely rash to some who are aware that the cones and craters whence the lavas issue, are often as fresh in their aspect as the majority of the cones of the forest zone of Etna.

The brim of the crater of the Puy de Pariou, near Clermont, is so sharp, and has been so little blunted by time, that it scarcely affords room to stand upon. This and other cones in an equally remarkable state of integrity have stood, we conceive, uninjured, not *in spite of* their loose porous nature, as some geologists might think, but in consequence of it. No rills can collect where all the rain is instantly absorbed by the sand and scoriæ, as we have shown to be the case on Etna (see above, p. 102), and nothing but a waterspout breaking directly upon the Puy de Pariou could carry away a portion of the hill, so long as it is not rent by earthquakes or engulphed.

Attempt to divide Volcanos into ante-diluvian, and post-diluvian.—The opinions above expressed are entirely at variance with the doctrines of those writers who have endeavoured to arrange all the volcanic cones of Europe under two divisions,

those of ante-diluvian and those of post-diluvian origin. To
the former they attribute such hills of sand and scoriæ as
exhibit on their surface evident signs of aqueous denudation ;
to the latter, such as betray no marks of having been exposed
to such aqueous action.   According to this classification almost
all the minor cones of Central France must be called post-
diluvian ; although, if we receive this term in its ordinary
acceptation as denoting posteriority of date to the Noachian
deluge, we are forced to suppose that all the volcanic eruptions
occurred within a period of little more than twenty centuries,
or between the era of the flood, which happened about 4000
years ago, and the earliest historical records handed down
to us respecting the former state of Central France. Dr.
Daubeny has justly observed, that had any of these French
volcanos been in a state of activity in the age of Julius Cæsar,
that general, who encamped upon the plains of Auvergne, and
laid siege to its principal city, (Gergovia, near Clermont,)
could hardly have failed to notice them.   Had there been
even any record of their existence in the time of Pliny or
Sidonius Apollinaris, the one would scarcely have omitted
to make mention of it in his Natural History, nor the other to
introduce some allusion to it among the descriptions of this
his native province.   This poet's residence was on the borders
of the Lake Aidat, which owed its very existence to the damming up of a river by one of the most modern lava currents*.

The ruins of several Roman bridges and of the Roman
baths at Royat confirm the conclusion that no sensible
alteration has taken place in the physical geography of the
district, not even in the chasms excavated through the newest
lavas since ages historically remote. We have no data at
present for presuming that any one of the Auvergne cones has
been produced within the last 4000 or 5000 years; and the

---

* Daubeny on Volcanos, p. 14.

same may be said of those of Velay. Until the bones of men or articles of human workmanship are found buried under some of their lavas, instead of the remains of extinct animals, which alone have hitherto been met with, we shall consider it probable, as we before hinted, that the latest of the volcanic eruptions may have occurred during the Miocene period.

Supposed effects of the Flood.

They who have used the terms ante-diluvian and post-diluvian in the manner above adverted to, proceed on the assumption that there are clear and unequivocal marks of the passage of a general flood over all parts of the surface of the globe. It had long been a question among the learned, even before the commencement of geological researches, whether the deluge of the Scriptures was universal in reference to the whole surface of the globe, or only so with respect to that portion of it which was then inhabited by man. If the latter interpretation be admissible, the reader will have seen, in former parts of this work, that there are two classes of phenomena in the configuration of the earth's surface, which might enable us to account for such an event. First, extensive lakes elevated above the level of the ocean ; secondly, large tracts of dry land depressed below that level. When there is an immense lake, having its surface, like Lake Superior, raised 600 feet above the level of the sea, the waters may be suddenly let loose by the rending or sinking down of the barrier during earthquakes, and hereby a region as extensive as the valley of the Mississippi, inhabited by a population of several millions, might be deluged *. On the other hand, there may be a country placed beneath the mean level of the ocean, as we have shown to be the case with part of Asia †, and such a region must be entirely laid under water should the tract which separates it from the ocean

* See vol. i. p. 89, and Second Edition, p. 101.
† Vol. ii. p. 163, and Second Edition, p. 169.

be fissured or depressed to a certain depth.   The great cavity of western Asia is 18,000 square leagues in area, and is occupied by a considerable population.   The lowest parts, surrounding the Caspian Sea, are 300 feet below the level of the Euxine,—here, therefore, the diluvial waters might overflow the summits of hills rising 300 feet above the level of the plain ; and if depressions still more profound existed at any former time in Asia, the tops of still loftier mountains may have been covered by a flood.

But it is undeniable, that the great majority of the older commentators have held the deluge, according to the brief account of the event given by Moses, to have consisted of a rise of waters over *the whole earth,* by which the summits of the loftiest mountains on the globe were submerged.   Many have indulged in speculations concerning the instruments employed to bring about the grand cataclysm ; and there has been a great division of opinion as to the effects which it might be expected to have produced on the surface of the earth.   According to one school, of which De Luc in former times, and more recently Dr. Buckland, have been zealous and eloquent supporters, the passage of the flood worked a considerable alteration in the external configuration of our continents.   By the last-mentioned writer the deluge is represented as a violent and transient rush of waters which tore up the soil to a great depth, excavated valleys, gave rise to immense beds of shingle, carried fragments of rock and gravel from one point to another, and, during its advance and retreat, strewed the valleys, and even the tops of many hills, with alluvium *.

But we agree with Dr. Fleming †, that in the narrative of Moses there are no terms employed that indicate the impetuous rushing of the waters, either as they rose or when they re-

* Buckland, Reliquiæ Diluvianæ.

† See a Memoir by the Rev. John Fleming, D.D., on the Geological Deluge, Edin. Phil. Journ., vol. xiv. p. 205, whose opinions were reviewed by the author in Oct. 1827, in an article in the Quarterly Review, No. lxxii. p. 481.

treated, upon the restraining of the rain and the passing of a
wind over the earth.   On the contrary, the olive-branch,
brought back by the dove, seems as clear an indication to us
that the vegetation was not destroyed, as it was then to Noah
that the dry land was about to appear.

We have been led with great reluctance into this digression,
in the hope of relieving the minds of some of our readers from
groundless apprehension respecting the bearing of many of the
views advocated in this work.   They have been in the habit of
regarding the diluvial theory above controverted as alone
capable of affording an explanation of geological phenomena in
accordance with Scripture, and they may have felt disapproba-
tion at our attempt to prove, in a former chapter *, that the
minor volcanos on the flanks of Etna may, some of them, be
more than 10,000 years old.   How, they would immediately
ask, could they have escaped the denuding force of a diluvial
rush of waters?   The same objection may have presented
itself when we quoted, with so much respect, the opinion of a
distinguished botanist, that some living specimens of the Baobab tree of Africa, or the Taxodium of Mexico, may be five
thousand years old †.   Our readers may also have been
astonished at the high antiquity assigned by us to the greater
part of the European alluviums, and the many different
ages to which we refer them ‡, as they may have been taught
to consider the whole as the result of one *recent* and *simultaneous* inundation.   Lastly, they may have felt some dis-
appointment at observing, that we attach no value whatever
to the hypothesis of M. Elie de Beaumont, adopted by Professor Sedgwick, that the sudden elevation of mountain-chains
' has been followed again and again by mighty waves deso-
lating whole regions of the earth §,' a phenomenon which,
according to the last-mentioned of these writers, has ' taken

* Chap. viii. p. 100.   † See above, p. 99.
‡ P. 147.   § P. 101.

away all anterior incredibility from the fact of a recent deluge*.'

For our own part, we have always considered the flood, if we are required to admit its universality·in the strictest sense of the term, as a preternatural event far beyond the reach of philosophical inquiry, whether as to the secondary causes employed to produce it, or the effects most likely to result from it. At the same time, it is evident that they who are desirous of pointing out the coincidence of geological phenomena with the occurrence of such a general catastrophe, must neglect no one of the circumstances enumerated in the Mosaic history, least of all so remarkable a fact as that the olive remained standing while the waters were abating.

Recapitulation.—We shall now briefly recapitulate some of the principal conclusions to which we have been led by an examination of the volcanic districts of Central France.

1st. Some of the volcanic eruptions of Auvergne took place during the Eocene period, others at an era long subsequent, probably during the Miocene period.

2ndly. There are no proofs as yet discovered that the most recent of the volcanos of Auvergne and Velay are subsequent to the Miocene period, the integrity of many cones and craters not opposing any sound objection to the opinion that they may be of indefinite antiquity.

3rdly. There are alluviums in Auvergne of very different ages, some of them belonging to the Miocene period. Many of these have been covered by lava-currents which have been poured out in succession while the excavation of valleys was in progress.

4thly. There are a multitude of cones in Auvergne, Velay, and the Vivarais, which have never been subjected to the action of a violent rush of waters capable of modifying considerably the surface of the earth.

5thly. If, therefore, the Mosaic deluge be represented as universal, and as having exercised a violent denuding force, all

* Anniv. Address to the Geol. Soc., Feb. 18th, 1831.

these cones, several hundred in number, must be post-dilu-vian.

6thly. But since the beginning of the historical era, or the invasion of Gaul by Julius Cæsar, the volcanic action in Au-vergne has been dormant, and there is nothing to countenance the idea that, between the date usually assigned to the Mosaic deluge and the earliest traditional and historical records of Central France (a period of little more than twenty centuries), all or any one of the more entire cones of loose scoriæ were thrown up.

Lastly, it is the opinion of some writers, that the earth's surface underwent no great modification at the era of the Mosaic deluge, and that the strictest interpretation of the scriptural narrative does not warrant us in expecting to find any geological monuments of the catastrophe, an opinion which is consistent with the preservation of the volcanic cones, however high their antiquity.

CHAPTER XX.

Eocene formations, *continued*—Basin of the Cotentin, or Valognes—Rennes—Basin of Belgium, or the Netherlands—Aix in Provence—Fossil insects—Tertiary strata of England—Basins of London and Hampshire—Different groups—Plastic clay and sand—London clay—Bagshot sand—Fresh-water strata of the Isle of Wight—Palæotherium and other fossil mammalia of Binstead—English Eocene strata conformable to chalk—Outliers on the elevated parts of the chalk—Inferences drawn from their occurrence—Sketch of a theory of the origin of the English tertiary strata.

HAVING in the last three chapters treated of the Eocene formations of different parts of France, we now propose to examine those which are found in the south-eastern division of England; but before we pass from the continent to our own island, we may briefly advert to several other spots where Eocene deposits have been observed. Their geographical position will be found delineated on the annexed map*.

MAP OF THE PRINCIPAL TERTIARY BASINS OF THE EOCENE
PERIOD.

No. 62.

| ▨ Primary rocks and | ▨ Eocene formations. |

strata older than the carboniferous series.

N.B. The space left blank is occupied by secondary formations, from the old red sandstone to the chalk inclusive.

* This map is copied from one given by M. Desnoyers, Mem. de la Soc. d'Hist. Nat. de Paris, 1825, pl. 9 ; compiled partly from that author's observations, and partly from Mr. Webster's map, Geol. Trans., 1st series, vol. ii. plate 10.

Basin of the Cotentin, or Valognes.—The strata in the environs of Valognes, in the department of La Manche, consist chiefly of a coarse limestone resembling the calcaire grossier of Paris, of which M. Desnoyers has given an elaborate description. It is occasionally covered with a compact fresh-water limestone alternating with fresh-water marls. In these Eocene strata more than 300 species of fossil shells have been discovered, almost all identical with species of the Paris basin. (See Tables, Appendix I.) Superimposed upon the Eocene strata of this basin is a newer marine deposit, extending over a limited area, the fossils of which agree with those of the Faluns of the Loire *. Here, therefore, the geologist has an opportunity of observing the superposition of the Miocene deposits upon those of the age of the Paris basin.

Rennes.—Several small patches, also, of marine strata, have been found by M. Desnoyers, in the neighbourhood of Rennes, which are characterised by Eocene fossils and repose on ancient rocks, as will be seen in the map.

Basin of Belgium, or the Netherlands.—The greater part of the tertiary formations of the Low Countries consist of clay and sand, much resembling those of the basin of London, afterwards to be described. The fossil shells, also, are of the same species, 49 of which will be found referred to by M. Deshayes, in the tables, Appendix I.

Aix, in Provence.—The tertiary strata of Aix and Fuveau in Provence are of great thickness and extent, the lower members being remarkable for containing coal grit and beds of compact limestone, such as we only find in England in ancient secondary groups. Yet these strata are for the most part of fresh-water origin, and contain several species of Eocene shells, together with many which are peculiar to this basin. It would require a fuller comparison than has yet been made of the fossil remains of Aix and Fuveau, before we can determine with accuracy the relative age of that formation. Some of the plants seem to agree with those of the Paris basin, while many

---

* Desnoyers, Mem. de la Soc. d'Hist. Nat. de Paris, 1825.

of the insects have been supposed identical with species now living *. These insects have been almost exclusively procured from a thin bed of grey calcareous marl, which passes into an argillaceous limestone found in the quarries of gypsum near Aix. The rock in which they are imbedded is so thinly laminated that there are sometimes more than 70 layers in the thickness of an inch. The insects are for the most part in an extraordinary state of preservation, and an impression of their form is seen both on the upper and under laminæ, as in the case of the Monte Bolca fishes. M. Marcel de Serres enumerates 62 genera belonging chiefly to the orders Diptera, Hemiptera and Coleoptera. On reviewing a collection brought from Aix, Mr. Curtis observes that they are all of European forms and most of them referrible to existing genera †. With the single exception of an Hydrobius, none of the species are aquatic. The antennæ, tarsi, and trophi are generally very obscure, or distorted, yet in a few the claws are visible, and the sculpture, and even some degree of local colouring, are preserved. The nerves of the wings, in almost all the Diptera, are perfectly distinct, and even the pubescence on the head of one of them. Several of the beetles have the wings extended beyond the elytra, as if they had made an effort to escape by flying, or had fallen into the water while on the wing ‡.

BASINS OF LONDON AND HAMPSHIRE.

The reader will see in the small map above given (No. 62, p. 275,) the position of the two districts usually called the basins of London and Hampshire, to which the Eocene formations of England are confined. These tracts are bounded by rising grounds composed of chalk, except where the sea intervenes. That the chalk passes beneath the tertiary strata, we can not only infer from geological data, but can prove by numerous artificial sections at points where wells have been sunk, or borings made through the overlying beds. The

---

* M. Marcel de Serres, Géog. des Ter. Tertiaires du Midi de la France.
† Murchison and Lyell. Ed. New Phil. Journ., Oct. 1829.
‡ Curtis, ibid., where figures of some of the insects are given.

Eocene deposits are chiefly marine, and have generally been divided into three groups: 1st, the Plastic clay and sand, which is the lowest group; 2dly, the London clay; and, 3rdly, the Bagshot sand. Of all these the mineral composition is very simple, for they consist almost entirely of clay, sand, and shingle, the great mass of clay being in the middle, and the upper and lower members of the series being more arenaceous.

Plastic clay and sand.—The lowest formation, which sometimes attains a thickness of from 400 to 500 feet, consists principally of an indefinite number of beds of sand, shingle, clay, and loam, irregularly alternating, some of the clay being used in potteries, in reference to which the name of Plastic clay has been given to the whole formation. The beds of shingle are composed of perfectly rolled chalk flints, with here and there small pebbles of quartz. Heaps of these materials appear sometimes to have remained for a long time covered by a tranquil sea. Dr. Buckland mentions that he observed a large pebble in part of this formation at Bromley, to which five full-grown oyster-shells were affixed, in such a manner as to show that they had commenced their first growth upon it, and remained attached through life *.

In some of the associated clays and sand, perfect marine shells are met with, which are of the same species as those of the London clay. The line of separation, indeed, between the superincumbent blue clay last alluded to, and the Plastic clay and sand, is quite arbitrary, as any geologist may be convinced who examines the celebrated section in Alum Bay, in the Isle of Wight †, where a distinct alternation of the two groups is observable, each marked with their most characteristic peculiarities. In the midst of the sands of the lower series a mass of clay occurs 200 feet thick, containing septaria, and replete with the usual fossils of the neighbourhood of London ‡.

* Geol. Trans., First Series, vol. iv. p. 300.

† See Mr. Webster's Memoir, Geol. Trans., vol. ii., First Series, and his Letters in Sir H. Englefield's Isle of Wight.

‡ See Mr. Webster's sections, plate 11. Geol. Trans., vol. ii., First Series.

The *arenaceous* beds are chiefly laid open on the confines of the basins of London and Hampshire, in following which we discover at many places great beds of perfectly rounded flints. Of this description, on the southern borders of the basin of London, are the hills of Comb Hurst and the Addington hills, which form a ridge stretching from Blackheath to Croydon. Here they have much the appearance of banks of sand and shingle formed near the shores of the tertiary sea; but whether they were really of littoral origin cannot be determined for want of a sufficient number of sections which might enable us to compare the tertiary strata at the edges with those in the central parts of each basin.

We have ample opportunities in the basin of Paris of examining steep cliffs of hard rock which bound many of the valleys, and innumerable excavations have been made for building-stone, limestone, and gypsum; but when we attempt to obtain a connected view of any considerable part of the tertiary series in the basin of London, we are almost entirely limited to a single line of coast-section; for in the interior the regular beds are much concealed by an alluvial covering of flint gravel spread alike over the summits and gentle slopes of the hills, and over the bottoms of the valleys.

Organic remains are extremely scarce in the Plastic clay; but when any shells occur they are of Eocene species. Vegetable impressions and fossil wood sometimes occur, and even beds of lignite, but none of the *species* of plants have, we believe, as yet been ascertained.

London clay.—This formation consists of a bluish or blackish clay, sometimes passing into a calcareous marl, rarely into a solid rock. Its thickness is very great, sometimes exceeding 500 feet *. It contains many layers of ovate or flattish masses of argillaceous limestone, which, in their interior, are generally traversed in various directions by cracks, partially or wholly filled by calcareous spar. These masses, called septaria, are sometimes continued through a thickness of 200 feet †.

---

* Con. and Phil. Outlines of Geol., p. 33.   † Outlines of Geol., p. 27.

A great number of the marine shells of this clay have been identified with those of the Paris basin, and are mentioned by name in Appendix I. It is quite evident, therefore, that these two formations belong to the same epoch.

No remains of terrestrial mammalia have as yet been found in this clay, but the occurrence of bones and skeletons of crocodiles and turtles prove, as Mr. Conybeare justly remarks, the existence of neighbouring dry land. The shores, at least, of some islands were accessible, whither these creatures may have resorted to lay their eggs. In like manner, we may infer the contiguity of land from the immense number of ligneous seed-vessels of plants, some of them resembling the cocoa-nut, and other spices of tropical regions, which have been found fossil in great profusion in the Isle of Sheppey. Such is the abundance of these fruits, that they have been supposed to belong to several hundred distinct species of plants.

Bagshot sand.—The third and uppermost group, usually termed the Bagshot sand, rests conformably upon the London clay, and consists of siliceous sand and sandstone devoid of organic remains, with some thin deposits of marl associated. From these *marls* a few marine shells have been obtained which are in an imperfect state, but appear to belong to Eocene species common to the Paris basin *.

Fresh-water strata of the Hampshire basin.—In the northern part of the Isle of Wight, and part of the opposite coast of Hampshire, fresh-water strata occur resting on the London clay. They are composed chiefly of calcareous and argillaceous marls, interstratified with some thick beds of siliceous sand, and a few layers of limestone sometimes slightly siliceous. The marls are often green, and bear a considerable resemblance to the green marls of Auvergne and the Paris basin. The shells and gyrogonites also agree specifically with some of those most common in the French deposits. Mr. Webster, who first described the fresh-water formation of Hampshire, divided it into an upper and lower series separated by intervening beds of marine

---

* Warburton, Geol. Trans., vol. i. Second Series.

origin. There are undoubtedly certain intercalated strata, both in the Isle of Wight and coast of Hampshire, marked by a slight intermixture of marine and fresh-water shells, sufficient to imply a temporary return of the sea, before and after which the waters of a lake, or rather, perhaps, some large river, prevailed *. The united thickness of the fresh-water and intercalated upper marine beds, exposed in a vertical precipice in Headen Hill, in the Isle of Wight, is about 400 feet, the marine series appearing about half way up in the cliff.

Eocene mammiferous remains.—Very perfect remains of tortoises and the teeth of crocodiles have been procured from the fresh-water strata, but a still more interesting discovery has recently been made. The bones of mammalia corresponding to those of the celebrated gypsum of Paris, have been disinterred at Binstead, near Ryde, in the Isle of Wight. In the ancient quarries near this town a limestone, belonging to the lower fresh-water formation, is worked for building. Solid beds alternate with marls, wherein a tooth of an Anoplotherium, and two teeth of the genus Palæotherium, were found. These remains were accompanied not only by several other fragments of the bones of Pachydermata (chiefly in a rolled and injured state), but also by the jaw of a new species of Ruminantia, apparently closely allied to the genus Moschus †. Mr. T. Allan of Edinburgh had several years before found the tooth of an Anoplotherium at the same spot, and when we alluded to this in our first volume ‡, we threw out some doubts as to the authenticity of his specimen, stating at the same time, that in the Binstead beds, if anywhere in our island, we should expect such remains to be found. Although we carried our scepticism too far, it has been attended with good results, for it induced Mr. Pratt to visit Binstead, where he verified and extended the discovery of Mr. Allan.

* See Memoirs of Mr. Webster, Geol. Trans., vol. ii., First Series, vol. i. part i., Second Series, and Englefield's Isle of Wight.—Professor Sedgwick, Ann. of Phil., 1822, and Lyell, Geol. Trans., vol. ii. Second Series.

† Pratt, Proceedings of Geol. Soc., No. 18, p. 239.

‡ First Edition, p. 153.

These newer strata of the Isle of Wight bear a certain degree of resemblance to some of the green marls and limestones in the Paris basin, yet, as a whole, no formations can be more dissimilar in mineral character than the Eocene deposits of England and Paris. In our own island the tertiary strata are more exclusively marine, and it might be said that the Parisian series differs chiefly from that of London in the very points in which it agrees with the formations of Auvergne, Cantal, and Velay. The tertiary formations of England are, in fact, almost exclusively of mechanical origin, and their composition bespeaks the absence of those mineral and thermal waters to which we have attributed the origin of the compact and siliceous limestones, the gypsum, and beds of pure flint, common to the Paris basin and Central France.

English tertiary strata conformable to the chalk.—The British Eocene strata are nearly conformable to the chalk on which they rest, being horizontal where the strata of the chalk are horizontal, and vertical where they are vertical. On the other hand, there are evident signs that the surface of the chalk had, in many places, been furrowed by the action of the waves and currents, before the Plastic clay and its sands were superimposed. In the quarries near Rochester and Gravesend, for instance, fine examples are seen of deep indentations on the surface of the chalk, into which sand, together with rolled and angular pieces of chalk-flint, have been swept *. But these appearances may be referred to the action of water when the chalk began to emerge during the Eocene period, and they by no means warrant the conclusion, that the chalk had undergone any considerable change of position before the tertiary strata were superimposed.

In this respect there is a marked difference between the reciprocal relations of our secondary and tertiary rocks and those which exist between the same groups throughout the greater part of the continent, especially in the neighbourhood of mountain-chains. Near the base, for example, of the Alps,

* Con. and Phil., Outlines of Geol., p. 62.

Apennines, and Pyrenees, we find the newer formations reposing unconformably upon the truncated edges of the older beds, and it is clear that, in many cases, the latter had been subjected to a complicated series of movements before the more modern strata were formed. The latter rise only to a certain height on the flanks of the mountains which usually tower above them, and are recognized at once by the geologist as having been upraised into land when the tertiary formations were still forming in the sea. The ancient borders also of that sea can often be defined with certainty, and the outline of some of its bays and sea-cliffs traced.

In England, although undoubtedly the greater portion of the tertiary strata is confined to certain spaces, we find outlying patches here and there at great distances beyond the general limits, and at great heights upon the chalk which separates the basins of London and Hampshire[*]. I have seen masses of clay extending in this manner to near the edge of the western escarpment of the chalk in Wiltshire, and Mr. Mantell has pointed out the same to me in the South Downs. Near the escarpment at Lewes, for example, there is a fissure in the chalk filled with sand, and with a ferruginous breccia, such as usually marks the lower members of the Plastic clay formation. From the fact of these tertiary outliers Dr. Buckland inferred, 'that the basins of London and Hants were originally united together in one continuous deposit across the now intervening chalk of Salisbury Plain in Wilts, and the plains of Andover and Basingstoke in Hants, and that the greater integrity in which the tertiary strata are preserved within the basins has resulted from the protection which their comparatively low position has afforded them from the ravages of diluvial denudation[†].'

We agree so far with this conclusion as to believe that the basins of London and Hampshire were not separated until part of the tertiary strata were deposited, but we do not think it probable that the tertiary beds ever extended continuously over

* Dr. Buckland, Geol. Trans., Second Series, vol. ii. p. 125.   † Ibid., p. 126.

those spaces where the outliers above mentioned occur, nor that the comparative thinness of those deposits in the higher chalk countries should be attributed chiefly to the greater degree of denudation which they have there suffered.

Origin of the English tertiary strata.—In explanation of the phenomena above described, we shall endeavour, in the two next chapters, to lay before the reader a view of the series of events which may have produced the leading geological and geographical features of the south-east of England.

A preliminary outline of these views may be useful in this place. We conceive that the chalk, together with many subjacent rocks, may have remained undisturbed and in horizontal stratification until after the commencement of the Eocene period. When at length the chalk was upheaved and exposed to the action of the waves and currents, it was rent and shattered, so that the subjacent secondary strata were exposed at the same time to denudation. The waste of these rocks, composed chiefly of sandstone and clay, supplied materials for the tertiary sands and clays, while the chalk was the source of flinty shingle, and of the calcareous matter which we find intermixed with the Eocene clays. The tracts now separating the basins of London and Hampshire were those first elevated, and which contributed by their gradual decay to the production of the newer strata. These last were accumulated in deep submarine hollows, formed probably by the subsidence of certain parts of the chalk, which sank while the adjoining tracts were rising.

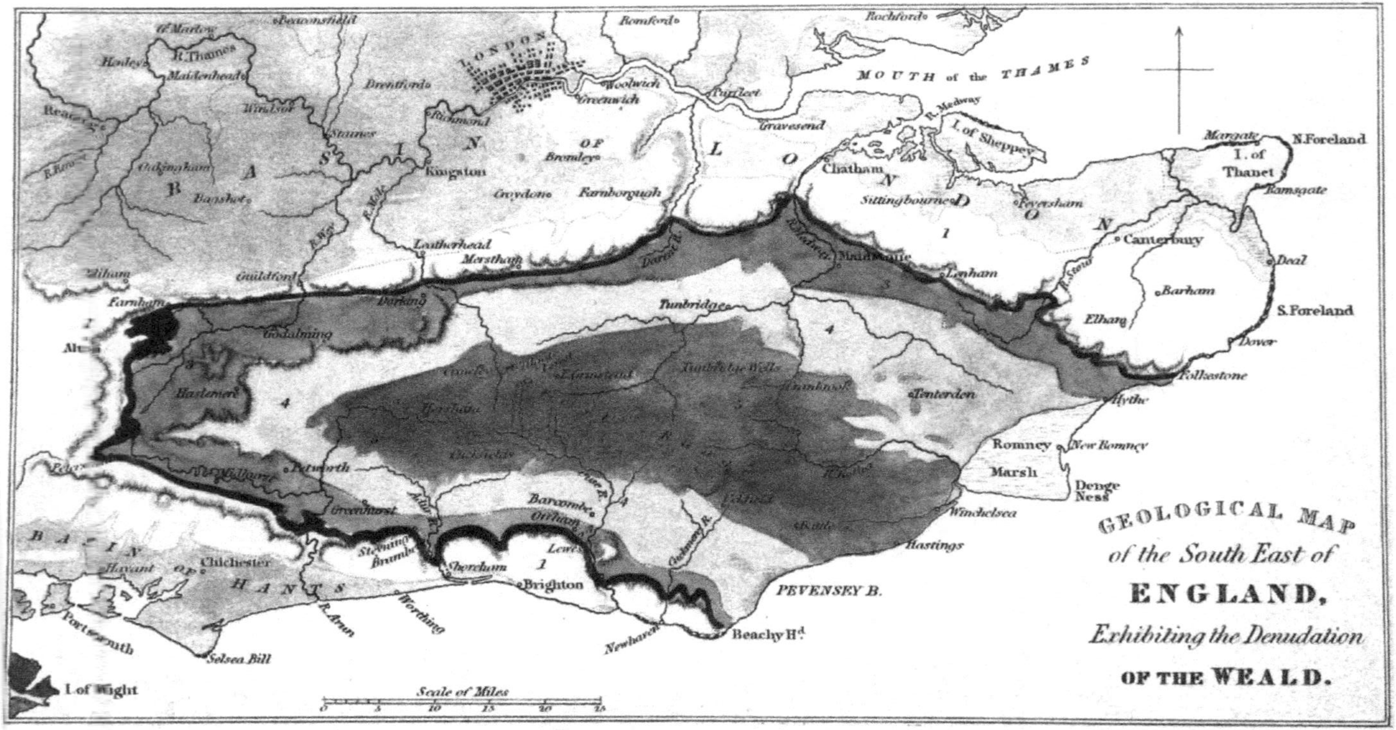

Plate 5.

GEOLOGICAL MAP
of the South East of
ENGLAND.
Exhibiting the Denudation
OF THE WEALD.

Scale of Miles

1	Chalk and Firestone
2	Gault
3	Lower Green Sand
4	Weald Clay
5	Hastings Sand

Tertiary Strata

I. of Wight

CHAPTER XXI.

Denudation of secondary strata during the deposition of the English Eocene
formations—Valley of the Weald between the North and South Downs.—Map
—Secondary rocks of the Weald divisible into five groups—North and South
Downs—Section across the valley of the Weald—Anticlinal axis—True scale
of heights—Rise and denudation of the strata gradual—Chalk escarpments
once sea-cliffs—Lower terrace of ' firestone,' how caused—Parallel ridges and
valleys formed by harder and softer beds—No ruins of the chalk on the central
district of the Weald—Explanation of this phenomenon—Double system of
valleys, the longitudinal and the transverse—Transverse how formed—Gorges
intersecting the chalk—Lewes Coomb—Transverse valley of the Adur.

Denudation of the Valley of the Weald.—In order to understand the theory of which we sketched an outline at the close
of the last chapter, it will be necessary that the reader should
be acquainted with the phenomena of denudation exhibited by
the chalk and some of the older secondary rocks in parts of
England most nearly contiguous to the basins of London and
Hampshire. It will be sufficient to consider one of the denuded districts, as the appearances observable in others are
strictly analogous; we shall, therefore, direct our attention to
what we may call *the Valley of the Weald,* or the region intervening between the North and South Downs.

Map.—In the coloured map given in Plate V. *, the district
alluded to is delineated, and it will be there seen that the
southern portion of the basin of London, and the north-east-
ern limits of that of Hampshire, are separated by a tract of
secondary rocks, between 40 and 50 miles in breadth, comprising within it the whole of Sussex and parts of the counties
of Kent, Surrey, and Hampshire.

There can be no doubt that the tertiary deposits of the
Hampshire basin formerly extended much farther along our
southern coast towards Beachy Head, for patches are still

---

* This map has been chiefly taken from Mr. Greenough's Map of England.

found near Newhaven, and at other points, as will be seen by the map. These are now wasting away, and will in time disappear, as the sea is constantly encroaching and undermining the subjacent chalk.

The secondary rocks, depicted on the map, may be divided into five groups :—

1. *Chalk and Upper green-sand.*—This group is the uppermost of the series ; it includes the white chalk with and without flints, and an inferior deposit called, provincially, ' Firestone,' and by English geologists the ' Upper green-sand.' It sometimes consists of loose siliceous sand, containing grains of silicate of iron, but often of firm beds of sandstone and chert.

2. Blue clay or calcareous marl, called provincially *Gault.*

3. *Lower green-sand,* a very complex group consisting of grey, yellowish, and greenish sands, ferruginous sand and sandstone, clay, chert, and siliceous limestone.

4. *Weald clay,* composed for the most part of clay without intermixture of calcareous matter, but sometimes including thin beds of sand and shelly limestone.

5. *Hastings sands,* composed chiefly of sand, sandstone, clay, and calcareous grit, passing into limestone *.

The first three formations above enumerated are of marine origin, the last two, Nos. 4 and 5, contain almost exclusively the remains of fresh-water and amphibious animals. But it is not our intention at present to enlarge upon the organic remains of these formations, as we have merely adverted to the rocks in order that we may describe the changes of position which they have undergone, and the denudation to which they have been exposed since the commencement of the Eocene period,—mutations which, if our theory be well founded, belong strictly to the history of *tertiary* phenomena.

By a glance at the map, the reader may trace at once the

* For an account of these strata in the south-east of England, see Mantell's Geology of Sussex, and Dr. Fitton's Geology of Hastings, where the memoirs of all the writers on this part of England are referred to.

superficial area occupied by each of the five formations above mentioned.　On the west will be seen a large expanse of chalk, from which two branches are sent off; one through the hills of Surrey and Kent to Dover, forming the ridge called the North Downs, the other through Sussex to the sea at Beachy Head, constituting the South Downs.　The space comprised between the North and South Downs, or ' the Valley of the Weald,' consists of the formations Nos. 2, 3, 4, 5, of the above table. It will be observed that the chalk terminates abruptly, and with a well-defined line towards the country occupied by those older strata.　Within that line is a narrow band coloured blue, formed by the gault, and within this again is the Lower green sand, next the Weald clay, and then, in the centre of the district, a ridge formed by the Hastings sands.

Section of the Valley of the Weald.—It has been ascertained by careful investigation, that if a line be drawn from any part of the North to the South Downs, which shall pass through the central group, No. 5, the beds will be found arranged in the order described in the annexed section (No. 63, p. 288).

We refer the reader at present to the dark lines of the section, as the fainter lines represent portions of rock supposed to have been carried away by denudation.

At each end of the diagram the tertiary strata *a* are exhibited reposing on the chalk.　In the centre are seen the Hastings sands (No. 5), forming an anticlinal axis, on each side of which the other formations are arranged with an opposite dip.　It has been necessary however, in order to give a clear view of the different formations, to exaggerate the proportional height of each in comparison to its horizontal extent, and we have subjoined a true scale in another diagram (No. 64) in order to correct the erroneous impression which might otherwise be made on the reader's mind.　In this section the distance between the North and South Downs is represented to exceed 40 miles; for we suppose the valley of the Weald to be here intersected in its longest diameter, in the direction of a line between Lewes and Maidstone.

No. 63.

S. *a*

N. *a*

1 2 3 4 5 4 3 2 1

Section from the London to the Hampshire basin across the Valley of the Weald.

a, Tertiary strata.   1, Chalk and firestone.   2, Gault.   3, Lower green-sand.   4, Weald clay.   5, Hastings sands.

No. 64.

Anticlinal axis of the Weald.

Highest point of South Downs, 858 feet.   Crowborough Hill, 804 feet.   Highest point of North Downs, 880 feet *.

Miles

0 2 4 6 8 10 20 30 40 50 Miles

Section of the country from the confines of the basin of London to that of Hants, with the principal heights above the level of the sea on a true scale.†

* Lieutenant H. Murphy, R. E., informs me that Botley Hill, near Godstone, in Surrey, was found by trigonometrical measurement to be 880 feet above the level of the sea; and Wrotham Hill, near Maidstone, which appears to be next in height of the North Downs, 795 feet.

† My friend Mr. Mantell, of Lewes, has kindly drawn up this scale at my request.

In attempting to account for the manner in which the five secondary groups above mentioned may have been brought into their present position, the following hypothesis has been very generally adopted. Suppose the five formations to lie in horizontal stratification at the bottom of the sea; then let a movement from below press them upwards into the form of a flattened dome, and let the crown of this dome be afterwards cut off, so that the incision should penetrate to the lowest of the five groups. The different beds would then be exposed on the surface in the manner exhibited in the map, plate 5 *.

It will appear from former parts of this work, that the amount of elevation here supposed to have taken place is not greater than we can prove to have occurred in other regions within geological periods of no great duration. On the other hand, the quantity of denudation or removal by water of vast masses which are assumed to have once reached continuously from the North to the South Downs is so enormous, that the reader may at first be startled by the boldness of the hypothesis. But he will find the difficulty to vanish when once sufficient time is allowed for the gradual and successive rise of the strata, during which the waves and currents of the ocean might slowly accomplish an operation, which no sudden diluvial rush of waters could possibly have effected.

Escarpments of the chalk once sea-cliffs.—In order to make the reader acquainted with the physical structure of the Valley of the Weald, we shall suppose him first to travel southwards from the London basin. On leaving the tertiary strata he will first ascend a gently-inclined plane, composed of the upper flinty portion of the chalk, and then find himself on the summit of a declivity consisting, for the most part, of different members of the chalk formation, below which the upper green-sand, and sometimes also the gault *crop out* †. This steep declivity is called by geologists ' the escarpment of the chalk,' which overhangs a

---

* See illustrations of this theory by Dr. Fitton, Geol. Sketch of Hastings.

† We use this term, borrowed from our miners, to express the coming up to the surface of one stratum from beneath another.

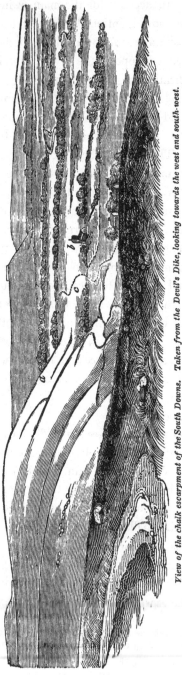

No. 65.

View of the chalk escarpment of the South Downs. Taken from the Devil's Dike, looking towards the west and south-west.

a, The town of Steyning is hidden by this point. *b*, Edburton church. *c*, Road. *d*, River Adur.

valley excavated chiefly out of the argillaceous or marly bed, termed Gault (No. 2). The escarpment is continuous along the southern termination of the North Downs, and the reader may trace it from the sea at Folkstone, westward to Guildford and the neighbourhood of Petersfield, and from thence to the termination of the South Downs at Beachy Head. In this precipice or steep slope the strata are cut off abruptly, and it is evident that they must originally have extended farther. In the accompanying wood-cut (No. 65), part of the escarpment of the South Downs is faithfully represented, where the denudation at the base of the declivity has been somewhat more extensive than usual, in consequence of the upper and lower green-sand being formed of very incoherent materials, the former, indeed, being extremely thin and almost wanting.

The geologist cannot fail to recognize in this view the exact likeness of a sea-cliff, and if he turns and looks in an opposite direction, or eastward, towards Beachy Head, he will see the same line of height prolonged. Even those who are not accustomed to speculate on the former changes which the surface has undergone, may fancy the broad and level plain to resemble the flat sands which were laid dry by the receding tide, and the different projecting masses of chalk to be the headlands of a coast which separated the different bays from each other.

No. 66.

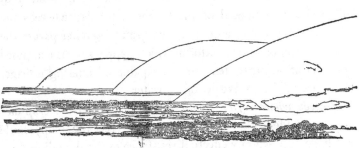

Chalk escarpment as seen from the hill above Steyning, Sussex. The castle and village of Bramber in the fore-ground.

Lower terrace of firestone.—We have said that the upper

green-sand ('firestone,' or 'malm rock,' as it is sometimes called) is almost absent in the tract here alluded to. It is, in fact, seen at Beachy Head to thin out to an inconsiderable stratum of loose green-sand; but farther to the westward it is of great thickness, and contains hard beds of blue chert and limestone. Here, accordingly, we find that it produces a corresponding influence on the scenery of the country, for it runs out like a step beyond the foot of the chalk-hills, and constitutes a lower

No. 67.

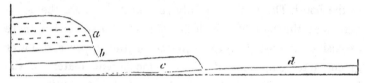

a, Chalk with flints.　　　　　b, Chalk without flints.
c, Upper green sand, or firestone.　　　d, Gault.

terrace varying in breadth from a quarter of a mile to three miles, and following the sinuosities of the chalk escarpment*.

It is impossible to desire a more satisfactory proof that the escarpment is due to the excavating power of water during the gradual rise of the strata. For we have shown, in our account of the coast of Sicily †, in what manner the encroachments of the sea tend to efface that succession of terraces which must otherwise result from the successive rises of a coast preyed upon by the waves. During the interval between two elevatory movements, the lower terrace will usually be destroyed, wherever it is composed of incoherent materials; whereas the sea will not have time entirely to sweep away another part of the same terrace, or lower platform, which happens to be composed of rocks of a harder texture and capable of offering a firmer resistance to the erosive action of water.

Valleys where softer strata, ridges where harder crop out. —It is evident that the Gault No. 2 (see the map) could not have opposed any effectual resistance to the denuding force

* Mr. Murchison, Geol. Sketch of Sussex, &c., Geol. Trans., 2nd Series, vol. ii. p. 98.
† See p. 111; and wood-cut No. 24.

of the waves; its outcrop, therefore, is marked by a valley, the breadth of which is often increased by the loose incoherent nature of the uppermost beds of the lower green-sand, which lie next to it, and which have often been removed with equal facility.

The formation last mentioned has been sometimes entirely smoothed off like the gault; but in those districts where chert, limestone, and other solid materials enter largely into its composition, it forms a range of hills parallel to the chalk, which sometimes rival the escarpment of the chalk itself in height, or even surpass it, as in Leith Hill. This ridge often presents a steep escarpment towards the Weald clay which crops out from under it. (See the strong lines in diagram No. 63, p. 288.)

The clay last mentioned forms, for the most part, a broad valley, separating the lower green-sand from the Hastings sands, or Forest ridge; but where subordinate beds of sandstone of a firmer texture occur, the uniformity of the plain is broken by waving irregularities and hillocks *.

In the central region, or Forest ridge, the strata have been considerably disturbed and are greatly fractured and shifted. One fault is known where the vertical shift of a bed of calcareous grit is no less than 60 fathoms †. It must not be supposed that the anticlinal axis, which we have described as running through the centre of the weald, is by any means so simple as is usually represented in geological sections. There are, on the contrary, a series of anticlinal and synclinal ‡

* Martin, Geol. of Western Sussex.   Fitton, Geol. of Hastings, p. 31.
† Fitton, ibid., p. 55.
‡ We adopt this term, first used, we believe, by Professor Sedgwick; its signification will best be understood by reference to the accompanying diagram.

No. 68.

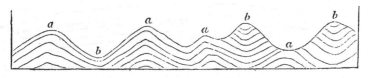

a, Anticlinal lines.   b, Synclinal lines.

lines, which form ridges and troughs running nearly parallel to each other.

Much of the picturesque character of the scenery of this district arises from the depth of the narrow valleys and ridges to which the sharp bends and fractures of the strata have given rise; but it is also in part to be attributed to the excavating power exerted by water, especially on the interstratified argillaceous beds.

From the above description it will appear that, in the tract intervening between the North and South Downs, there are a series of parallel valleys and ridges; the valleys appearing evidently to have been formed principally by the removal of softer materials, while the ridges are due to the resistance offered by firmer beds to the destroying action of water.

Rise and denudation of the strata gradual.—Let us then consider how far these phenomena agree with the changes which we should naturally expect to occur during the gradual rise of the secondary strata. Suppose the line of the most violent movements to have coincided with what is now the central ridge of the Weald Valley; in that case, the first land which emerged must have been situated where the Forest ridge is now placed. Here a number of reefs may have existed, and islands of chalk, which may have been gradually devoured by the ocean in the same manner as Heligoland and other European isles have disappeared in modern times, as related in our first volume *.

Suppose the ridge or dome first elevated to have been so rent and shattered on its summit as to give more easy access to the waves, until at length the masses represented by the fainter lines in the annexed diagram were removed. Two strips of land might

No. 69.

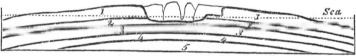

then remain on each side of a channel, in the same manner as

* Page 289, and Second Edition, page 330.

the opposite coasts of France and England, composed of chalk, present ranges of white cliffs facing each other.   A powerful current might then rush, like that which now ebbs and flows through the straits of Dover, and might scoop out a channel in the gault. We must bear in mind that the intermittent action of earthquakes would accompany this denuding process, fissuring rocks, throwing down cliffs, and bringing up, from time to time, new stratified masses, and thus greatly accelerating the rate of waste.   If the lower bed of chalk on one side of the channel should be harder than on the other, it would cause an under terrace, as represented in the above diagram, resembling that presented by the upper green-sand in parts of Sussex and Hampshire.   When at length the gault was entirely swept away from the central parts of the channel, the lower green-sand (3, diagram No. 70,) would be laid bare, and portions of it would

No. 70.

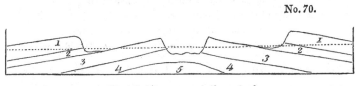

The dotted line represents the sea-level.

become land during the continuance of the upheaving earthquakes.   Meanwhile the chalk cliffs would recede farther from one another, whereby four parallel strips of land, or perhaps rows of islands, would be caused.

The edges of the argillaceous strata, No. 2, are still exposed to erosion by the waves, and a portion of the clay, No. 4, is already removed.   This clay, as it gradually rises, will be swept off from part of the subjacent group, No. 5, which will then be laid bare, and may afterwards become land by subsequent elevation.

Why no ruins of chalk on central district.—By this theory of the successive emergence and denudation of the groups, 1, 2, 3, 4, 5, we may account for an alluvial phenomenon which seems inexplicable on any other hypothesis.   The summits of the chalk downs are covered everywhere with flint gravel, which

is often entirely wanting on the surface of the clay at the foot
of the chalk escarpment, and no traces of chalk flint have ever
been found in the alluvium of the central district, or Forest
ridge. It is rare, indeed, to see any wreck of the chalk, even
at the distance of two or three miles from the escarpments of
the North and South Downs. To this general rule, however,
an exception occurs near Barcombe, about three miles to the
north of Lewes, where we obtain the accompanying section *.

No. 71.

Section from the North escarpment of the South Downs to Barcombe.

1, Gravel composed of partially-rounded chalk flints.
2, Chalk with and without flints.
3, Lowest chalk or chalk marl (upper green-sand wanting).
4, Gault.   5, Lower green-sand.   6, Weald clay.

It will be seen that the valley at the foot of the escarpment
extends, in this case, not only over the gault, but over the
'lower green-sand' to the Weald clay. On this clay a thick
bed of flints, evidently derived from the waste of chalk, remains in the position above described.

We say that there is no detritus of the chalk and its flints on
the central ridge of the Weald. I have sought in vain for a
vestige of such fragments, and Mr. Mantell, who has had
greater opportunities of minute investigation, assures me that
he has never been able to detect any. Now whether we embrace
or reject the theory of the former continuity of the chalk and
other groups over the whole space intervening between the
North and South Downs, we cannot certainly imagine that
any transient and tumultuous rush of waters could have swept
over this country, which should not have left some fragments

* The author visited this locality with Mr. Mantell, to whom he is indebted for
this section.

of the chalk and its flints in the deep valleys of the Forest ridge. Indeed, if we adopted the diluvial hypothesis of Dr. Buckland, we should expect to find vast heaps of broken flints drifted more frequently into the valleys of the Gault and Weald clay, instead of being so frequently confined to the summit of the chalk downs. On the other hand, it is quite conceivable that the slow agency of oceanic currents may have cleared away, in the course of ages, the matter which fell into the sea from wasting cliffs. The reader will recollect our account of the manner in which the sea has advanced, within the last century, upon the Norfolk coast at Sherringham *.

No. 72.

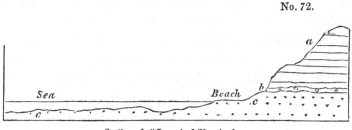

Section of cliffs west of Sherringham.

a, Crag.   b, Ferruginous flint breccia on the surface of the chalk.
c, Chalk with flints.

The beach, at the foot of the cliff, is composed of bare chalk with flints, as is the bed of the sea near the shore. No one would suspect, from the appearance of the beach at low water, that a few years ago beds of solid chalk, together with sand and loam of the superincumbent crag, formed land on the very spot where the waves are now rolling; still less that these same formations extended, within the last 50 years, to a considerable distance from the present shore, over a space where the sea has now excavated a channel 20 feet deep.

As in this recent instance the ocean has cleared away part of the chalk, and its capping of crag, so the tertiary sea may have swept away not only the chalk, but the layer of broken flints on its surface, which was probably a marine alluvium of the

* Vol. i. p. 268, and Second Edition, p. 307.

Eocene period. Hence these flints might naturally occur on the downs, and be wanting in the valleys below.

If the reader will refer to the preceding diagrams (Nos. 69 and 70), and reflect not only on the successive states of the country there delineated, but on all the intermediate conditions which the district must have passed through during the process of elevation and denudation before supposed, he will understand why no wreck of the chalk (No. 1) should occur at great distances from the chalk escarpments. For it is evident that when the ruins of the uppermost bed (No. 1, diagram 69) had been thrown down upon the surface of the bed immediately below, those ruins would subsequently be carried away when this inferior stratum itself was destroyed. And in proportion to the number and thickness of the groups, thus removed in succession, is the probability lessened of our finding any remnants of the highest group strewed over the bared surface of the lowest.

Transverse valleys.—There is another peculiarity in the geographical features of the south-east of England which must not be overlooked when we are considering the action of the denuding causes. By reference to the map (Plate 5), the reader will perceive that the drainage of the country is not effected by water-courses following the great valleys excavated out of the argillaceous strata (Nos. 2 and 4), but by valleys which run in a transverse direction, passing through the chalk to the basin of the Thames on the one side, and to the English channel on the other.

In this manner the chain of the North Downs is broken by the rivers Wey, Mole, Darent, Medway, and Stour; the South Downs by the Arun, Adur, Ouse, and Cuckmere *.

If these transverse hollows could be filled up, all the rivers, observes Mr. Conybeare, would be forced to take an easterly course, and to empty themselves into the sea by Romney Marsh and Pevensey levels †.

---

* Conybeare, Outlines of Geol., p. 81,   † Ibid., p. 145.

Mr. Martin has suggested that the great cross fractures of the chalk which have become river channels have a remarkable correspondence on each side of the valley of the Weald; in several instances the gorges in the North and South Downs appearing to be directly opposed to each other. Thus, for example, the defiles of the Wey, in the North Downs, and of the Arun in the South, seem to coincide in direction; and, in like manner, the Ouse is opposed to the Darent, and the Cuckmere to the Medway *. But we think it very possible that these coincidences may be accidental. It is, however, by no means improbable, as hinted by the author above mentioned, that the great amount of elevation towards the centre of the Weald district gave rise to transverse fissures. And as the longitudinal valleys were connected with that linear movement which caused the anticlinal lines running east and west, so the cross fissures might have been occasioned by the intensity of the upheaving force towards the centre of

No. 73.

Transverse valley of the Adur in the South Downs.

a, Town of Steyning.   b, River Adur.   c, Old Shoreham.

* Geol. of Western Sussex, p. 61.

the line, whereby the effect of a double axis of elevation was in some measure produced.

In order to give a clearer idea of the manner in which the chalk-hills are intersected by these transverse valleys, we subjoin a sketch (No. 73) of the gorge of the river Adur, taken from the summit of the chalk-downs, at a point in the bridle-way leading from the towns of Bramber and Steyning to Shoreham. If the reader will refer again to the view given in a former wood-cut (No. 65, p. 290), he will there see the exact point where the gorge, of which we are now speaking, interrupts the chalk escarpment. A projecting hill, at the point *a*, hides the town of Steyning, near which the valley commences where the Adur passes directly to the sea at Old Shoreham. The river flows through a nearly level plain, as do most of the others which intersect the hills of Surrey, Kent, and Sussex ; and it is evident that these openings, so far at least as they are due to aqueous erosion, have not been produced by the rivers, many of which, like the Ouse near Lewes, have filled up arms of the sea, instead of deepening the hollows which they traverse.

In regard to the origin of the transverse ravines, there can be no doubt that they are connected with lines of fracture, and perhaps, in some places, there may be an anticlinal dip on both sides of the valley, as suggested by a local observer *. But this notion requires confirmation.

No. 74.

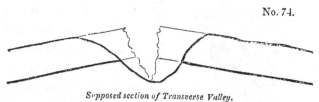

Supposed section of Transverse Valley.

The ravine, called the Coomb, near Lewes, affords a beautiful example of the manner in which narrow openings in the chalk may have been connected with shifts and dislocations in the strata. This coomb is seen on the eastern side of the valley of the Ouse, in the suburbs of the town of Lewes. The steep

* Martin, Geol. of Western Sussex, p. 64, plate III. fig. 3.

declivities on each side are covered with green turf, as is the bottom, which is perfectly dry. No outward signs of disturbance are visible, and the connexion of the hollow with subterranean movements would not have been suspected by the geologist, had not the evidence of great convulsions been clearly exposed in the escarpment of the valley of the Ouse, and in the numerous chalk pits worked at the termination of the

No. 75.

The Coomb, near Lewes.

Coomb. By aid of these we discover that the ravine coincides precisely with a line of fault, on one side of which the chalk with flints *a*, appears at the summit of a hill, while it is thrown down to the bottom on the other.

No. 76.

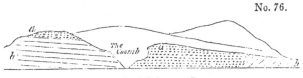

Fault in the cliff-hills near Lewes.

a, Chalk with flints.   b, Lower chalk *.

The fracture here alluded to is one of those which run east

* I examined this spot in company with Mr. Mantell, to whom I am indebted for the above section.

and west, and of which there are many in the Weald district, parallel to the central axis of the Forest ridge.

In whatever manner the transverse gorges originated, they must evidently have formed ready channels of communication between the submarine longitudinal valleys and those deep parts of the sea wherein we imagine the tertiary strata to have been accumulated. If the strips of land which first rose had been unbroken, and there had been no free passage through the cross fractures, the currents would not so easily have drifted away the materials detached from the wasting cliffs, and it would have been more difficult to understand how the wreck of the denuded strata could have been so entirely swept away from the base of the escarpments.

In the next chapter we shall resume the consideration of these subjects, especially the proofs of the former continuity of the chalk of the North and South Downs, and the probable connexion of the denudation of the Weald valley with the origin of the Eocene strata.

Denudation of the Valley of the Weald, *continued*—The alternative of the proposition that the chalk of the North and South Downs were once continuous,
considered—Dr. Buckland on the Valley of Kingsclere—Rise and denudation
of secondary rocks gradual—Concomitant deposition of tertiary strata gradual
—Composition of the latter such as would result from the wreck of the secondary rocks—Valleys and furrows on the chalk how caused—Auvergne, the
Paris basin, and south-east of England one region of earthquakes during the
Eocene period—Why the central parts of the London and Hampshire basins
rise nearly as high as the denudation of the Weald—Effects of protruding
force counteracted by the levelling operations of water—Thickness of masses
removed from the central ridge of the Weald—Great escarpment of the chalk
having a direction north-east and south-west—Curved and vertical strata in the
Isle of Wight—These were convulsed after the deposition of the fresh-water
beds of Headen Hill—Elevations of land posterior to the crag—Why no Eocene
alluviums recognizable—Concluding remarks on the intermittent operations of
earthquakes in the south-east of England, and the gradual formation of valleys
—Recapitulation.

Extent of denudation in the Valley of the Weald.—' It would
be highly rash,' observes Mr. Conybeare, speaking of the denudation of the Weald, ' to assume that the chalk at any period
actually covered the whole space in which the inferior strata
are now exposed, although the truncated form of its escarpment
evidently shows it to have once extended much farther than at
present *.'

We believe that few geologists who have considered the
extent of country supposed to have been denuded, and who
have explored the hills and valleys of the central, or Forest
ridge, without being able to discover the slightest vestige of chalk
in the alluvium †, will fail to participate, at first, in the doubts
here expressed as to the original continuity of the upper secondary formations over the anticlinal axis of the Weald. For our
own part, we never traversed the wide space which separates the
North and South Downs, without desiring to escape from the
conclusions advocated in the last chapter; and yet we have

* Outlines, p. 144.   † See above, p. 295.

been invariably brought back again to the opinion, that the chalk was originally continuous, on a more deliberate review of the whole phenomena.

It may be useful to consider the only other alternative of the hypothesis before explained. If the marine groups, Nos. 1, 2, 3, were not originally continuous, it is necessary to imagine

No. 77.

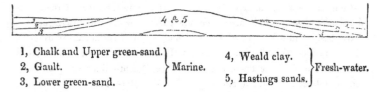

1, Chalk and Upper green-sand.			4, Weald clay.	
2, Gault.	} Marine.		5, Hastings sands.	} Fresh-water.
3, Lower green-sand.				

that they each terminated at some point between their present outgoings and the secondary strata of the Forest ridge. Thus we might suppose them to have thinned out one after the other, as in the above diagram, and never to have covered the entire area occupied by the fresh-water strata, Nos. 4 and 5.

We grant that had such been the original disposition of the different groups, they might, as they gradually emerged from the sea, have become denuded in the manner explained in the last chapter, so that the country might equally have assumed its present configuration. But, although we know of no invincible objection to such an hypothesis, there are certainly no appearances which favour it. If the strata Nos. 4 and 5 had been unconformable to the Lower green-sand No. 3, then, indeed, we might have imagined that the older groups had been disturbed by a series of movements antecedently to the deposition of No. 3, and, in that case, some parts of them might be supposed to have emerged or formed shoals in the ancient sea, interrupting the continuity of the newer marine deposits. But the group No. 4 is *conformable* to No. 3, and the only change which has been observed to take place at the junction, is an occasional intermixture of the Weald clay with the superior marine sand, such as might have been caused by a slight superficial movement in the waters when the sea first overflowed the fresh-water strata.

On the other hand, the green-sand and chalk, as they approach the central axis of the Weald, are not found to contain littoral shells, or any wreck of the fresh-water strata, such as might indicate the existence of an island with its shores or wasting cliffs. Had any such signs been discovered, we might have been inclined to suppose the geography of the region to have once borne some resemblance to that exhibited in the diagram No. 77.

Dr. Buckland on Valleys of Elevation.—We are indebted to Dr. Buckland for an able memoir in illustration of several districts of similar form and structure to the Weald, which occur at no great distance in the south of England. His paper is intitled, ' On the formation of the Valley of Kingsclere and other valleys by the elevation of the strata which enclose them *.'

The valley of Kingsclere, situate a few miles south of New-

No. 78.

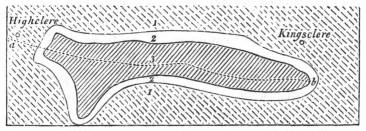

Valley of Kingsclere.

a, b, Anticlinal line marking the opposite dip of the strata on each side of it.

No. 79.

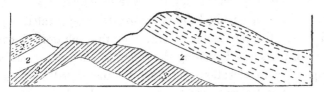

Section across the Valley of Kingsclere from north to south.

1, Chalk with flints.   2, Lower chalk without flints.
3, Upper green-sand, or firestone, containing beds of chert.

* Geol. Trans., 2nd Series, vol. ii. p. 119.

bury, in Berkshire, is about five miles long and two in breadth.
The upper and lower chalk, as will be seen in the accompanying section *, and the upper green-sand dip in opposite directions from an anticlinal axis which passes through the middle
of the valley along the line *a*, *b*, of the ground-plan (No. 78).

We subjoin an additional wood-cut, as conveying a scale of

N.   No. 80.   S.

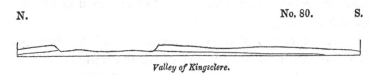

Valley of Kingsclere.

heights more nearly approaching to that of nature, although
the altitudes, in proportion to the horizontal extent, are even
in this, perhaps, somewhat in excess. On each side of the valley we find escarpments of chalk, the strata of which dip in opposite directions, in the northern escarpment to the north, and
in the southern to the south. At the eastern and western extremities of the valley, the two escarpments become confluent,
precisely in the same manner as do those of the North and
South Downs, at the eastern end of the Weald district, near
Petersfield. And as, a few miles east of the town last mentioned (see Map, plate V.), the firestone, or upper green-sand,
is laid open in the sharp angle between the escarpment of the
Alton Hills and the western termination of the South Downs† ;
so in the valley of Kingsclere the same formation is seen to
crop out from beneath the chalk.

The reader might imagine, on regarding Dr. Buckland's
section (No. 79), where, for the sake of elucidating the geological phenomena, the heights are greatly exaggerated in proportion to the horizontal extent, that the solution of continuity of the strata bounding the valley of Kingsclere had been
simply due to elevation and fracture, unassisted by aqueous
causes; but by reference to the true scale (No. 80), it will

* Copied by permission from Dr. Buckland's plate XVII., Geol. Trans., 2nd
Series, vol. ii.

† See Mr. Murchison's map, plate XIV., ibid.

immediately appear, that a considerable mass of chalk must have been removed by denudation.

If the anticlinal dip had been confined to the valley of Kingsclere, we might have supposed that the upheaving force had acted on a mere point, forcing upwards the superincumbent strata into a small dome-shaped eminence, the crown of which had been subsequently cut off. But Dr. Buckland traced the line of opposite dip far beyond the confluence of the chalk escarpments, and found that it was prolonged in a more north-west direction far beyond the point a, diagram No. 78. In following the line thus extended, the strata are seen in numerous chalk-pits to have an opposite dip on either side of a central axis, from which we may clearly infer the linear direction of the movement. Perhaps the intensity of the disturbing force was greatest where the denudation of the valley of Kingsclere took place; but this cannot be confidently inferred, for the quantity of matter removed by aqueous agency must depend on the set of the tides and currents at the period of emergence, and not solely on the amount of elevation and derangement of the strata.

Many of the valleys enumerated by Dr. Buckland as having a similar conformation to that of Kingsclere, run east and west, like the anticlinal ridge of the Weald valley. Several of these occur in Wiltshire and Dorsetshire, and they are all circumscribed by an escarpment whose component strata dip outwards from an anticlinal line running along the central axis of the valley. One of these, distant about seven miles to the north-east of Weymouth, is nearly elliptical in shape, and in size does not much exceed that of the Coliseum at Rome. Their drainage is generally effected by an aperture in one of their lateral escarpments, and not at either extremity of their longer axis, as would have happened had they been simply excavated by the sweeping force of rapid water *.

'It will be seen,' continues Dr. Buckland, 'if we follow on Mr. Greenough's map the south-western escarpment of the

* Dr. Buckland, Geol. Trans., 2nd Series, vol. ii, p 122.

chalk in the counties of Wilts and Dorset, that, at no great distance from these small elliptical valleys of elevation, there occur several longer and larger valleys, forming deep notches, as it were, in the lofty edge of the chalk. These are of similar structure to the smaller valleys we have been considering, and consist of green-sand, inclosed by chalk at one extremity, and flanked by two escarpments of the same, facing each other with an opposite dip; but they differ in the circumstance of their other and broader extremity being without any such inclosure, and gradually widening till it is lost in the expanse of the adjacent country.

The cases I now allude to are the Vale of Pewsey, to the east of Devizes, that of the Wily, to the east of Warminster, and the valley of the Nadder, extending from Shaftsbury to Barford, near Salisbury; in which last not only the strata of green-sand are brought to the surface, but also the still lower formations of Purbeck and Portland beds, and of Kimmeridge clay.

It might at first sight appear that these valleys are nothing more than simple valleys of denudation; but the fact of the strata composing their escarpments having an opposite and out-ward dip from the axis of the valley, and this often at a high angle, as near Fonthill and Barford, in the Vale of the Nadder, and at Oare, near the base of Martinsell Hill, in the Vale of Pewsey, obliges us to refer their inclination to some antecedent violence, analogous to that to which I have attributed the position of the strata in the inclosed valleys near Kingsclere, Ham, and Burbage. Nor is it probable that, without some pre-existing fracture or opening in the lofty line of the great chalk escarpment, which is here presented to the north-west, the power of water alone would have forced open three such deep valleys as those in question, without causing them to maintain a more equable breadth, instead of narrowing till they end in a point in the body of the chalk *.

Rise and denudation of the secondary rocks gradual.—To

---

* Dr. Buckland, Geol. Trans., 2nd Series, vol. ii. p. 123.

return to the valley of the Weald, the strata of the North Downs are inclined to the north, at an angle of from 10° to 15°, and in the narrow ridge of the Hog's back, west of Guildford, in Surrey, about 45°; those in the South Downs dip to the south at a slight angle. It is superfluous to dwell on the analogy which in this respect the two escarpments bear to those which flank the valleys above alluded to ; and in regard to the greater distance which separates the hills of Surrey from those of Sussex, the difficulty may be reduced simply to a question of time. If the rise of the land and its degradation by aqueous causes was accomplished by an indefinite number of minor convulsions, during an immense lapse of ages, we behold in the ocean a power fully adequate to perform the work of demolition. If, on the other hand, we embrace the hypothesis of paroxysmal elevation, or, in other words, suppose a submarine tract to have been converted instantaneously into high land, we may seek in vain for any known cause capable of sweeping away even those portions of chalk and other rocks which, all are agreed, must once have formed the prolongation of the existing escarpments. It is common in such cases to call in one arbitrary hypothesis to support another, and as the upheaving force operated with sudden violence, so a vast diluvial wave is introduced to carry away, with almost equal celerity, the mountain mass of strata assumed to have been stripped off.

Materials of the tertiary strata whence derived.—If, then, we conclude that the wreck of the denuded district was *removed* gradually, it follows that it was *deposited* by degrees elsewhere. If any part of the sea immediately adjacent to the district which was then emerging, was of considerable depth, the drift matter would be consigned to that submarine region, since every current charged with sediment must purge itself in the first deep cavity which it traverses, as does a turbid river in a lake. Suppose that while the wave sand currents were excavating the longitudinal valleys, D and C (No. 81, p. 310), the deposits *a* were thrown down to the bottom of the contiguous deep

No. 81.

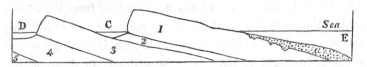

water E, the sediment being drifted through transverse fissures, as before explained.   In this case, the rise of the formations Nos. 1, 2, 3, 4, 5, may have been going on contemporaneously with the excavation of the valleys C and D, and with the accumulation of the strata *a*.   There must be innumerable points on our own coast where the sea is of great depth near to islands and cliffs now exposed to rapid waste, and in all these the denuding and reproductive processes must be going on in the immediate proximity of each other.   Such may have been the case during the rise of the Valley of the Weald, and the deposition of the beds of the London and Hampshire basins.

The theory above proposed requires that the deposits *a* should be composed, for the most part, of a mixture of such mineral ingredients as would result from the degradation of the secondary groups, Nos. 1, 2, 3, 4, 5.   Now the tertiary strata answer extremely well to these conditions.   They consist, as we have before seen, of alternations of variously-coloured sands and clays, as do the secondary strata from the group No. 5 to No. 2 inclusive, the principal difference being, that the latter are more consolidated.

If it be asked, where do we find the ruins of the *white chalk* among our Eocene strata?   We reply, that the flint pebbles which are associated in such immense abundance with the sands of the plastic clay, are derived evidently from the destruction of chalk ; and as to the soft white calcareous matrix, we may suppose it to have been reduced easily to fine sediment, and to have contributed, when in a state of perfect solution, to form the shells of Eocene testacea; or when mixed with the waste of the argillaceous groups, Nos. 2 and 4, which have been peculiarly exposed to denudation, it may have entered into the composition of the London clay, which contains no slight proportion of calcareous matter.   In the crag of Norfolk, undoubtedly, we find great heaps of broken pieces of white chalk, with slightly-worn and angular flints; but in this case we may infer that the attrition was not continued for a long time; whereas the large accumulations of perfectly-rolled shingle, which are interstratified with our Eocene formations, proves that they were acted upon for a protracted period by the waves.   We have many opportunities of witnessing the entire demolition of the chalk on our southern coast, as at Seaford, for example, in Sussex, where large masses are, year after year, detached from the cliffs, and soon disappear, leaving nothing behind but a great bank of flint shingle *.

Valleys and furrows in the chalk how caused.—The furrows which occur on the surface of the chalk, filled with sand and pebbles of the plastic clay, may be easily explained if we suppose the English Eocene strata to have been formed during a period of local convulsion.   For if portions of the secondary rocks emerged from the sea in the south-east of our island during that period, it is probable that the chalk underwent many oscillations of level, and that certain tracts became land and then sea, and then land again, so that parts of the surface, first excavated by currents or rivers, were occasionally submerged, and, after being covered by tertiary deposits, upraised again.   We must also remember, that almost every part of the chalk must have been exposed for some time to the action of the waves, if we assume the elevation to have been slow and by successive movements.   The valleys seen everywhere on the surface, and the layers of partially-rolled and broken flints which very generally overspread it, may be referred to the sea breaking upon the reefs and shoals when the rocks were about to emerge.   We apprehend, indeed, that no formidable difficulty will be encountered in explaining the position of the tertiary sand which sometimes fills rents and furrows in the chalk, or the occurrence of banks of shingle at the junction of the tertiary strata and the chalk, if we once admit that the

* Vol. i. p. 279, and Second Edition, p. 319.

Eocene deposits originated while the chalk and other secondary rocks were rising from the deep and wasting away.

Earthquakes during the Eocene period.—We have pointed out, in a former chapter, our reasons for concluding that the Paris basin was a theatre of subterranean convulsions during the Eocene period, the older beds of the calcaire grossier having been raised and exposed to the action of the waves before, or at least during, the deposition of the upper or second marine group *. These convulsions were doubtless connected with that depression which let in the sea upon the second fresh-water formation, and gave rise to the superposition of the upper marine beds. We have also demonstrated, in a preceding chapter, that some of the earlier volcanic eruptions in Auvergne happened before the Eocene species of animals were extinct, and we suggested that the great lakes of Central France may have been drained by alterations of level which accompanied the outbreak of those earlier Eocene volcanos of Auvergne. We ought not, therefore, to be surprised if we discover proofs that the south-east of England participated in the earthquakes which seem to have extended at that time over a considerable part of the neighbouring continent; and we may refer the alternation of marine and fresh-water beds in the Isle of Wight and coast of Hampshire, to changes of level analogous to those which gave rise to the intercalation of the upper marine formation in the Paris basin.

Why the English Eocene strata rise nearly as high as the denuded secondary districts. — Those geologists who have hitherto regarded the rise and denudation of the lands in the south-east of England as events posterior in date to the deposition of the London clay, will object to the foregoing reasoning, that not only certain outlying patches of tertiary strata, but even the central parts of the London and Hampshire basins, attain very considerable altitudes above the level of the sea. Thus the London clay at Highbeach, in Essex, reaches the height of 750 feet, an elevation exceeding that

* See above, p. 248.

of large districts of the chalk and other denuded secondary rocks. But these facts do not, we think, militate against the theory above proposed, for we have assumed a long-continued series of elevatory movements in a region where the degradation and reproduction of strata were in progress.

If this be granted, it is evident that the great antagonist powers, the igneous and the aqueous, would, throughout the whole period, be brought into play in their fullest energy, the igneous labouring continually to produce the greatest inequality of surface, by uplifting certain lines of country and depressing others ; the aqueous no less incessantly engaged in reducing the whole to a level, by cutting off the summits of the upraised tracts, and throwing the materials thence removed into the adjoining hollows. If the volcanic forces eventually prevail, so as to convert the whole region into land, we must expect that some of the materials drifted into the hollows, and forming the newer strata, will be brought up to view, while the denuded districts are raised at the same time. If these last continue, *in general*, to occupy a higher position above the level of the sea, it is all that can be expected after the levelling operations before alluded to.

Now the tracts occupied by our Eocene formations are low, not so much with reference to the secondary rocks which remain, as to these masses which must be supposed by our theory to have disappeared, having been carried away by denudation. Let the portions removed from the space intervening between the North and South Downs, and which are expressed by faint lines in our section, wood-cut No. 63, be restored, and we may readily conceive that those masses may have formed shoals and dry land for ages before any part of our tertiary basins emerged.

The estimate of Mr. Martin is not, perhaps, exaggerated, when he computes the probable thickness of strata removed from the highest part of the Forest ridge to be about 1900 feet. So that if we restore to Crowborough Hill, in Sussex, the beds of Weald clay, Lower green-sand, Gault, and chalk, which have been removed by denudation, that hill, instead of rising to the height of 800 feet, would be more than trebled in altitude *, and be about 2700 feet high.   It would then tower far above the highest outlyers of tertiary strata which are scattered over our chalk, for Inkpen Hill, the greatest elevation of chalk in England, rises only 1011 feet above the level of the sea.

Some geologists who have thought it necessary to suppose all the strata of the London and Hampshire basins to have been once continuous, have estimated the united thickness of the three marine Eocene groups before described, as amounting to 1300 feet, and have been bold enough to imagine a mass of this height to have been once superimposed upon the chalk which formerly covered the axis of the Weald †.   Hence they were led to infer that Crowborough Hill was once 4000 feet high, and was then cut down from 4000 to 800 feet by *diluvial action*.

We, on the contrary, deem it wholly unnecessary to suppose any removal of rocks newer than the secondary from the central parts of the valley of the Weald ; and we suppose the waste of the older rocks to have been caused gradually during the emergence of the country.   The small strips of land which were first protruded in an open sea above the level of the waves, may have been entirely carried away, again and again, in the intervals between successive movements, until at last a great number of reefs and islands rising at once, afforded protection to each other against the attacks of the waves, and the lands began to increase.   We do not conceive, therefore, that a mountain ridge first rose to the height of more than 2000 feet, and was then lowered to less than half that elevation ; but that a stratified mass, more than 2000 feet thick, was, by the continual stripping off of the uppermost beds as they rose, diminished to a thickness of about 800 feet.

It is not our intention, at present, to point out the applica-

* Phil. Mag. and Annals, No. 26, New Series, p. 117.
† Martin, ibid.

tion of the above theory to the region immediately westward
of the great line of chalk escarpment which runs through the
central parts of England from Dorsetshire to Cambridgeshire.
The denudation in that country has doubtless been on a great
scale, and was, perhaps, effected during the Eocene period;
for we know of no reason why one line of movements should
not have been in progress in a direction north-east and south-
west, while others were heaving up the strata in lines running
east and west. We may remark, at the same time, that if the
chalk in the interior of England, in those tracts from which it
has been extensively swept away, began to rise during the tertiary epoch, and before the emergence of the chalk which once
extended over the central axis of the Weald, some tertiary
deposits may, in that case, have been thrown down upon that
central ridge. We have at present, however, no data to lend
countenance to such conjectures.

Vertical strata of the Isle of Wight.—A line of vertical and
inclined strata running east and west, or parallel to the central
axis of the Weald, extends through the isles of Wight and
Purbeck, and through Dorsetshire, and has been observed by
Dr. Fitton to reappear in France, north of Dieppe. The
same strata which are elevated in the Weald Valley are upheaved also on this line in the centre of the Isle of Wight,
where all the tertiary strata appear to have partaken in the
same movement *.

From the horizontality of the fresh-water series in Alum
Bay, as contrasted with the vertical position of the marine tertiary beds, Mr. Webster was at first led very naturally to
conclude, that the latter had undergone great derangement
before the deposition of the former. It appears, however,
from the subsequent observations of Professor Sedgwick †,
that these appearances are deceptive, and that at the eastern

---

* See Mr. Webster's section, Geol. Trans., vol. ii. First Series, plate XI.

† Anniv. Address to the Geol. Soc., Feb. 1831, p. 9. Professor Sedgwick informs me that his observations, made six years ago, have recently been confirmed
by Professor Henslow.

extremity of the Isle of Wight, part of the fresh-water series is vertical, like the marine. Hence it is now ascertained that as the chalk is horizontal at the southern extremity of the Isle of Wight, while it is vertical in the centre of that island, so the Eocene strata are horizontal in the north of the island, and vertical in the centre. We have only to imagine that the great flexure' of the secondary and tertiary beds, so ingeniously suggested by Mr. Webster in his theoretical section *, extended to the fresh-water formations, in order to comprehend how a very simple series of movements may have brought the whole of the Isle of Wight groups into their present position.

We are unable to assign a precise date to the convulsions which produced this great curve in the stratified rocks of the Isle of Wight; but we may observe that, although subsequent to the deposition of the fresh-water beds, it does not follow that it was not produced in the Eocene period. It may have been contemporaneous with those movements which raised the central parts of the London and Hampshire basins, which, as we before explained, were subsequent to the principal elevation and denudation of the central axis of the Weald.

Land has certainly been elevated on our eastern coast since the commencement of the older Pliocene period, as is attested by the moderate height attained by the crag strata †. But these changes of level may have been partial, and if the crag does not extend farther over the Eocene formations, and into the Weald Valley, it is probably because those regions were dry land when the strata of crag were forming in the sea.

The first land that rose in the south-eastern extremity of England may have been placed, as we before hinted, where we now find the central axis of elevation in the Weald. Perhaps the chalk islands there formed may have supported that tropical vegetation whereof we find memorials in the fossil

---

* Englefield's Isle of Wight, plate XLVII. fig. 1.

† We alluded, at p. 182, to the supposed discovery of recent marine shells at the height of 140 feet above the sea in Sheppey; but we have since learnt from Professor Sedgwick, that the information communicated to the Geological Society on this subject was erroneous.

plants of Sheppey; and the shores of those islands may have
been frequented, during the ovipositing season, by the turtles
and crocodiles, of which the teeth and skeletons are imbedded
in the London clay *.

Eocene alluviums.—The river which produced that body of
water in which the fresh-water strata of Hampshire originated,
must have drained some contiguous lands which may have
emerged during the Eocene period. On these lands we may
suppose the Paleothere, Anoplothere, and Moschus of Binstead
to have lived. The discovery of the two former genera, associated as they are with well-known Eocene species of testacea,
is most interesting. It shows that in England, or rather on
the space now occupied by part of our island, as well as in the
Paris basin, Auvergne, Cantal, and Velay, there were mum-
malia of a peculiar type during the Eocene period. Yet we
have never found a single fragment of the bones of any of these
quadrupeds in our alluviums or cave breccias. In these
formations we find the bones of the mastodon and mammoth,
of the rhinoceros, hippopotamus, lion, hyæna, bear, and other
quadrupeds, all of extinct species. They are accompanied by
recent fresh-water shells, or by the marine fossils of the crag,
and evidently belong to an epoch posterior to the Eocene.
Where, then, are the terrestrial alluviums of that surface which
was inhabited by the Paleothere and its congeners? Have the
remains which were buried at so remote a period decomposed,
so that they no longer afford any zoological characters which
might enable us to distinguish the Eocene from more modern
alluviums?

It seems clear that a peculiar and rare combination of favourable circumstances is required to preserve mammiferous or
other remains in terrestrial alluviums in sufficient quantity to
afford the geologist the means of assigning the date of such
deposits. For this reason we are scarcely able, at present, to
form any conjecture as to the relative ages of the numerous

---

* We have introduced these islands into the map of Europe, in the 2nd volume,
which may be supposed to relate to the commencement of the Eocene period.

alluviums which cover the surface of Scotland, a country which probably became land long before the commencement of the tertiary epochs.

Elevation of land gradual.—As we have assumed, throughout this and the preceding chapter, that the elevatory force was developed in a succession of minor convulsions in the southeast of England, we may seem called upon to answer an objection which has been drawn from the verticality of the strata in the Isles of Wight and Purbeck. Mr. Conybeare has remarked, that the vertical strata are traced through a district nearly 60 miles in length, so that '*if* their present position were the effect of a *single convulsion,* no disturbance in the least comparable with it has occurred in modern times *.' As we can by no means dissent from this proposition, we only ask where is the evidence that a *single* effort of the subterranean force, rather than reiterated movements, produced that sharp flexure of which we suppose the vertical strata of the Isle of Wight to form a part, the remainder of the arc having been carried away by denudation.

It appears extremely probable that the Cutch earthquake of 1819, so often alluded to by us †, may have produced an incipient curve, running in a linear direction through a tract at least 60 miles in length. The strata were upraised in the Ullah Bund, and depressed below the level of the sea in the adjoining tract, where the fort of Sindree was submerged. It would be impossible, if the next earthquake should raise the Bund still higher, and sink to a lower depth the adjoining tract, to discriminate, by any geological investigations, the different effects of the two earthquakes, unless a minute survey of the effects of the first shock had been made and put on record. In this manner we may suppose the strata to be bent, again and again, in the course of future ages, until parts of them become perpendicular.

To some it may appear, that there is a unity of effect in the

---

* Phil. Mag. and Annals, No. 49, new Series, p. 21.
† Vol. i., Second Edition, p. 465, and vol. ii., First Edition, p. 265.

line of deranged strata in the isles of Wight and Purbeck, as also in the central axis of the Weald, which is inconsistent with the supposition of a great number of separate movements recurring after long intervals of time. But we know that earthquakes are repeated throughout a long series of ages, in the same spots, like volcanic eruptions. The oldest lavas of Etna were poured out many thousand, perhaps myriads, of years, before the newest, and yet they have produced a symmetrical mountain; and if rivers of melted matter thus continue to flow in the same direction, and towards the same points, for an indefinite lapse of ages, what difficulty is there in conceiving that the subterranean volcanic force, occasioning the rise or fall of certain parts of the earth's crust, may, by reiterated movements, produce the most perfect unity of result ?

Excavation of Valleys.—In our attempt to explain the origin of the existing valleys in the south-east of England, it will be seen that we refer their excavation chiefly to the ocean. We are aware that we cannot generalize these views and apply them to the valleys of all parts of the world. In Central France, for example, rivers and land-floods, co-operating with earthquakes, have deeply intersected the lacustrine and volcanic deposits, and have hollowed out valleys as deep, perhaps, as any in our Weald district. In what manner these effects may be brought about in the course of time, by the action of running water, even without the intervention of the sea, may be understood by what we have said of the removal of rock by aqueous agency during the Calabrian earthquakes *.

Those geologists who contend that the valleys in England are not due to what they term ' modern causes,' are in the habit of appealing to the fact, that the rivers in the interior of England are working no sensible alterations, and could never, in their present state, not even in millions of years, excavate the valleys through which they now flow. We suspect that a false theory is involved even in the term ' modern causes,' as

* Vol. i. chap. xxiv.

if it could be assumed that there were *ancient causes* differing from those which are now in operation. But if we substitute the phrase, existing causes, we shall find that the argument now controverted amounts to little more than this, ' that in a country free from subterranean movements, the action of running water is so trifling that it could never hollow out, in any lapse of ages, a deep system of valleys, and, *therefore,* no known combination of existing causes could ever have given rise to our present valleys ! '

The advocates of these doctrines, in their anxiety to point out the supposed absurdity of attributing to ordinary causes those inequalities of hill and dale, which now diversify the earth's surface, have too often kept entirely out of view the many recorded examples of elevations and subsidences of land during earthquakes, the frequent fissuring of mountains, and opening of chasms, the damming up of rivers by landslips, the deflection of streams from their original courses, and more important, perhaps, than all these, the denuding power of the ocean, during the rise of our continents from the deep. Few of the ordinary causes of change, whether igneous or aqueous, can be observed to act with their full intensity in any one place at the same time; hence it is easy to persuade those who have not reflected long and profoundly on the working of the numerous igneous and aqueous agents, that they are entirely inadequate to bring about any important fluctuations in the configuration of the earth's surface.

Recapitulation.—We shall now briefly recapitulate the conclusions to which we have arrived respecting the geology of the south-east of England, in reference to the nature and origin of the Eocene formations considered in this and the two preceding chapters.

1. In the first place, it appears that the tertiary strata rest exclusively upon the chalk, and consist, with some trifling exceptions, of alternations of clay and sand.

2. The organic remains agree with those of the Paris basin, but the *mineral character* of the deposit is extremely different,

those rocks in particular, which are common to the Paris basin and Central France, being wanting, or extremely rare, in the English tertiary formations.

3. The Eocene deposits of England are generally conformable to the chalk, being horizontal where the beds of chalk are horizontal, and vertical where they are vertical; so that both series of rocks appear to have participated in nearly the same movements.

4. It is not possible to define the limits of the ancient borders of the tertiary sea in the south-east of England, in the same manner as can be frequently done in those countries where the secondary rocks are unconformable to the tertiary.

5. Although the tertiary deposits are chiefly confined to the tracts called the basins of London and Hampshire, insulated patches of them are, nevertheless, found on some of the highest summits of the chalk intervening between these basins.

6. These outliers, however, do not necessarily prove that the great mass of tertiary strata was once continuous between the basins of London and Hampshire, and over other parts of the south-east of England now occupied by secondary rocks.

7. On the contrary, it is probable that these secondary districts were gradually elevated and denuded when the basins of London and Hampshire were still submarine, and while they were gradually becoming filled up with tertiary sand and clay.

8. If, in illustration of this theory, we examine one of the districts thus supposed to have been denuded, we find in the Valley of the Weald decided proofs, that since the emergence of the secondary rocks, an immense mass of chalk and subjacent formations has been removed by the force of water.

9. We infer from the existence of large valleys along the outcrop of the softer beds, and of parallel chains of hills where harder rocks come up to the surface, that water was the removing cause; and from the shape of the escarpments presented by the harder rocks, and the distribution of alluvium over different parts of the surface of the Weald district, we

conclude that the denudation was successive and gradual during the rise of the strata.

10. We may suppose that the materials carried away from the denuded district were conveyed into the depths of the contiguous sea, through channels produced by cross fractures which have since become river-channels, and which now intersect the chalk in a direction at right angles to the general axis of elevation of the country.

11. The analogous structure of the Valley of Kingsclere, and other valleys which run east and west, like the Valley of the Weald, but are much narrower, accord also with the hypothesis, that they were all produced by the denuding power of water co-operating with elevatory movements.

12. The mineral composition of the materials thus supposed to have been removed in immense abundance from the Valley of the Weald, are precisely such as would, by degradation, form the English Eocene strata.

13. It is probable that there were many oscillations of level during the Eocene period, so that some tracts were alternately land and then sea, and then land again. These fluctuations may account for the furrowed surface of the chalk on which the tertiary strata sometimes repose, for the valleys on its surface, for the banks of shingle associated with the Plastic clay, for the partial deposits of sand and clay on elevated tracts of chalk, and for the alternations of marine and fresh-water strata in the Hampshire basin.

14. The volcanic eruptions of the Eocene period in Auvergne, the changes of level which took place at the same time in the Paris basin, and those above alluded to in the south-east of England, may all have belonged to one theatre of subterranean convulsion.

15. The basins of London and Hampshire may have been partly formed by subsidence in the bed of the sea, contemporaneously with the elevation and emergence of the Weald district.

16. The movements which threw the chalk and the tertiary strata of the isles of Wight and Purbeck into a vertical position, were subsequent to the formation of the Eocene freshwater strata of the Isle of Wight, but may possibly have occurred during the Eocene period.

17. The masses of secondary rock which have been removed by denudation from the central axis of the Weald would, if restored, rise to more than double the height now attained by any patches of tertiary strata in England.

18. If, therefore, the Eocene strata do not appear to occupy a much lower level than the secondary rocks, from the destruction of which they have been formed, it is because the highest summits of the latter have been cut off during the rise of the land, and thrown into those troughs where we now find the tertiary deposits.

19. The upheaving of the strata of the London and Hampshire basins may have been in great part effected towards the close of the Eocene period ; but it must also have been in some part due to the movements which raised the crag.

CHAPTER XXIII.

Secondary formations—Brief enumeration of the principal groups—No species common to the secondary and tertiary rocks—Chasm between the Eocene and Maestricht beds—Duration of secondary periods—Former continents placed where it is now sea—Secondary fresh-water deposits why rare—Persistency of mineral composition why apparently greatest in older rocks—Supposed universality of red marl formations—Secondary rocks why more consolidated—Why more fractured and disturbed—Secondary volcanic rocks of many different ages.

SECONDARY FORMATIONS.

As we have already exceeded the limits originally assigned to this work, it is not our intention to enter, at present, upon a detailed description of the formations usually called 'Secondary,' the elucidation of which might well occupy another volume. By 'secondary,' we mean those stratified rocks older than the tertiary, which contain distinct organic remains, and which sometimes pass into the strata called 'Primary,' to be described in our concluding chapters.

The observations which we are about to offer have chiefly for their object to show that the rules of interpretation adopted by us for the tertiary formations, are equally applicable to the phenomena of the secondary series. This last has been divided into several groups, and we shall briefly enumerate some of the principal of these for the convenience of reference, without pretending to offer to the student a systematic classification, founded on a full comparison of fossil remains.

PRINCIPAL SECONDARY GROUPS. (*Descending Series.*)

1. *Strata from the chalk of Maestricht to the lower green-sand inclusive.*

The number of species of testacea already procured from the different members of this division amount to about 1000. The principal subdivisions are the Maestricht beds, the chalk with and without flints, the upper green-sand, the gault, and

the lower green-sand. The first of these groups is seen at
St. Peter's Mount, Maestricht, reposing upon the upper flinty
chalk of England and France. It is characterized by a peculiar assemblage of organic remains, perfectly distinct from those
of the tertiary period. M. Deshayes, after a careful comparison, and after making drawings of more than 200 species of
the Maestricht shells, has been unable to identify any one of
them with the numerous tertiary fossils in his collection. On
the other hand, there are several shells which are decidedly
common to the calcareous beds of Maestricht and the white
chalk. The names of twelve of these, communicated by M.
Deshayes, will be found in Appendix II., p. 60.

But the fossils of the Maestricht beds extend not merely
into the white chalk of the French geologists, but into their
' green-sand,' which appears to correspond very nearly with
the upper green-sand of the English geologists. A list of
five species of shells, common to the Maestricht beds and
the upper green-sand of France, will be found in Appendix
II., p. 60.

It will be seen by the above lists, that the belemnite, one of
the cephalopodes not found in any tertiary formation, occurs
in the Maestricht beds; an ammonite has also been discovered
in this group by Dr. Fitton, and is now in the collection of the
Geological Society of London.

That gigantic species of reptile, the Mososaurus of Maestricht,
has also been found by Mr. Mantell in the English chalk.

2. *The Wealden, or the strata from the Weald clay to the Purbeck limestone inclusive.*

The numerous fossil-shells of this group are referrible to fresh-
water genera, which are associated with many remains of
fluviatile and terrestrial reptiles and land-plants. We believe
that no species, whether animal or vegetable, in this group, has
been distinctly identified with any found either in the superincumbent marine beds of the first division, or in the subjacent
rocks of the group No. 3, which are also of marine origin.

3. *Oolite, or Jura limestone formation.*

This division, in which we do not include the lias, contains a great number of subordinate members, several of which may relate, perhaps, to periods as important as our subdivisions of the great tertiary epoch. The shells, even of the uppermost part of the series, appear to differ entirely from the species found in the division No. 1.

4. *The Lias.*

The shells of the argillaceous limestone, termed lias, and other associated strata, differ considerably from those of the preceding group, as do the greater number of species of vertebrated animals.

5. *Strata intervening between the Lias and the Carboniferous group.*

The formations which are referrible to the interval which separated the great coal formations from the division last mentioned, are very various, and some of them, like the new red sandstone, contain few organic remains.   One group, however, belonging to this period, the Muschelkalk of the Germans, which has no precise equivalent among the English strata, contains many organic remains belonging to species perfectly distinct from the fossils of the lias, and equally so from those of the carboniferous era next to be mentioned.

6. *Carboniferous group, comprising the coal-measures, the mountain limestone, the old red sandstone, the transition limestone, the coarse slates and slaty sandstones called graywacke by some writers, and other associated rocks*

The mountain and transition limestones of the English geologists contain many of the same species of shells in common, and we shall therefore refer them for the present to the same great period; and, consequently, the coal, which alternates in some districts with mountain limestone, and the old red sandstone which intervenes between the mountain and transition limestones, will be considered as belonging to the same period. The coal-bearing strata are characterized by several hundred species of plants, which serve very distinctly to mark the vegetation of part of this era. Some of the rocks, termed graywacke in Germany, are connected by their fossils with the mountain limestone.

With this group we shall conclude our enumeration for the present; for although other divisions may hereafter be requisite, we are not aware that any antecedent periods can yet be established on the evidence of a distinct assemblage of fossil remains. Traces of organization undoubtedly occur in rocks more ancient than the transition limestone, and its associated sandstones, called graywacke; but we cannot refer them to a distinct geological period, according to the principles laid down in this work, until we have obtained data for determining the specific characters of a considerable number of fossil remains.

In reviewing the above groups we may first call the reader's attention to the important fact stated on the authority of M. Deshayes, that no species of fossil shells has yet been found common to the secondary and tertiary formations*. This marked discordance in the organic remains of the two series is not confined to the testacea, but extends, so far as a careful comparison has yet been instituted, to all the other departments of the animal kingdom, and to the fossil plants. I am informed by M. Agassiz, whose great work on fossil fish is anxiously looked for by geologists, that after examining about 500 species of that class, in formations of all ages, he could discover no one common to the secondary and tertiary rocks; nay, all the secondary species hitherto known to him, belong to

---

* M. Deshayes assures me that he has seen no tertiary shells in the Gosau beds, supposed by some geologists to be intermediate between the secondary and tertiary formations; but that some of the most characteristic species of Gosau occur in the green-sand beneath the chalk, at Mons in Belgium.

genera distinct from those established for the classification of tertiary and recent fish.

Chasm between the Eocene and Maestricht formations.—
There appears, then, to be a greater chasm between the organic remains of the Eocene and Maestricht beds, than between the Eocene and Recent strata; for there are some living shells in the Eocene formations, while there are no Eocene fossils in the newest secondary group. It is not improbable that a greater interval of time may be indicated by this greater dissimilarity in fossil remains. In the 3rd and 4th chapters we endeavoured to point out that we have no right to expect, even when we have investigated a greater extent of the earth's surface, that we shall be able to bring to light an unbroken chronological series of monuments from the remotest eras to the present; but as we have already discovered a long succession of deposits of different ages, between the tertiary groups first known and the *recent* formations, so we may, perhaps, hereafter detect an equal, or even greater series, intermediate between the Maestricht beds and the Eocene strata.

Duration of secondary periods.—The different subdivisions of the secondary group No. 1, extending from the chalk of Maestricht to the lower green-sand inclusive, may, perhaps, relate to a lapse of ages as immense as the united tertiary periods, of which we have sketched the eventful history in this volume. Such a conjecture, at least, seems warranted, if we can form any estimate of the quantity of time, by comparing the amount of vicissitude in animal life which has occurred during its lapse.

Position of former continents.—The existence of sea as well as land, at every geological period, is attested by the remains of terrestrial plants imbedded in the deposits of all ages, even the most remote. We find fluviatile shells not unfrequently in the secondary strata, and here and there some fresh-water formations; but the latter are less common than in the tertiary series. For this fact we have prepared the reader's mind, by the views advanced in the third chapter respecting the different circumstances under which we conceive the secondary and tertiary strata to have originated. We have there hinted, that the former may have been accumulated in an ocean like the Pacific, where coralline and shelly limestone are forming, or in a basin like the bed of the western Atlantic, which may have received for ages the turbid waters of great rivers, such as the Amazon, and Orinoco, each draining a considerable extent of continent. The *tertiary* deposits, on the other hand, may have been accumulated during the growth of a continent, by the successive emergence of new lands, and the uniting together of islands. During such changes, inland seas and lakes would be caused, and afterwards filled up with sediment, and then raised above the level of the waters.

That the greater part of the space now occupied by the European continent was sea when some of the secondary rocks were produced, must be inferred from the wide areas over which several of the marine groups are diffused ; but we do not suppose that the quantity of land was less in those remote ages, but merely that its position was very different. In the above tabular view of the secondary rocks, we have shown that immediately below the division No. 1, or ' the chalk and green-sand,' is placed a fresh-water formation called, in the south-east of England, the Wealden. This group has been ascertained to extend from west to east (from Lulworth Cove to the boundary of the Lower Boulonnois) about 200 English miles, and from north-west to south-east (from Whitchurch to Beauvais), about 220 miles, the depth or total thickness of the beds, where greatest, being about 2000 feet *.

Now these phenomena most clearly indicate, that there was a constant supply in this region, for a long period, of a considerable body of fresh water, such as might be supposed to have drained a continent, or a large island, containing within it a lofty chain of mountains. Dr. Fitton, in speaking of these appearances, recalls to our recollection that the delta of the newly-discovered Quorra, or Niger, in Africa, stretches into the interior

* Fitton's Geology of Hastings, p. 58.

for more than 170 miles, and occupies, it is supposed, a space of more than 300 miles along the coast, thus forming a surface of more than 25,000 square miles, or equal to about one half of England *.

Now if this modern ' delta,' or, in other words, that part of the bed of the Atlantic which has been converted into land by matter deposited immediately at the river's mouth, be so extensive, how much larger may be the space over which the same kind of sediment may be distributed by the action of the tides and currents! If, then, groups like the Wealden may be formed near the mouths of great rivers, others, like the lias, may be produced by the wider dispersion of similar materials over larger submarine areas. For we may conceive that the Niger may carry out the remains of land plants, and the carcasses and bones of fluviatile reptiles, into places where they may be swept away by currents and afterwards mingled far and wide with the marine shells and corals of the Atlantic.

The reader will remember that we stated, in the first volume†, that the common crocodile of the Ganges frequents both fresh and salt water, the same species being sometimes seen far inland, many hundred miles from the sea, and at the same time swarming on the sand-banks in the salt and brackish water beyond the limits of the delta.

If we are asked where the continent was placed from the ruins of which the Wealden strata were derived, we are almost tempted to speculate on the former existence of the Atlantis of Plato, which may be true in geology, although fabulous as an historical event. We know that the present European lands have come into existence almost entirely since the deposition of the chalk, and the same period may have sufficed for the disappearance of a continent of equal magnitude, situated farther to the west.

Secondary fresh-water deposits why rare.—If there were extensive tracts of land in the secondary period, we may presume that there were lakes also; yet we are not aware of any

* Fitton's Geology of Hastings, p. 58, who cites Lander's Travels.
† Page 243; Second Edition, p. 279.

pure lacustrine formations interstratified with rocks older than the chalk. Perhaps their absence may be accounted for by the adoption of the theoretical views above set forth; for if the present ocean coincides for the most part with the site of the ancient continent, the places occupied by lakes must have been submerged.   It should also be recollected, that the area covered by lakes, at any one time, is very insignificant in proportion to the sea, and, therefore, we may expect that, after the earth's surface has undergone considerable revolutions in its physical geography, the lacustrine strata will be concealed, for the most part, under superimposed marine deposits.

Persistency of mineral character.—In the same manner as it is rare and difficult to find ancient lacustrine strata, so also we can scarcely expect to discover newer marine groups preserving the same lithological characters continuously throughout wide areas.   The chalk now seen stretching for thousands of miles over different parts of Europe, has become visible to us by the effect, not of one, but of many distinct series of movements. Time has been required, and a succession of geological periods, to raise it above the waves in so many regions; and if calcareous rocks of the Eocene or Miocene periods have been formed, preserving an 'homogeneous mineral composition throughout equally extensive regions, it may require convulsions as numerous as all those which have occurred since the origin of the chalk, to bring them up within the sphere of human observation.   Hence the rocks of more modern periods may appear of partial extent, as compared to those of remoter eras, not because there was any original difference of circumstances throughout the globe when they were formed, but because there has not been sufficient time for the development of a great series of subterranean volcanic operations since their origin.

At the same time, the reader should be warned not to place implicit reliance on the alleged persistency of the same mineral characters in secondary rocks *.   When it was first ascertained that an order of succession could be traced in the principal

* See some remarks on this subject, vol. i. p. 90, and Second Edition, p. 102.

groups of strata above enumerated by us, names were given to
each, derived from the mineral composition of the rocks in
those parts of Germany, England or France, where they happened to be first studied. When it was afterwards acknowledged that the zoological and phytological characters of the
same formations were far more persistent than their mineral
peculiarities, the old names were still retained, instead of being
exchanged for others founded on more constant and essential
characters. The student was given to understand, that the
terms chalk, green-sand, oolite, red marl, coal, and others,
were to be taken in a liberal and extended sense; that chalk
was not always a cretaceous rock, but, in some places, as on the
northern flanks of the Pyrenees, and in Catalonia, a saliferous
red marl. Green-sand, it was said, was rarely green, and frequently not arenaceous, but represented in parts of the south
of Europe by a hard dolomitic limestone. In like manner, it
was declared that the oolitic texture was rather an exception
to the general rule in rocks of the oolitic period, and that no
particle of carbonaceous matter could often be detected in the
true *coal* formation of many districts where it attains great
thickness. It must be obvious to every one, that inconvenience
and erroneous prepossessions could hardly fail to arise from
such a nomenclature, and accordingly a fallacious mode of reasoning has been widely propagated, chiefly by the influence of
a language so singularly inappropriate.

After the admission that the identity or discordance of
mineral character was by no means a sure test of agreement or
disagreement in the age of rocks, it was still thought, by many
geologists, that if they found a rock at the antipodes agreeing
precisely in mineral composition with another well known in
Europe, they could fairly presume that both are of the same
age, *until the contrary could be shown.*

Now it is usually difficult or impossible to combat such an
assumption, on geological grounds, so long as we are imperfectly acquainted with the geology of a distant country, inasmuch as there are often no organic remains in the foreign

stratum, and even if these abound and are specifically different from the fossils of the supposed European equivalent, it may be objected, that we cannot expect the same species to have inhabited very distant quarters of the globe at the same time.

Supposed universality of red marl.—We shall select a remarkable example of the erroneous mode of generalizing now alluded to. A group of red marl and sandstone, sometimes containing salt and gypsum, is found in England interposed between the lias and the carboniferous strata. For this reason, other red marls and sandstones, associated some of them with salt and others with gypsum, and occurring not only in different parts of Europe, but in Peru, India, the salt deserts of Asia, those of Africa, in a word, in every quarter of the globe, have been referred to one and the same period. The burden of proof is not supposed to rest with those who insist on the identity of age of all these groups, so that it is in vain to urge as an objection, the improbability of the hypothesis which would imply that all the moving waters on the globe were once simultaneously charged with sediment of a red colour.

But the absurdity of pretending to identify, in age, all the red sandstones and marls in question, has at length been sufficiently exposed, by the discovery that, even in Europe, they belong decidedly to many different epochs. We have already ascertained, that the red sandstone and red marl with which the rock-salt of Cardona is associated, may be referred to the period of our chalk and green-sand *. We have pointed out that in Auvergne there are red marls and variegated sandstones, which are undistinguishable in mineral composition, from the new red sandstone of English geologists, but which were deposited in the Eocene period ; and, lastly, the gypseous red marl of Aix in Provence, formerly supposed to be a marine secondary group, is now acknowledged to be a tertiary freshwater formation.

---

* I was led to this opinion when I visited Cardona in 1830, and before I was aware that M. Dufrénoy had arrived at the same conclusions. Ann. des Sci. Nat., Avril, 1831, p. 449.

Secondary rocks why more consolidated.—One of the points where the analogy between the secondary and tertiary formations has been supposed to fail is the greater degree of solidity observable in the former. Undoubtedly the older rocks, in general, are more stony than the newer; and most of the tertiary strata are more loose and incoherent in their texture than the secondary. Many exceptions, however, may be pointed out, especially in those calcareous and siliceous deposits which have been precipitated in great part from the waters of mineral springs, and have been originally compact. Of this description are a large proportion of the Parisian Eocene rocks, which are more stony than most of the English secondary groups.

But a great number of strata have evidently been consolidated *subsequently to their deposition* by a slow lapidifying process. Thus loose sand and gravel are bound together by waters holding carbonate and oxide of iron, carbonate of lime, silica, and other ingredients, in solution. These waters percolate slowly the earth's crust in different regions, and often remove gradually the component elements of fossil organic bodies, substituting other substances in their place. It seems, moreover, that the draining off of the waters during the elevation of land may often cause the *setting* of particular mixtures, in the same manner as mortar hardens when desiccated, or as the recent soft marl of Lake Superior becomes highly indurated when exposed to the air *. The conversion of clay into shale, and of sand into sandstone, may, in many cases, be attributed to simple pressure, produced by the weight of superincumbent strata, or by the upward heaving of subjacent masses during earthquakes. Heat is another cause of a more compact and crystalline texture, which will be considered when we speak of the strata termed ' primary.' All the changes produced by these various means require *time* for their completion; and this may explain, in a satisfactory manner, why the older rocks are most consolidated, without

* Vol. i. p. 226, and Second Edition, p. 259.

entitling us to resort to any hypothesis respecting an *original* distinctness in the degree of lapidification of the secondary strata.

Secondary rocks why more disturbed.—As the older formations are generally more stony, so also they are more fractured, curved, elevated, and displaced, than the newer. Are we, then, to infer, with some geologists, that the disturbing forces were more energetic in remoter ages? No conclusion can be more unsound; for as the moving power acts from below, the newer strata cannot be deranged without the subjacent rocks participating in the movement; while we have evidence that the older have been frequently shattered, raised, and depressed, again and again, before the newer rocks were formed. It is evident that if the disturbing power of the subterranean causes be exerted with *uniform* intensity in each succeeding period, the quantity of convulsion undergone by different groups of strata will generally be great in proportion to their antiquity. But exceptions will occur, owing to the partial operation of the volcanic forces at particular periods, so that we sometimes find tertiary strata more elevated and disturbed, in particular countries, than are the secondary rocks in others.

Some of the enormous faults and complicated dislocations of the ancient strata may probably have arisen from the continued repetition of earthquakes in the same place, and sometimes from two distinct series of convulsions, which have forced the same masses in different, and even opposite directions, sometimes by vertical, at others by horizontal movements.

Secondary volcanic rocks of different ages.—The association of volcanic rocks with different secondary strata is such as to prove, that there were igneous eruptions at many distinct periods, as also that they were confined during each epoch, as now, to limited areas. Thus, for example, igneous rocks contemporaneous with the carboniferous strata abound in some countries, but are wanting in others. So it is evident that the bottom of the sea, on which the oolite and its contemporary deposits were thrown down, was, for the most part, free from submarine eruptions; but at some points, as in the Hebrides, it seems that the same ocean was the theatre of volcanic action. We have mentioned in the first volume *, that as the ancient eruptions occurred in succession, sufficient time usually intervening between them to allow of the accumulation of many subaqueous strata, so also should we infer that subterranean movements, which are another portion of the volcanic phenomena, occurred separately and in succession.

* Chap. v. p. 88; and Second Edition, p. 100.

CHAPTER XXIV.

On the relative antiquity of different mountain-chains—Theory of M. Elie de Beaumont—His opinions controverted—His method of proving that different chains were raised at distinct periods—His proof that others were contemporaneous—His reasoning, why not conclusive—His doctrine of the parallelism of contemporaneous lines of elevation—Objections—Theory of parallelism at variance with geological phenomena as exhibited in Great Britain—Objections of Mr. Conybeare—How far anticlinal lines formed at the same period are parallel—Difficulties in the way of determining the relative age of mountains.

RELATIVE ANTIQUITY OF MOUNTAIN-CHAINS.

THAT the different parts of our continents have been elevated, in succession, to their present height above the level of the sea, is an opinion which has been gradually gaining ground with the progress of science; but no one before M. Elie de Beaumont had the merit even of attempting to collect together the recorded facts which bear on this subject, and to reduce them to one systematic whole. The above-mentioned geologist was eminently qualified for the task, as one who had laboured industriously in the field of original observation, and who combined a considerable knowledge of facts with an ardent love of generalization.

But he has been ambitious, we think unfortunately, of anticipating the march of discovery in reference to the comparative antiquity of different mountain-chains and their supposed connexion with changes in the animate world. His speculations differ entirely from the conclusions to which we have arrived, and we therefore think it necessary to explain fully the reasons of our dissent. In order to put the reader in possession of the principal points of M. de Beaumont's theory, we shall first offer a brief sketch of them, and then proceed to analyze the data on which they are founded.

Theory of M. Elie de Beaumont.

1st. He supposes 'that in the history of the earth there have been long periods of comparative repose, during which the deposition of sedimentary matter has gone on in regular continuity, and there have also been short periods of paroxysmal violence during which that continuity was broken.

'2ndly. At each of these periods of violence or "revolution" in the state of the earth's surface, a great number of mountain-chains have been formed suddenly.

'3rdly. All the chains thrown up by a particular revolution have one uniform direction, being parallel to each other within a few degrees of the compass, even when situated in remote regions; but the chains thrown up at different periods have, for the most part, different directions.

'4thly. Each "revolution," or, as it is sometimes termed, "frightful convulsion," has coincided in date with another geological phenomenon, namely, "the passage from one independent sedimentary formation to another," characterized by a considerable difference in "organic types."

'5thly. There has been a recurrence of these paroxysmal movements from the remotest geological periods, and they may still be reproduced, and the repose in which we live may hereafter be broken by the sudden upthrow of another system of parallel chains of mountains.

'6thly. We may presume that one of these revolutions has occurred within the historical era when the Andes were upheaved to their present height, for that chain is the best defined and least obliterated feature observable in the present exterior configuration of the globe, and was probably the last elevated.

'7thly. The instantaneous upheaving of great mountain masses must cause a violent agitation in the waters of the sea, and the rise of the Andes may, perhaps, have produced that transient deluge which is noticed among the traditions of so many nations.

'Lastly. The successive revolutions above mentioned cannot

be referred to ordinary volcanic forces, but may depend on the secular refrigeration of the heated interior of our planet*.'

It will at once be seen, that the greater number of the above propositions are directly opposed to that theory which we have endeavoured to deduce, partly from the study of the earth's structure, and partly from the analogy of changes now in progress in the animate and inanimate world.

Our opinions respecting the alternation of periods of general repose and disorder have been explained in former chapters † ; and we have pointed out our objections to the hypothesis which substitutes paroxysmal violence for the reiterated recurrence of minor convulsions ‡.

The speculation of M. de Beaumont concerning the ' secular refrigeration' of the internal nucleus of the globe, considered as a cause of the instantaneous rise of mountain-chains, appears to us mysterious in the extreme, and not founded upon any induction from facts; whereas the intermittent action of subterranean volcanic heat is a known cause capable of giving rise to the elevation and subsidence of the earth's crust without interruption to the *general* repose of the habitable surface.

We have shown, in the second volume, that we believe the changes in physical geography, which are unceasingly in progress, to be among the causes which contribute, in the course of ages, to the extermination of certain species of animals and plants; but the influence of these causes is slow and, for the most part, indirect, and has no analogy with those sudden catastrophes which are introduced into the theory now under review. What have appeared to us to be the true causes of the abrupt transitions from one set of strata to another, containing distinct organic remains, have been explained at length in the third and fourth chapters of this volume §.

* Ann. des Sci. Nat., Septembre, Novembre, et Décembre, 1829. Revue Française, No. 15, May, 1830. The last version by M. de B. which I have seen is in the Phil. Mag. and Annals, No. 58, new series, p. 241.

† Vol. i. pp. 64 and 88, and Second Edition, pp. 73 and 100; vol. ii. p. 196, and Second Edition, p. 203.

‡ Vol. i. p. 79, and Second Edition, p. 90.

§ See particularly from p. 26 to p. 34.

The notion of deluges accompanying the protrusion of mountain-chains is founded on a belief of the instantaneousness of the movement which we are prepared to controvert, and on other assumptions which we have discussed in a former part of this volume *. On these topics, therefore, it will be unnecessary for us to dilate at present, and we shall merely address ourselves to the analysis of that evidence whereby M. de Beaumont endeavours to establish the successive elevation of different mountain-chains, and the supposed law of parallelism in the lines of contemporaneous elevation.

M. de Beaumont's proofs that different chains were raised at different epochs.—' We observe,' says M. Elie de Beaumont, ' along nearly all mountain-chains, when we attentively examine them, that the most recent rocks extend horizontally up to the foot of such chains, as we should expect would be the case if they were deposited in seas or lakes of which these mountains have partly formed the shores ; whilst the other sedimentary beds tilted up, and more or less contorted on the flanks of the mountains, rise in certain points even to their highest crests †.' There are, therefore, in each chain two classes of sedimentary rocks, the ancient or inclined beds, and the newer or horizontal. It is evident that the first appearance of the chain itself was an event ' intermediate between the period when the beds, now upraised, were deposited, and that when the strata were produced horizontally at its feet.'

No. 82.

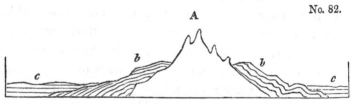

Thus the chain A received its present form after the deposition of the strata *b*, which have undergone great movements, and before the deposition of the group *c*, in which the strata have not suffered derangement.

* See above, p. 148.
† Phil. Mag. and Annals, No. 58, new series, p. 242.

If we then discover another chain, B, in which we find not only the formation *b*, but the group *c* also, disturbed and

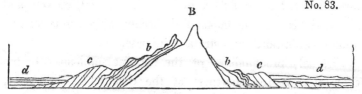

B

No. 83.

thrown on its edges, we may infer that the latter chain is of subsequent date to A; for B was elevated *after* the deposition of *c*, and before that of the group *d ;* whereas A originated *before* the strata *c* were formed.

In order to ascertain whether other mountain ranges are of contemporaneous date with A and B, or whether they are referrible to *distinct* periods, we have only to inquire whether the geological phenomena are identical, namely, whether the inclined and undisturbed sets of strata correspond to those in the types above mentioned.

Objections to M. de Beaumont's theory.—Now all this reasoning is perfectly correct, so long as the particular groups of strata *b* and *c* are not confounded with the geological periods to which they may belong, and provided due latitude is given to the term contemporaneous ; for it should be understood to allude not to a moment of time, but to the interval, whether brief or protracted, which has elapsed between two events, namely, between the accumulation of the inclined and that of the horizontal strata.

But, unfortunately, the distinct import of the terms 'formation' and 'period' has been overlooked, or not attended to by M. de Beaumont, and hence the greater part of his proofs are equivocal, and his inferences uncertain; and even if no errors had arisen from this source, the length of some of his intervals is so immense, that to affirm that all the chains raised in such intervals were *contemporaneous,* is an abuse of language.

In order to illustrate our argument, let us select the Pyrenees as an example. This range of mountains, says M. de Beaumont, rose suddenly (à un seul jet) to its present

elevation at a certain epoch in the earth's history, namely,
between the deposition of the chalk and that of the tertiary
formations; for the former are seen in vertical, curved, and
distorted beds on the flanks of the chain, while the latter rest
upon them in horizontal strata at its base.

The only proof offered of the extreme suddenness of the
convulsion is the shortness of the time which intervened
between the formation of the chalk and that of the tertiary
strata. 'For it follows,' we are told, 'from the unconformable
position of two systems of beds, the inclined and the horizontal,
that the elevation of the former has not been effected in a
continuous and progressive manner, but that it has been produced in a space of time comprised between the periods of
deposition of the two consecutive rocks, and during which no
regular series of beds was formed; in a word, that it was sudden
and of short duration *.'

We are prepared to show that the Pyrenees cannot be
assumed to have risen, as M. de Beaumont imagines, in the interval between the period of the chalk and that of the tertiary
strata; for we can only say that the movement took place after
the commencement of the chalk epoch, and before the close
of the Miocene tertiary period. But, first, let us suppose the
premises of our author to be correct, and let us permit him to
exclude the whole period of the chalk, on one hand, and of the
tertiary formations in contact with it on the other; what will
then be the duration of the interval? We can only estimate
its importance by ascertaining what description of chalk is
found on the flanks of the Pyrenees, and what horizontal tertiary formations at their base.

Now the beds called chalk, although they differ widely in
mineral composition from the white chalk with flints of
England and France, contain the same species of fossil shells,
and may, therefore, on that evidence, be referred to the same
age †. On the other hand, the horizontal tertiary strata at the

* Phil. Mag. and Annals, No. 58, new series, p. 243.
† The fossils which I collected in company with Captain S. E. Cook, R. N.,

western end of the Pyrenees, near Bayonne, are certainly of
the Miocene period.

Such, then, being the age of the strata, and granting even
that the movement occurred after the period of the white
chalk, and before the beginning of the Miocene era, there still
remains ample scope for conjecture as to the date of the event.
For the upheaving of the Pyrenees may have been going on
when the animals of the Maestricht beds flourished, or during
the indefinite ages which may have elapsed between their extinction and the introduction of the Eocene tribes, or during
the Eocene epoch, or between that and the Miocene.   Or the
rise may have been going on continuously throughout several
or all of these periods

But this is not all; we must include within the possible space
of time wherein the convulsions may have happened, part of the
epochs both of the chalk and of the Miocene species.   We have
stated, that the newer Pliocene beds in Sicily have been raised
during the newer Pliocene epoch, partly, perhaps, in the
Recent, but this latter supposition will lend equal support to
our present argument.   Now, it is evident that the greater part
of the species of testacea which pre-existed in the Mediterranean
have survived the elevation of the newer Pliocene beds in
Sicily, and in the same manner there is no reason to conclude
that the rise of the chalk in the Pyrenees exterminated the
animals which lived in the sea wherein the chalk was formed.
In that case, a series of convulsions may not only have begun,
but may even have been completed before the era when the
Maestricht beds originated.

In like manner the sea may have been inhabited by Miocene
testacea for ages before the deposition of those particular
Miocene strata which occur at the foot of the Pyrenees, and
the disturbing forces may have operated in the Miocene period,

from the newest secondary beds on the flanks of the Pyrenees, near Bayonne, were
examined by M. Deshayes, and found identical with species of the chalk near
Paris.

notwithstanding the horizontality of the tertiary formations of that age.

In order to illustrate the grave objections above advanced, which are aimed at the validity of the whole of de Beaumont's reasoning, let the reader suppose, that in some country three styles of architecture had prevailed in succession, each for a period of 1000 years; first the Greek, then the Roman, and then the Gothic; and that a tremendous earthquake was known to have occurred in the same district during some part of the three periods,—a shock of such violence as to have levelled to the ground every building. If an antiquary, desirous of discovering the date of the catastrophe, should first arrive at a city where several Greek temples were lying in ruins and half engulphed in the earth, while many Gothic edifices were standing uninjured, could he determine on these data the era of the shock? Certainly not. He could merely affirm that it happened at some period after the introduction of the Greek style, and before the Gothic had fallen into disuse. Should he pretend to define the date of the convulsion with greater precision, and decide that the earthquake must have occurred in the interval between the Greek and Gothic periods, that is to say, when the Roman style was in use, the fallacy in his reasoning would be too palpable to escape detection for a moment.

Yet such is the nature of the erroneous induction which we are now exposing. For, in the example above proposed, the erection of a particular edifice is not more distinct from the period of architecture in which it may have been raised, than is the deposition of chalk, or any other set of strata, from the geological epoch to which they may belong. Yet, if on these grounds we are compelled to include in the interval in which the elevation of each chain may have happened, the periods of those two classes of formations before alluded to, the deranged and the horizontal, it follows that, even if all the facts appealed to by de Beaumont are correct, his intervals are of indefinite extent. He is not even warranted in asserting that the chain

A (p. 340) is older than B (p. 341), if he means that it was
elevated at a different *geological period,* for both may have been
upheaved during the same period, namely, that when the strata
c were formed.

Supposed parallelism of contemporaneous lines of elevation.—
So, also, when he infers that two chains were simultaneously
upraised, the proof fails, since the close of the period of the
disturbed strata and the commencement of the era of the undisturbed must be added to the lapse of time during which the
two chains may have originated, and in separate parts of which
each may have been produced. With the insufficiency of the
above evidence the whole force of the argument in support
of the parallelism of lines of contemporaneous movement is
annihilated.

This hypothesis, indeed, of parallelism appears, even as
stated by the author, in some degree at variance with itself.
When certain European chains had been assumed to have been
raised at the same time on the data already impugned, it was
found that several of these contemporaneous chains had a
parallel direction. Hence it was presumed to be a general law
in geological dynamics that the chains upheaved at the same
time are parallel. For example, it was said that the Pyrenees
and other coetaneous chains, such as the northern Apennines,
have a direction about W. N. W. and E. S. E., and to this line
the Alleghanies in North America conform, as also the ghauts
of Malabar, and certain chains in Egypt, Syria, northern
Africa, and other countries; and from this mere conformity in
direction it was presumed that all these mountain-ranges were
thrown up simultaneously.

To select another example, the principal chain of the Alps,
differing in age and direction from the Pyrenees, is parallel to
the Sierra Morena, the Balkan, the chain of Mount Atlas, the
central chain of the Caucasus, and the Himalaya. All these
ridges, therefore, were probably heaved up by the same paroxysmal convulsion! The western Alps, on the other hand, rose at
a still earlier period, when the parallel chains of Kiöl, in Scan-

dinavia, certain chains in Morocco, and the littoral Cordillera of Brazil, were formed!

Not only do these speculations refer to mountains never touched, as M. Boué remarks, by the hammer of the geologist, but they proceed on the supposition, that in these distant chains the geological and geographical axes always coincide. Now we know that in Europe the *strike** of the beds is not always parallel to the direction of the chain. As an exception, we may instance that pointed out by Von Dechen †, who states that in the Hartz the direction or *strike* of the strata of slate and grey-wacke is sometimes from E. and W. and frequently N. E. and S. W.; whereas the geographical direction of the mountain-chain is decidedly from E. S. E. to W. N. W.

In addition to these uncertainties, which should, in the present state of science, have deterred a geologist even from speculating on the phenomena of unexplored regions, the important admission is made by M. de Beaumont himself, that the elevating forces, whose activity must be referred to *different* epochs, have sometimes acted in Europe in *parallel* lines. ' It is worthy of remark, says that author, that the directions of three systems of mountains, namely, first, that of the Pilas and the Côte d'Or; secondly, that of the Pyrenees; and thirdly, that of the islands of Corsica and Sardinia, are respectively parallel to three other systems, namely, first, that of Westmoreland and the Hunsdruck, secondly, that of the Ballons (or Vosges) and the hills of the Bocage, in Calvados; and thirdly, the system of the north of England. The corresponding directions only differ in a few degrees, and the two series have succeeded each other in the same order, leading to the supposition, that there has been *a kind of periodical*

* The term ' strike ' has been recently adopted by some of our most eminent geologists from the German ' streich,' to signify what our miners call the ' line of bearing' of the strata. Such a term was much wanted, and as we often speak of *striking off* in a given direction, the expression seems sufficiently consistent with analogy in our language.

† Trans. of De la Beche's Geol. Manual, p. 41.

recurrence of the same, or nearly the same, directions of elevation *.'

Here then we have three systems of mountains, A, B, C, which were formed at successive epochs, and have each a different direction ; and we have three other systems, D, E, F, which, although they are assumed to have the same strike, as the series first mentioned (D corresponding with A, E with B, and F with C), are nevertheless declared to have been formed at different periods. On what principle, then, is the age of an Indian or transatlantic chain referred to one of these European lines rather than another? why is the age of the Alleghanies, or the ghauts of Malabar, determined by their parallelism to B rather than to E, to the Pyrenees rather than to the Ballons of the Vosges?

The substance of the last objection has been anticipated by M. Boué†, who, at the same time, disputes the accuracy of many of the *facts* appealed to by M. de Beaumont. Other errors in fact have also been pointed out by MM. Keferstein, Von Dechen, and De la Beche‡. But the incorrectness of some of these data might not have affected the validity of the general theory if it had been founded on a solid basis. In regard to the Alps, MM. Necker and Studer have informed me, that on re-examining that chain since de Beaumont's memoirs were published, they have been unable to reconcile the phenomena there exhibited with his views relating to the strike and dip of that great chain.

Professor Sedgwick has declared his adhesion to the opinions of de Beaumont; but we are not aware that he had maturely considered them in all their bearings; and he has stated some important objections to the doctrine of ' parallelism §.' Among others, he has remarked, that in consequence of the spheroidal figure of the earth, different mountain-chains, running north and

---

* Phil. Mag. and Annals, No. 58, new series, pp. 255, 256.
  † Journ. de Géologie, tome iii. p. 338.
  ‡ Geol. Manual, p. 501, and Second Edition, p. 519.
  § Anniv. Address to the Geol. Soc., Feb., 1831.

south, cannot be strictly said to be parallel, since they would, if prolonged, cross each other at the poles.

Objections of Mr. Conybeare.—An inquiry was proposed, in 1831, by the British Association for the Promotion of Science, ' whether the theory of M. Elie de Beaumont, concerning the parallelism of lines of elevation of the same geological era, is agreeable to the phenomena as exhibited in Great Britain ?' Mr. Conybeare, in the first part of his report, in answer to this inquiry *, points out many lines of *distinct* ages in England which are exactly *parallel*, and others which, according to the rules laid down by M. de Beaumont, ought to agree in age with certain continental chains, and yet do not, having an entirely *different* direction.   He imagines that the general strike of the secondary strata of our island, from N. E. to S. W., has been the result, not of any violent or single convulsion, but, on the contrary, of ' a gradual, gentle, and protracted upheaving, continued without interruption *during the whole period of the formation of all these strata.*'

The same author has also adverted to some of the difficulties attending the exact determination of the geological epochs of the elevation of each chain, especially where the disturbed and undisturbed strata in contact are not very nearly of the same age, or, as he expresses it, ' where they are not terms immediately following one another in the regular geological series †.' We were forcibly struck with the uncertainty arising from this cause during a late tour, when we discovered that at the eastern end of the Pyrenees, on the side of France, tertiary strata of the older Pliocene epoch abut against vertical mica-schist ; while at the western extremity of the same mountain-range we find the disturbed series to consist of chalk, the undisturbed of Miocene strata.   The chain is then lost in the sea, and we are precluded from pushing our investigations farther to the westward ; but

* Phil. Mag. and Journ. of Sci., No. 2, third series, p. 118.   The second part, I believe, is not yet published.

† Ibid., p. 120.

if we could follow the strike of the beds in their submarine prolongation, who shall say that the *tilted* group might not be
found to include strata newer than the chalk, the *horizontal*
beds older than the Miocene ?

Supposed instantaneous rise of a mountain-chain.—' Every-
thing shows, says M. Elie de Beaumont, that the *instantaneous*
elevation of the beds of a whole mountain-chain is an event of
a different order from those which we *daily* witness [*].'

We observe with pleasure the rejection, by Mr. Conybeare,
of the hypothesis that the disturbances affecting large geographical districts have been produced *at one blow*, rather than by
a series of shocks which may have occurred at intervals through
a long period of ages, and that he contends for the greater
probability of successive convulsions, on the ground that such
an hypothesis is most conformable to the only analogy presented by actual causes—' the operations of volcanic forces [†].'

Modern volcanic lines not parallel.—By that analogy we are
led to suppose that the lines of convulsion, at former epochs,
were far from being uniform in direction, for the trains of
active volcanos are not parallel, as every one is aware who has
studied Von Buch's masterly survey of the general range of
volcanic lines over the globe [‡], and the elevations and subsidences caused by modern earthquakes, although they may
sometimes run in parallel lines within limited districts, have not
been observed to have a common direction in distant and independent theatres of volcanic action.

We do not doubt that in many regions the ridges, troughs,
and fissures caused by modern earthquakes, are, to a certain
extent, parallel to each other, but only within a limited range
of country ; and such appears to have been the case in many
districts at former eras. The anticlinal lines of the Weald
Valley, before alluded to, and of the Isle of Wight, may, in
this manner, have been contemporaneous, that is to say, both

---

[*] Phil. Mag. and Annals, No. 58, new series, p. 243.

[†] Phil. Mag. and Journ. of Sci., No. 2, third series, p. 121.

[‡] Physical. Besch. der Canarischen Inseln. Berlin, 1825.

may have been formed in some part of the Eocene period,—an hypothesis which does not involve the theory of their having been due to paroxysmal convulsions during one part of that vast period.

It should be observed, that as some trains of burning volcanos are parallel to each other, so at all periods some independent lines of elevation may be parallel *accidentally*, or not in obedience to any known law of parallelism ; but, on the contrary, as exceptions to the general rule. We hope that the speculations of M. de Beaumont will be useful in inducing geologists to inquire how far the uniformity in the direction of the beds, in a region which has been agitated at any particular period, may extend ; but we trust that travellers will not be led away with the idea that, on arriving in India, America, or New Holland, they have only to use the compass and examine the strike of the beds in order to discover the relative era of the movement by which they were upraised. Such problems can in truth be only solved by a patient and laborious investigation of the sedimentary formations occurring in each region, and especially by the study of their organic remains.

Difficulties attending the determination of the relative age of mountains.—If we are asked whether we cherish no expectation of fixing a chronological succession of epochs of elevation of different mountain-chains, we reply, that in the present state of our science we have no hope of making more than a loose approximation to such a result. The difficulty depends chiefly on the broken and interrupted nature of the series of sedimentary formations hitherto brought to light, which appears so imperfect that we can rarely be sure that the memorials of some great interval of time are not wanting between two groups now classed as consecutive. Another great source of ambiguity arises from the small progress which we have yet made in identifying strata in countries somewhat distant from each other.

There may be instances where the same set of strata, preserving throughout a perfect identity of mineral character, may be traced continuously from the flanks of one independent mountain-chain to the base of another, the beds being vertical or inclined in one chain, and horizontal in the other. We might then decide with confidence, according to the method proposed by M. de Beaumont, on the relative eras when these chains had undergone disturbance; and from one point thus securely established, we might proceed to another, until we had determined the dates of many neighbouring lines of convulsion.

We fear that the cases are rare where such evidence can be obtained; and, for the most part, we can identify the age of strata, not by their continuity and homogeneous mineral character, but by organic remains. When by their aid we prove strata to be contemporaneous, we must generally speak with great latitude, merely intending that they were deposited in the same geological epoch during which certain animals and plants flourished.

CHAPTER XXV.

On the rocks usually termed ' Primary '—Their relation to volcanic and sedimentary formations—The ' primary ' class divisible into stratified and unstratified—Unstratified rocks called Plutonic—Granite veins—Their various forms and mineral composition—Proofs of their igneous origin—Granites of the same character produced at successive eras—Some of these newer than certain fossiliferous strata—Difficulty of determining the age of particular granites—Distinction between the volcanic and the plutonic rocks—Trappean rocks not separable from the volcanic—Passage from trap into granite—Theory of the origin of granite at every period from the earliest to the most recent.

ON THE ROCKS COMMONLY CALLED PRIMARY.

We shall now treat of the class of rocks usually termed ' primary,' a name which, as we shall afterwards show, is not always applicable, since the formations so designated sometimes belong to different epochs, and are not, in every case, more ancient than the secondary strata.   In general, however, this division of rocks may justly be regarded as of higher antiquity than the oldest secondary groups before described, and they may, therefore, with propriety be spoken of in these concluding chapters, for we have hitherto proceeded in our retrospective survey of geological monuments from the newer to those of more ancient date.

In order to explain to the reader the relation which we conceive the rocks termed 'primary' to bear to the tertiary and secondary formations, we shall resume that general view of the component parts of the earth's crust of which we gave a slight sketch in the preliminary division of our subject in the 2nd chapter *.

We there stated that sedimentary formations, containing organic remains, occupy a large part of the surface of our continents, but that here and there volcanic rocks occur, breaking through, alternating with, or covering the sedimentary deposits,

* See above, p. 8.

so that there are obviously two orders of mineral masses formed
at the surface which have a distinct origin, the aqueous and
the volcanic.

No. 84.

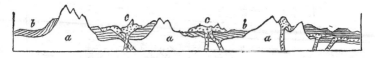

a, Formations called primary (stratified and unstratified).
b, Aqueous formations.   c, Volcanic rocks.

Besides these, however, there is another class, which cannot
be assimilated precisely to either of the preceding, and which is
often seen underlying the sedimentary, or breaking up to the
surface in the central parts of mountain-chains, constituting
some of the highest lands, and, at the same time, passing down
and forming the inferior parts of the crust of the earth.   This
class, usually termed 'primary,' is divisible into two groups, the
stratified and the unstratified.   The stratified consists of the
rocks called gneiss, mica-schist, argillaceous-schist (or clay-
slate), hornblende-schist, primary limestone, and some others.
The unstratified, or Plutonic, is composed in great measure of
granite, and rocks closely allied to granite.   Both these groups
agree in having, for the most part, a highly crystalline texture,
and in not containing organic remains.

Plutonic rocks.—The unstratified crystalline rocks have been
very commonly called Plutonic, from the opinion that they
were formed by igneous action at great depths, whereas the
volcanic, although they also have risen up from below, have
cooled from a melted state upon or near to the surface.   The
theory conveyed by the name Plutonic is, we believe, correct.
Granite, porphyry, and other rocks of the same family, often
occur in large amorphous masses, from which small veins and
dikes are sent off, which traverse the stratified rocks called
'primary,' precisely in the manner in which lava is seen in
some places to penetrate the secondary strata.

Granite Veins.—We find also one set of granite veins
intersecting another, and granitiform porphyries intruding

themselves into granite, in a manner analogous to that of the volcanic dikes of Etna and Vesuvius, where they cut and shift each other, or pass through alternating beds of lava and tuff.

No. 85.

Granite veins traversing stratified rocks.

The annexed diagram will explain to the reader the manner in which these granite veins often branch off from the principal mass. Those on the right-hand side, and in the middle, are taken from Dr. Macculloch's representation of veins passing through the gneiss at Cape Wrath, in Scotland *. The veins on the left are described, by Captain Basil Hall, as traversing the argillaceous schist of the Table-Mountain at the Cape of Good Hope †.

No. 86.

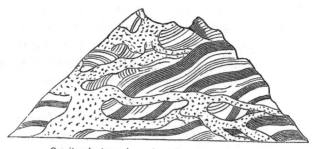

Granite veins traversing gneiss at Cape Wrath, in Scotland.

We subjoin another sketch from Dr. Macculloch's interesting

* Western Islands, plate 31.

† Account of the structure of the Table-Mountain, &c., Trans. Roy. Soc. Edin., vol. vii.

representations of the granite veins in Scotland, in which
the contrast of colour between the vein and some of the dark
varieties of hornblende-schist associated with the gneiss renders
the phenomena more conspicuous.

The following sketch of a group of granite veins in Cornwall
is given by Messieurs Von Oeynhausen and Von Dechen*.

No. 87.

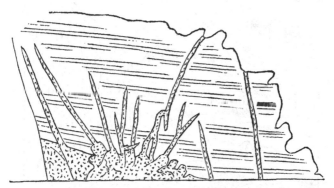

Granite veins passing through Hornblende slate, Carnsilver Cove, Cornwall.

The main body of the granite here is of a porphyritic appearance
with large crystals of felspar; but in the veins it is fine-grained
and without these large crystals. The general height of the
veins is from 16 to 20 feet, but some are much higher.

The vein-granite of Cornwall very generally assumes a finer
grain, and frequently undergoes a change in mineral composition, as is very commonly observed in other countries.
Thus, according to Professor Sedgwick, the main body of the
Cornish granite is an aggregate of mica, quartz, and felspar; but
the veins are sometimes without mica, being a granular aggregate of quartz and felspar. In other varieties quartz prevails
to the almost entire exclusion both of felspar and mica; in
others, the mica and quartz both disappear, and the vein is
simply composed of white granular felspar †.

Changes are sometimes caused in the intersected strata very

* Phil. Mag. and Annals, No. 27, new Series, March, 1829.
† On Geol. of Cornwall. Trans. of Cambridge Soc., vol. i. p. 124.

analogous to those which the contact of a fused mass might be supposed to produce.

No. 88.

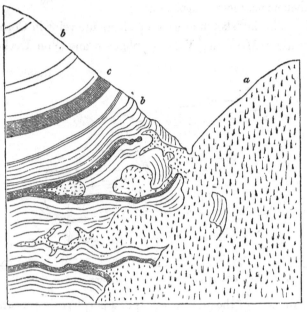

Junction of granite and limestone in Glen Tilt.

a, Granite.   b, Limestone.   c, Blue argillaceous schist.

The above diagram from a sketch of Dr. Macculloch, represents the junction of the granite of Glen Tilt in Perthshire, with a mass of stratified limestone and schist. The granite, in this locality, often sends forth so many veins as to reticulate the limestone and schist, the veins diminishing towards their termination to the thickness of a leaf of paper or a thread. In some places fragments of granite appear entangled as it were in the limestone, and are not visibly connected with any larger mass, while sometimes, on the other hand, a lump of the limestone is found in the midst of the granite. The ordinary colour of the limestone of Glen Tilt is lead blue, and its texture large grained and highly crystalline; but where it approximates to the granite, particularly where it is penetrated by the smaller veins, the crystalline texture disappears, and it

assumes an appearance exactly resembling that of horn-stone.
The associated argillaceous schist often passes into hornblende
slate, where it approaches very near to the granite *.

In the plutonic, as in the volcanic rocks, there is every
gradation from a tortuous vein to the most regular form of a
dike, such as we have described as intersecting the tuffs and
lavas of Vesuvius and Etna. In these dikes of granite, which
may be seen, among other places, on the southern flank of Mount
Battoch, one of the Grampians, the opposite walls sometimes
preserve an exact parallelism for a considerable distance. It is
not uncommon for one set of granite veins to intersect another,
and sometimes there are three sets, as in the environs of
Heidelborg, where the granite of the right bank of the Rhine
is seen to consist of three varieties differing in colour, grain, and
various peculiarities of mineral composition. One of these,
which is evidently the second in age, is seen to cut through an
older granite, and another, still newer, traverses both the
second and the first. These phenomena were lately pointed
out to me by Professor Leonhard at Heidelberg.

In Shetland there are two kinds of granite. One of these,
composed of hornblende, mica, felspar, and quartz, is of a dark
colour, and is seen underlying gneiss. The other is a red
granite which penetrates the former everywhere in veins †.

Granites of different ages.—It was formerly supposed that
granite was the oldest of rocks, the mineral product of a particular period or state of the earth formed long antecedently to
the introduction of organic beings into the planet. But it
is now ascertained that this rock has been produced again
and again, at successive eras, with the same characters, penetrating the stratified rocks in different regions, but not always
associated with strata of the same age. Nor are organic remains
always entirely wanting in the formations invaded by granite,
although their absence is more usual. Many well authenticated
exceptions to the rule are now established on the authority of

* Macculloch, Geol. Trans., vol. iii. p. 259.
† Macculloch, Syst. of Geol., vol. i. p. 58,

numerous observers, amongst the earliest of whom we may cite
Von Buch, who discovered in Norway a mass of granite overlying an ancient secondary limestone, containing orthocerata
and other shells and zoophytes *.

A considerable mass of granite in Sky is described by
Dr. Macculloch as incumbent on limestone and shale, which
are of the age of the English lias †. The limestone, which,
at a greater distance from the granite, contains shells, exhibits no traces of them near the junction of the igneous rock,
where it has been converted into a pure crystalline marble ‡.

This granite of Sky was at first termed ' Syenite,' by which
name many geologists have denominated the more modern
granites; but authors have entirely failed in their attempt to
establish a distinction between granites and syenites on mineralogical characters. The latter have sometimes been defined to
consist of a triple compound of felspar, quartz, and hornblende,
but the oldest granites are very commonly composed of these
ingredients only. In his later publications Dr. Macculloch has
with great propriety, we think, called the plutonic rock of Sky
a granite §.

In different parts of the Alps a comparatively modern granite
is seen penetrating through secondary strata, which contain
belemnites, and other fossils, and are supposed to be referrible
to the age of the English lias. According to the observations of
M. Elie de Beaumont and Hugi, masses of this granite are sometimes found partially overlying the secondary beds, and altering
them in a manner which we shall describe more particularly
when we treat of the changes in composition and structure superinduced upon sedimentary deposits in contact with Plutonic
rocks‖ (see wood-cut, No. 90, p. 371).

In such examples we can merely affirm, that the granite is

* Travels through Norway and Lapland, p. 45. London, 1813.
† See Murchison, Geol. Trans., 2nd Series, vol. ii. part ii. p. 311—321.
‡ Western Islands, vol. i. p. 330.
§ Syst. of Geol., vol. i. p. 150.
‖ Elie de Beaumont, Sur les Montagnes de l'Oisans, Mem. de la Soc. d'Hist.
Nat. de Paris, tome v. Hugi, Natur. Historische Alpenreise, Soleure, 1830.

newer than a secondary formation containing belemnites, but we can form no conjecture when it originated, not even whether it be of secondary or tertiary date. It is, indeed, very necessary to be on our guard against the inference that a granite is usually of about the same age as the group of strata into which it has intruded itself, for in that case we shall be inclined to assume rashly that the granites found penetrating a more modern secondary rock, such as the lias for example, are much newer than those found invading strata older than the carboniferous series. The contrary may often be true, for the plutonic rock which was last in a melted state, may not have been forced up anywhere so near the surface as to enter into the newer groups of strata, and it may have been injected into a part of the earth's crust formed exclusively of the older sedimentary formations.

'In a deep series of strata,' says Dr. Macculloch, 'the superior or distant portions may have been but slightly disturbed, or have entirely escaped disturbance, by a granite which has not emitted its veins far beyond its immediate boundary. However certain, therefore, it may be, that any mass of granite is posterior to the gneiss, the micaceous schist, or the argillaceous schists, which it traverses, or into which it intrudes, we are unable to prove that it is not also posterior to the secondary strata that lie above them *.'

There can be no doubt, however, that some granites are more ancient than any of our regular series which we identify by organic remains, because there are rounded pebbles of granite, as well as gneiss, in the conglomerates of the oldest fossiliferous groups.

Distinction between volcanic and plutonic rocks—Trap.— The next point to consider is the distinction between the plutonic and volcanic rocks. When geologists first began to examine attentively the structure of the northern parts of Europe, they were almost entirely ignorant of the phenomena of existing volcanos, and when they met with basalt and other

---

* Syst. of Geol., vol. i. p. 136.

rocks composed chiefly of augite, hornblende, and felspar, which are now admitted by all to have been once in a state of fusion, they were divided in opinion whether they were of igneous or of aqueous origin. We have shown in our sketch of the history of geology in the first volume, how much the polemical controversies on this subject retarded the advancement of the science, and how slowly the analogy of the rocks in question to the products of burning volcanos was recognized.

Most of the igneous rocks first investigated in Germany, France, and Scotland, were associated with marine strata, and in some places they occurred in tabular masses or platforms at different heights, so as to form on the sides of some hills a succession of terraces or *steps*, from which circumstance they were called ' trap' by Bergman (from *trappa*, Swedish for a staircase), a name afterwards adopted very generally into the nomenclature of the science.

When these trappean rocks were compared with lavas produced in the atmosphere, they were found to be in general less porous and more compact ; but in this instance the terms of comparison were imperfect, for a set of rocks, formed almost entirely under water, was contrasted with another which had cooled in the open air.

Yet the ancient volcanos of Central France were classed, in reference probably to their antiquity, with the trap rocks, although they afford perfect counterparts to existing volcanos, and were evidently formed in the open air. Mont Dor and the Plomb du Cantal, indeed, may differ in many respects from Vesuvius and Etna in the mineral constitution and structure of their lavas ; but it is that kind of difference which we must expect to discover when we compare the products of any two active volcanos, such as Teneriffe and Hecla, or Hecla and Cotopaxi.

The amygdaloidal structure in many of the trap formations proves that they were originally cellular and porous like lava, but the cells have been subsequently filled up with silex, carbonate of lime, zeolite, and other ingredients which form the nodules. Dr. Macculloch, after examining with great attention the igneous rocks of Scotland, observes ' that it is a mere dispute about terms to refuse to the ancient eruptions of trap the name of submarine volcanos, for they are such in every essential point, although they no longer eject fire and smoke *.'

The same author also considers it not improbable that some of the volcanic rocks of the same country may have been poured out in the open air†.

The recent examination of the igneous rocks of Sicily, especially those of the Val di Noto, has proved that all the more ordinary varieties of European trap have been produced under the waters of the sea in the Newer Pliocene period, that is to say, since the Mediterranean has been inhabited by a great proportion of the existing species of testacea. We are, therefore, entitled to feel the utmost confidence, that if we could obtain access to the existing bed of the ocean, and explore the igneous rocks poured out within the last 5000 years beneath the pressure of a sea of considerable depth, we should behold formations of modern date scarcely distinguishable from the most ancient trap rocks of our island. We cannot, however, expect the identity to be perfect, for time is ever working some alteration in the composition of these mineral masses, as, for example, by converting porous lava into amygdaloids.

Passage from trap into granite.—If a division be attempted between the trappean and volcanic rocks, it must be made between different parts of the same volcano,—nay even the same rock, which would be called ' trap,' where it fills a fissure and has assumed a solid crystalline form on slow cooling, must be termed volcanic, or lava, where it issues on the flanks of the mountain. Some geologists may perhaps be of opinion that melted matter, which has been poured out in the open air, may be conveniently called volcanic, while that which appears to have cooled at the bottom of the sea, or under pressure, but at no great depth from the surface, may be termed ' trap ;' but we believe that such distinctions will lead only to confusion, and that we must consider trap and volcanic as synonymous. On the other hand, the difficulty of discrimi-

* Syst. of Geol., vol. ii. p. 114.   † Ibid.

nating the volcanic from the plutonic rocks is sufficiently
great; for we must draw an arbitrary line between them, there
being an insensible passage from the most common forms of
granite into trap or lava.

' The ordinary granite of Aberdeenshire,' says Dr. Macculloch, ' is the usual ternary compound of quartz, felspar, and
mica, but sometimes hornblende is added to these, or the
hornblende is substituted for the mica.   But in many places
a variety occurs which is composed simply of felspar and
hornblende, and in examining more minutely this duplicate
compound, it is observed in some places to assume a fine grain,
and at length to become undistinguishable from the greenstones
of the trap family.   It also passes in the same uninterrupted
manner into a basalt, and at length into a soft claystone, with a
schistose tendency on exposure, in no respect differing from
those of the trap islands of the western coast*.'   The same
author mentions, that in Shetland a granite composed of hornblende, mica, felspar, and quartz, graduates in an equally
perfect manner into basalt †.

It would be easy to multiply examples to prove that the
granitic and trap-rocks pass into each other, and are merely
different forms which the same elements have assumed according
to the different circumstances under which they have consolidated from a state of fusion.   What we have said respecting the
mode of explaining the different texture of the central and external parts of the Vesuvian dikes may enable the reader in some
measure to comprehend how such differences may originate ‡.

The same lava which is porous where it has flowed over
from the crater, and where it has cooled rapidly and under
comparatively slight pressure, is compact and porphyritic in
the dike.   Now these dikes are evidently the channels of communication between the crater and the volcanic foci below; so
that we may suppose them to be continuous to the depth of
several hundred fathoms, or perhaps two or three miles, or even
more; and the fluid matter below, which cools and consoli-

* Syst. of Geol., vol. i. p. 157.   † Ibid., p. 158.
‡ See above, p. 124.

dates slowly under so enormous a pressure, may be supposed
to acquire a very distinct texture and become granite.

If it be objected that we do not find in mountain-chains volcanic dikes passing upwards into lava, and downwards into
granite, we may answer that our vertical sections are usually
of small extent, and it is enough that we find in certain localities a transition from trap to porous lava, and in others a
passage from granite to trap. It should also be remembered,
that a large proportion of the igneous rocks, when first formed,
cannot be supposed to reach the surface, and these may assume
the usual granitic texture without graduating into trap, or into
ouch lava and scoriæ as are found on the flanks of a volcanic
cone.

Theory of the origin of granite at all periods.—It is not
uncommon for lava-streams to require more than ten years to
cool in the open air, and a much longer period where they
are of great depth. The melted matter poured out from
Jorullo, in Mexico, in the year 1759, which accumulated in
some places to the height of 550 feet, was found to retain a
high temperature half a century after the eruption [*]. For
what immense periods, then, must we not conclude that great
masses of subterranean lava in the volcanic foci may remain in
a red hot or incandescent state, and how gradual must be the
process of refrigeration! This process may be sometimes
retarded for an indefinite period, by the accession of fresh
supplies of heat, for we find that the lava in the crater of Stromboli, one of the Lipari islands, has been in a state of constant
ebullition for the last 2000 years, and we must suppose this
fluid mass to communicate with some cauldron or reservoir of
fused matter below. In the Isle of Bourbon, also, where there
has been an emission of lava once in every two years for a long
period, we may infer that the lava below is permanently in a
state of liquefaction.

When melted matter is injected into the fissures of a contiguous rock at a considerable depth, it may cool rapidly if that
rock has not acquired a high temperature; but suppose, on he

[*] See vol. i. p. 378, and Second Edition, p. 433.

contrary, that it has been heated, and still continues for centuries, or thousands of years, at a red heat, the vein may acquire a highly crystalline texture.

The great pressure of a superincumbent mass, and exclusion from contact with air or water, are probably the usual conditions necessary to produce the granitic texture; but the same may sometimes be superinduced at a slighter distance from the surface by slow refrigeration, when additional supplies of heat check, from time to time, the cooling process and cause it to be indefinitely protracted.

If, for the reasons above alluded to, we conceive it probable that plutonic rocks have originated in the nether parts of the earth's crust, as often as the volcanic have been generated at the surface, we may imagine that no small quantity of the former class has been forming in the recent epoch, since we suppose that about 2000 volcanic eruptions may occur in the course of every century, either above the waters of the sea or beneath them *.

We may also infer, that during each preceding period, whether tertiary or secondary, there have been granites and granitiform rocks generated, because we have already discovered the monuments of ancient volcanic eruptions at almost every period.

In the next chapter we shall endeavour to show, that in consequence of the great depths at which the plutonic rocks usually originate, and the manner in which they are associated with the older sedimentary strata of each district, it is rarely possible to determine with exactness their relative age. Yet there is reason to believe that the greater portion of the plutonic formations now visible are of higher antiquity than the oldest secondary strata. We shall also endeavour to point out, that this opinion is by no means inconsistent with the theory that *equal* quantities of granite may have been produced in succession, during *equal periods* of time, from the earliest to the most modern epochs.

* See vol. i. chap. xxii.

CHAPTER XXVI.

On the stratified rocks usually called 'primary'—Proofs from the disposition of their strata that they were originally deposited from water—Alternation of beds varying in composition and colour—Passage of gneiss into granite—Alteration of sedimentary strata by trappean and granitic dikes—Inference as to the origin of the strata called 'primary'—Conversion of argillaceous into hornblende schist—The term 'Hypogene' proposed as a substitute for primary—'Metamorphic' for 'stratified primary' rocks—No regular order of succession of hypogene formations—Passage from the metamorphic to the sedimentary strata—Cause of the high relative antiquity of the visible hypogene formations—That antiquity consistent with the hypothesis that they have been produced at each successive period in equal quantities—Great volume of hypogene rocks supposed to have been formed since the Eocene period—Concluding remarks.

ON THE STRATIFIED ROCKS CALLED 'PRIMARY.'

We stated in the last chapter, that the rocks usually termed 'primary' are divisible into two natural classes, the stratified and the unstratified. The propriety of the term stratified, as applied to the first-mentioned class, will not be questioned when the rocks so designated are carefully compared with strata known to result from aqueous deposition.

Mode of stratification.—If we examine gneiss, which consists of the same materials as granite, or mica-schist which is a binary compound of quartz and mica, or clay-slate, or any other member of the so-called primary division, we find that it is made up of a succession of beds, the planes of which are, to a certain extent, parallel to each other, but which frequently deviate from parallelism in a manner precisely analogous to that exhibited by sedimentary formations of all ages. The resemblance is often carried farther, for in the crystalline series we find beds composed of a great number of layers placed diagonally, as we have shown to be the case in the

Crag and other formations *.   This disposition of the layers

No. 89.

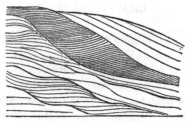

Lamination of clay-slate, Montagne de Seguinat, near Gavarnie, in the Pyrenees.

is illustrated in the accompanying diagram, in which I have represented carefully the stratification of a coarse argillaceous schist, which I examined in the Pyrenees, part of which approaches in character to a green and blue roofing slate, while part is extremely quartzose, the whole mass passing downwards into micaceous schist. The vertical section here exhibited is about three feet in height, and the layers are sometimes so thin that fifty may be counted in the thickness of an inch. Some of them consist of pure quartz.

The stratification now alluded to must not be confounded with that fissile texture sometimes observed in the older rocks, by virtue of which they divide in a direction different both from the general planes of stratification and from the planes of those transverse layers of which a single stratum may be made up.

Another striking point of analogy between the stratification of the crystalline formations and that of the secondary and tertiary periods is the alternation in each of beds varying greatly in composition, colour, and thickness. We observe, for instance, gneiss alternating with layers of black hornblende-schist, or with granular quartz or limestone, and the interchange of these different strata may be repeated for an indefinite number of times. In like manner, mica schist alternates with chlorite-schist, and with granular limestone in thin layers.

As we observe in the secondary and tertiary formations

* See above, p. 173.

strata of pure siliceous sand alternating with micaceous sand
and with layers of clay, so in the 'primary' we have beds of
pure quartz rock alternating with mica-schist and clay-slate.
As in the secondary and tertiary series we meet with limestone
alternating again and again with micaceous or argillaceous sand,
so we find in the 'primary' gneiss and mica-schist alternating
with pure and impure granular limestones.

Passage of gneiss into granite—If, then, reasoning from the
principle that like effects have like causes, we attribute the
stratification of gneiss, mica-schist, and other associated rocks,
to sedimentary deposition from a fluid, we encounter this difficulty, that there is often a transition from gneiss, one of the
stratified series, into granite, which, as we have shown, is of
igneous origin.   Gneiss is composed of the same ingredients as
granite, and its texture is equally crystalline.   It sometimes
occurs in thick beds, and in these the rock is often quite
undistinguishable, in hand specimens, from granite; yet the
lines of stratification are still evident.   These lines imply deposition from water, while the passage into granite would lead us
to infer an igneous origin.   In what manner can we reconcile
these apparently conflicting views?   The Huttonian hypothesis
offers, we think, the only satisfactory solution of this problem.
According to that theory, the materials of gneiss were originally
deposited from water in the usual form of aqueous strata, but
these strata were subsequently altered by their proximity to
granite, and to other plutonic masses in a state of fusion, until
they assumed a granitiform texture.   The reader will be prepared, by what we have said of granite, to conclude, that when
voluminous masses of melted rock have been for ages in an
incandescent state, in contact with sedimentary deposits, they
must produce some alteration in their texture, and this alteration
may admit of every intermediate gradation between that resulting from perfect fusion, and the slightest modification which
heat can produce.

The geologist has been conducted, step by step, to this

theory by direct experiments on the fusion of rocks in the laboratory, and by observation of the changes in the composition and texture of stratified masses, as they approach or come in contact with igneous veins and dikes. In studying the latter class of phenomena, we have the advantage of examining the condition of the rock at some distance from the dike where it has escaped the influence of heat, and its state where it has been near to, or in contact with, the fused mass. The changes thus exhibited may be regarded as the results of a series of experiments, made on a great scale by nature under every variety of condition, both as relates to the mineral ingredients of the rocks, the intensity of heat or pressure, the celerity or slowness of the cooling process, and other circumstances.

Strata altered by volcanic dikes—Plas Newydd.—We shall select a few examples of these alterations in illustration of our present argument. One of the most interesting is the modification of strata in the proximity of a volcanic dike near Plas Newydd, in Anglesea, described by Professor Henslow. The dike is 134 feet wide, and consists of basalt (dolerite of some authors), a compound of felspar and augite. Strata of shale and argillaceous limestone, through which it cuts perpendicularly, are altered to a distance of thirty, or even in some places to thirty-five feet, from the edge of the dike. The shale, as it approaches the basalt, becomes gradually more compact, and is most indurated where nearest the junction. Here it loses part of its schistose structure, but the separation into parallel layers is still discernible. In several places the shale is converted into hard porcellanous jasper. In the most hardened part of the mass the fossil shells, principally *Productæ*, are nearly obliterated, yet even here their impressions may frequently be traced. The argillaceous limestone undergoes analogous mutations, losing its earthy texture as it approaches the dike, and becoming granular and crystalline. But the most extraordinary phenomenon is the appearance in the shale of numerous crystals of analcime and garnet, which are distinctly confined to those portions of the rock affected by the dike*. Garnets have been observed, under very analogous circumstances, in High Teesdale, by Professor Sedgwick, where they also occur in shale and limestone, altered by a basaltic dike. This discovery is most interesting, because garnets often abound in mica-schist, and we see in the instances above cited, that they did not previously exist in the shale and limestone, and that they have evidently been produced by heat in rocks in which the marks of stratification have not been effaced.

Stirling Castle.—To select another example : we find in the rock of Stirling Castle, a calcareous sandstone fractured and forcibly displaced by a mass of green-stone, which has evidently invaded the strata in a melted state. The sandstone has been indurated, and has assumed a texture approaching to hornstone near the junction. So also in Arthur's Seat and Salisbury Craig, near Edinburgh, a sandstone is seen to come in contact with greenstone, and to be converted into a jaspideous rock †.

Antrim.—In the north of Ireland, in several parts of the county of Antrim, chalk, with flints, is traversed by basaltic dikes. The chalk is converted into granular marble near the basalt, the change sometimes extending eight or ten feet from the wall of the dike, being greatest at that point, and thence gradually decreasing till it becomes evanescent. ' The extreme effect,' says Dr. Berger, ' presents a dark brown crystalline limestone, the crystals running in flakes as large as those of coarse *primitive* limestone; the next state is saccharine, then fine-grained and arenaceous ; a compact variety having a por-cellanous aspect, and a bluish-grey colour succeeds ; this, towards the outer edge, becomes yellowish-white, and insen-sibly graduates into the unaltered chalk. The flints in the altered chalk usually assume a grey yellowish colour ‡.' All

* Trans. of Cambridge Phil. Soc., vol. i. p. 406.

† Illust. of Hutt. Theory, § 253 and 261. Dr. Macculloch, Geol. Trans., 1st series, vol. ii. p. 305.

‡ Dr. Berger, Geol. Trans., 1st series, vol. iii. p. 172.

traces of organic remains are effaced in that part of the limestone which is most crystalline.

As the carbonic acid has not been expelled, in this instance,
from that part of the rock which must be supposed to have
been melted, the change must have taken place under considerable pressure; for we know, by the experiments of Sir James
Hall, that it would require the weight of about 1700 feet of
sea-water, which would be equivalent to the pressure of a
column of liquid lava 600 feet high, to prevent this acid from
being given off.

Another of the dikes of the north-east of Ireland has converted a mass of red sandstone into hornstone *.   By another,
the slate-clay of the coal-measures has been indurated, and has
assumed the character of flinty slate † ; and in another place
the slate-clay of the lias has been changed into flinty slate,
which still retains numerous impressions of ammonites ‡.   One
of the greenstone dikes of the same country passes through a
bed of coal, which it reduces to a cinder for the space of nine
feet on each side §.

The secondary sandstones in Sky are converted into solid
quartz in several places where they come in contact with veins
or masses of trap; and a bed of quartz, says Dr. Macculloch,
has been found near a mass of trap, among the coal-strata of
Fife, which was in all probability a stratum of ordinary sandstone subsequently indurated by the action of heat ‖.

Alterations of strata in contact with granite. — Having
selected these from innumerable examples of mutations caused
by volcanic dikes, we may next consider the changes produced
by the contiguity of plutonic rocks.   To some of these , we
have already adverted, when speaking of granite veins, and
endeavouring to establish the igneous origin of granite.   We
mentioned that the main body of the Cornish granite sends
forth veins through the killas of that country ¶, a coarse
argillaceous schist, which is converted into hornblende-schist

---

* Rev. W. Conybeare, Geol. Trans., 1st series, vol. iii. p. 201.   Diﬀ

† Ibid., p. 205.   ‡ Ibid. p. 213, and Playfair, Illust. of Hutt. Theory, § 253.

§ Ibid., p. 206.   ‖ Syst. of Geol., vol. i. p. 206.   ¶ See diagram, No. 87.

near the contact with the veins.   These appearances are well
seen at the junction of the granite and killas in St. Michael's
Mount, a small island nearly 300 feet high, situated in the bay,
at the distance of about three miles from Penzance.

In the department of the Hautes Alpes, in France, near
Vizille, M. Elie de Beaumont traced a black argillaceous
limestone, charged with belemnites to within a few yards of a
mass of granite.   Here the limestone begins to put on a

No. 90.

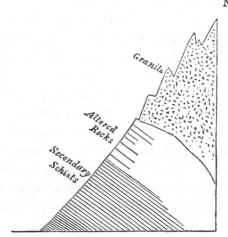

Junction of granite with Jurassic or oolite strata in the Alps, near Champoleon.

granular texture, but is extremely fine-grained.   When nearer
the junction it becomes grey and has a saccharoid structure.
In another locality, near Champoleon, a granite composed of
quartz, black mica, and rose-coloured felspar, is observed
partly to overlie the secondary rocks, producing an alteration
which extends for about thirty feet downwards, diminishing in
the inferior beds which lie farthest from the granite.   (See wood-
cut No. 90.)   In the altered mass the argillaceous beds are
hardened, the limestone is saccharoid, the grits quartzose, and in
the midst of them is a thin layer of an imperfect granite.   It is
also an important circumstance, that near the point of contact
both the granite and the secondary rocks become metalliferous,
and contain nests and small veins of blende, galena, iron, and
copper pyrites.   The stratified rocks become harder and more

2 B 2

crystalline, but the granite, on the contrary, softer and less perfectly crystallized near the junction *.

It will appear from sections described by M. Hugi, that some of the secondary beds of limestone and slate, which are in a similar manner overlaid by granite, have been altered into gneiss and mica-schist †.   Some of these altered sedimentary formations are supposed, by M. Elie de Beaumont, to be of the age of the lias of England, and others to be even as modern as the jurassic or oolite formations.

We can scarcely doubt, in these cases, that the heat communicated by the granitic mass reduced the contiguous strata to semi-fusion, and that on cooling slowly the rock assumed a crystalline texture.   The experiments of Gregory Watt prove, distinctly, that a rock need not be perfectly melted in order that a re-arrangement of its component particles should take place, and that a more crystalline texture should ensue. We may easily suppose, therefore, that all traces of shells and other organic remains may be destroyed, and that new chemical combinations may arise, without the mass being so fused as that the lines of stratification should be wholly obliterated.

In allusion to the passage from granite to gneiss before described, Dr. Macculloch remarks, that 'in numerous parts of Scotland, where the leading masses of gneiss are schistose, evenly stratified, and scarcely ever traversed by granite veins, they become contorted and irregular as they approach the granite; assuming also the granitic character, and becoming intersected by veins, numerous in proportion to the vicinity of the mass.   The conclusion,' he adds, 'is obvious; the fluid granite has invaded the aqueous stratum as far as its influence could reach, and thus far has filled it with veins, disturbed its regularity and generated in it a new mineral character, often absolutely confounded with its own.   And if the more remote beds, and those alternating with other rocks, are not thus

* Elie de Beaumont, Sur les Montagnes de l'Oisans, &c., Mem. de la Soc. d'Hist. Nat. de Paris, tome v.

† Natur. Historische Alpenreise, Soleure, 1830.

affected, it is not only that it has acted less on those, but that, if it had equally affected them, they never could have existed, or would have been all granitic and venous gneiss[*].

According to these views, gneiss and mica-schist may be nothing more than micaceous and argillaceous sandstones altered by heat, and certainly, in their mode of stratification and lamination, they correspond most exactly. Granular quartz may have been derived from siliceous sandstone, compact quartz from the same. Clay-slate may be altered shale, and shale appears to be clay which has been subjected to great pressure. Granular marble has probably originated in the form of ordinary limestone, having in many instances been replete with shells and corals now obliterated, while calcareous sands and marls have been changed into impure crystalline limestones.

Associated with the rocks termed primary we meet with anthracite, just as we find beds of coal in sedimentary formations, and we know that, in the vicinity of some trap dikes, coal is converted into anthracite. ' Hornblende schist,' says Dr. Macculloch, ' may at first have been mere clay, for clay or shale is found altered by trap into Lydian stone, a substance differing from hornblende-schist almost solely in compactness and uniformity of texture[†].' ' In Shetland,' remarks the same author, ' argillaceous schist (or clay-slate), when in contact with granite, is sometimes converted into hornblende-schist, the schist becoming first siliceous, and ultimately, at the contact, hornblende-schist [‡].'

This theory, if confirmed by observation and experiment, may enable us to account for the high position in the series usually held by clay slate relatively to hornblende-schist, as also to gneiss and mica-schist, which so commonly alternate with hornblende-schist. For we must suppose the heat which alters the strata to proceed, in almost all cases, from below upwards, and to act with greatest intensity on the inferior strata. If, therefore, several sets of argillaceous strata or shales be superimposed upon each other in a vertical series of beds in the same

* Syst. of Geol., vol. ii. p. 145.   † Ibid., vol. i. p. 210.   ‡ Ibid., p. 211.

district, the lowest of these will be converted into hornblende-schist, while the uppermost may continue in the condition of clay-slate.

The term 'Hypogene' proposed for Primary.—If our readers have followed us in the train of reasoning explained in this and the preceding chapter, they must already be convinced that the popular nomenclature of Geology, in reference to the so called 'primary' rocks, is not only imperfect, but in a great degree founded on a false theory; inasmuch as some granites and granitic schists are of origin posterior to many secondary rocks. In other words, some *primary* formations can already be shown to be newer than many *secondary* groups—a manifest contradiction in terms.

Yet granite and gneiss, and the families of stratified and unstratified rocks connected with each, belong to one great natural division of mineral masses, having certain characters in common, and it is therefore convenient that the class to which they belong should receive some common name—a name which must not be of chronological import, and must express, on the one hand, some peculiarity equally attributable to granite and gneiss (to the plutonic as well as the *altered* rocks), and which, on the other, must have reference to characters in which those rocks differ both from the volcanic and from the *unaltered* sedimentary strata. We propose the term 'hypogene' for this purpose, derived from ὑπο, *subter*, and γινομαι, *nascor*, a word implying the theory that granite and gneiss are both *nether-formed* rocks, or rocks which have not assumed their present form and structure at the surface. It is true that gneiss and all stratified rocks must have been deposited originally at the surface, or on that part of the surface of the globe which is covered by water; but according to the views explained in this and the foregoing chapter, they could never have acquired their crystalline texture, unless acted upon by heat under pressure in those regions, and under those circumstances where the plutonic rocks are generated.

The term 'Metamorphic' proposed for stratified primary.—

We divide the hypogene rocks, then, into the unstratified, or plutonic, and the *altered* stratified. For these last the term ' metamorphic' (from μετα, *trans*, and μορφη, *form*) may be used. The last-mentioned name need not, however, be often resorted to, because we may speak of hypogene *strata*, hypogene *limestone*, hypogene *schist*, and this appellation will suffice to distinguish the formations so designated from the plutonic rocks. By referring to the table (No. I.) at the close of this chapter, the reader will see the chronological relation which we conceive the two classes of hypogene rocks to bear to the strata of different ages.

No order of succession in hypogene formations.—When we regard the tertiary and secondary formations simply as mineral masses uncharacterized by organic remains, we perceive an indefinite series of beds of limestone, clay, marl, siliceous sand, sandstone, coal, and other materials, alternating again and again without any fixed or determinate order of position. The same may be said of the hypogene formations, for in these a similar want of arrangement is manifest, if we compare those occurring in different countries. Gneiss, mica-schist, hornblende-schist, quartz rock, hypogene limestone, and the rest, have no invariable order of superposition, although, for reasons above explained, clay-slate must usually hold a superior position relatively to hornblende schist.

We do not deny, that in a particular mountain-chain, a chronological succession of hypogene formations may be recognized, for the same reason that in a country of limited extent there is an order of position in the secondary and tertiary rocks, limestone predominating in one part of the series, clay in another, siliceous sand in a third, and so of other compounds. It is probable that a similar prevalence of a regular order of arrangement in the hypogene series throughout certain districts, led the earlier geologists into a belief, that they should be able to fix a definite order of succession for the various members of this great class throughout the world.

That expectation has not been realized ; yet was it more reasonable than the doctrine of the universality of certain rocks which were admitted to be of sedimentary origin ; for there is certainly a remarkable identity in the mineral character of the hypogene formations, both stratified and unstratified, in all countries ; although the notion of a uniform order of succession in the different groups must be abandoned.

The student may, perhaps, object to the views above given of the relation of the sedimentary and metamorphic rocks, on the ground that there is frequently, indeed usually, an abrupt passage from one to the other. This phenomenon, however, admits of the same explanation as the fact, that the beds of lakes and seas are now frequently composed of hypogene rocks. In these localities the hypogene formations have been brought up to the surface and laid bare by denudation. New sedimentary strata are thrown down upon them, and in this manner the two classes of rocks, the aqueous and the hypogene, come into immediate contact, without any gradation from one to the other. As we suppose the plutonic and metamorphic rocks to have been uplifted at all periods in the earth's history, so as to have formed the bottom of the ocean and of lakes, by the same operations which have carried up marine strata to the summits of lofty mountains, we must suppose the juxtaposition of the two great orders of rocks now alluded to, to have been a necessary result of all former revolutions of the globe.

But occasionally a transition is observable from strata containing shells, and displaying an evident mechanical structure, to others which are partially altered, and from these again we sometimes pass insensibly into the hypogene series. Some of the argillaceous-schists in Cornwall are of this description, being undistinguishable from the hypogene schists of many countries, and yet exhibiting, in a few spots, faint traces of organic remains. In parts of Germany, also, there are schists which, from their chemical condition, are identical with hypogene-schists, yet are interstratified with greywacke, a rock probably modified by heat, but which contains casts of shells, and often displays unequivocal marks of being an aggregate of fragments of pre-existing rocks.

Those geologists who shrink from the theory, that all the hypogene strata, so beautifully compact and crystalline as they are, have once been in the state of the ordinary mud, clay, marl, sand, gravel, limestone, and other deposits now forming beneath the waters, resort, in their desire to escape from such conclusions, to the hypothesis, that *chemical causes* once acted with intense energy, and that by their influence more crystalline strata were precipitated ; but this theory appears to us to be as mysterious and unphilosophical as the doctrine of a 'plastic virtue,' introduced by the earlier writers to explain the origin of fossil-shells and bones.

Relative age of the visible hypogene rocks.—We shall now return to the subject already in part alluded to at the close of the last chapter—the relative age of the hypogene rocks as compared to the secondary. How far are they entitled in general to the appellation of ' primary,' in the sense of being anterior in age to the period of the carboniferous strata, in which last we include the greywacke and many of the rocks commonly called transition ? It is undoubtedly true that we can rarely point out metamorphic or plutonic rocks which can be proved to have been formed in any secondary or tertiary period. We can, in some instances, demonstrate, as we have already shown, that there are granites of posterior origin to certain secondary strata, and that *secondary* strata have sometimes been converted into the *metamorphic.* But examples of such phenomena are rare, and their rarity is quite consistent with the theory, that the hypogene formations, both stratified and unstratified, have been always generated in equal quantities during periods of equal duration.

We conceive that the granite and gneiss, formed at periods more recent than the carboniferous era, are still for the most part concealed, and those portions which are visible can rarely be shown, by geological evidence, to have originated during

secondary periods.   It is very possible, for example, that con-
siderable¹ tracts of hypogene strata in the Alps may be altered
oolite, altered lias, or altered secondary rocks inferior to the
lias ; but we can scarcely ever hope to substantiate the fact,
because, whenever the change of texture is complete, no characters remain to afford us any insight into the probable age of
the mass.   Where granite happens to have intruded itself in
such a manner as partially to overlie a mass of lias or·other
strata, as in the case before alluded to (diagram No. 90, p. 371),
we may prove that *fossilliferous* strata have become gneiss,
mica-schist, clay-slate, or granular marble ; but if the action of
the heat upon the strata had been more intense, the same inferences could not have been drawn.   It might then have been
supposed that no Alpine hypogene strata were newer than the
carboniferous period.

The metamorphic strata of Scotland are certainly in·great
part older than the carboniferous, which are found incumbent
upon them in an unaltered state ; but it appears that secondary
deposits as new, or newer than the lias, have come in contact,
in the Western Islands, with granite, and have there assumed
the hypogene texture.

A considerable source of difficulty and misapprehension, in
regard to the antiquity of the metamorphic rocks, may arise
from the circumstance of their having been deposited at one
period, and having assumed their crystalline texture at another.
Thus, for example, if an Eocene granite should invade ·the
lias and superinduce a hypogene structure, to what period shall
we refer the altered strata ?   Shall we say that they are metamorphic rocks of the Eocene or Liassic eras ?   They assumed
their stratified form when the animals and plants of the lias
flourished; they became metamorphic during the Eocene period.
It would be preferable in such instances, we think, to consider
them as hypogene strata of the Eocene period, or of that in
which they were altered ; yet it would rarely be possible to
establish their true age.   We should know the granite, ·to
which the change of texture was due, to be newer than the lias

which it penetrated; but there would rarely be any date to show that it might not have been injected at the close of the Liassic period, or at some much later era.

The metamorphic rocks must be the oldest, that is to say, they must lie at the bottom of each series of superimposed strata, because the influence of the volcanic heat proceeds from below upwards; but the hypogene strata of one country may be, and frequently are, of a very different age from those of another. The greater part, however, of the visible hypogene rocks are, we believe, more ancient than the carboniferous formations. In the latter, we frequently discover pebbles of hypogene rocks, namely, granite, gneiss, mica-schist, and clay-slate; and the carboniferous rocks often rest unchanged upon the hypogene. According to our views of the operations of earth quakes, we ought not to expect plutonic and metamorphic rocks of the more modern eras to have reached the surface generally, for we must imagine many geological periods to elapse before a mass which has put on its particular form far below the level of the sea, can have been upraised and laid open to view above that level. Beds containing marine shells sometimes appear at the height of two or three miles in the principal mountain-chains, but they always belong to formations of considerable antiquity; still more should we be prepared to find the hypogene rocks now in sight to be of high relative antiquity, since, in order to be brought up to view, they must probably have risen from a position far inferior to the bottom of the ocean.

We shall endeavour to elucidate the cause of the great age of the plutonic and metamorphic rocks, *now in sight*, by a familiar illustration. Suppose two months to be the usual time required for passing from some tropical country to our island, and that an annual importation takes place of a certain tropical species of insect, the ordinary term of whose life is two months, and which can only be reared in the climate of that equatorial country. It is evident that no living individuals could ever be seen in England except in extreme old age. The young may come annually into the world in great numbers, but in order to see them, we must travel to lands near the equator.

In like manner, if the hypogene rocks can only originate at great depths in the regions of subterranean heat, and if it requires many geological epochs to raise them to the surface, they must be very ancient before they make their appearance in the superficial parts of the earth's crust.   They may still be forming in every century, and they may have been produced in equal quantities during each successive geological period of equal duration ; but in order to see them in a nascent state, slowly consolidating from a state of fusion, or semi-fusion, we must descend into the 'fuelled entrails' of the earth, into the regions described by the poets, where for ages the land has

——— ever burn'd
With solid, as the lake with liquid fire.

As the progress of decay and reproduction by aqueous agency is incessant on the surface of the continents, and in the bed of the ocean, while the hypogene rocks are generated below, or are rising gradually from the volcanic foci, thus there must ever be a remodelling of the earth's surface in the time intermediate between the origin of each set of plutonic and metamorphic rocks, and the protrusion of the same into the atmosphere or the ocean.   Suppose the principal source of the Etnean lavas to lie at the depth of ten miles, we may easily conceive that before they can be uplifted to the day several distinct series of earthquakes must occur, and between each of these there might usually be one or more periods of tranquillity. The time required for so great a development of subterranean elevatory movements might well be protracted until the deposition of a series of sedimentary rocks, equal in extent to all our secondary and tertiary formations, had taken place.   We conceive, therefore, that the relative age of the *visible* plutonic and metamorphic rocks, as compared to the unaltered sedimentary strata, must always be determined by the relations of two forces,—the power which uplifts the hypogene rocks, and that aqueous agency which degrades and renovates the earth's surface; or, in other words, it must depend on the quantity of aqueous action which takes place between two periods, that when the heated and melted rocks are cooled and consolidated in the nether regions, and that when the same emerge to the day.

Volume of hypogene rocks supposed to have been formed since the Eocene period.—If we were to indulge in speculations on the probable quantity of hypogene formations, both stratified and unstratified, which may have been formed beneath Europe and the European seas since the commencement of the Eocene period, we should conjecture, that the mass has equalled, if not exceeded in volume, the entire European continents. The grounds of this opinion will be understood by reference to what we have said of the causes which may have upheaved part of Sicily to a great height above the level of the sea since the beginning of the Newer Pliocene period*. If the theory which, in that instance, attributes the disturbance and upheaving of the superficial strata to the action of subterranean heat be deemed admissible, the same argument will apply with no less force to every other district, elevated or depressed, since the commencement of the tertiary period.

But we have shown, in our remarks on the map of Europe, in the second volume, that the conversion of sea into land, since the Eocene period, embraces an area equal to the greater part of Europe, and even those tracts which had in part emerged before the Eocene era, such as the Alps, Apennines, and other mountain-chains, have risen to the additional altitude of from 1000 to 4000 feet since that era. We have also stated the probability of a great amount of subsidence and the conversion of considerable portions of European land into sea during the same period—changes which may also be supposed to arise from the influence of subterranean heat.

From these premises we conclude, that the liquefaction and alteration of rocks by the operation of volcanic heat at suc-

* See above, p. 107.

cessive periods, has extended over a subterranean space equal at least in area to the present European continent, and often through a portion of the earth's crust 4000 feet or more in thickness.

The principal effect of these volcanic operations in the nether regions, during the tertiary periods, or since the existing species began to flourish, has been to heave up to the surface hypogene formations of an age anterior to the carboniferous. We imagine that the repetition of another series of movements, of equal violence, might upraise the plutonic and metamorphic rocks of many of the secondary periods ; and if the same force should still continue to act, the next convulsions might bring up the *tertiary* and *recent* hypogene rocks, by which time we imagine that nearly all the sedimentary strata now in sight would either have been destroyed by the action of water, or would have assumed the metamorphic structure, or would have been melted down into plutonic and volcanic rocks.

At the close of this chapter the reader will find a table of the chronological relations of the principal divisions of rocks according to the views above set forth. The sketch is con-fessedly imperfect, but it will elucidate our theory of the con-nexion which may exist between the hypogene rocks of different periods, and the alluvial, volcanic, and sedimentary formations. A second table is added, containing the names of some of the principal groups of sedimentary strata mentioned in this work, arranged in their order of superposition.

Concluding Remarks.—In our history of the progress of geology, in the first volume, we stated that the opinion originally promulgated by Hutton, ' that the strata called *primitive* were mere altered sedimentary rocks,' was vehe-mently opposed for a time, the main objection to the theory being its supposed tendency to promote a belief in the past eternity of our planet. Previously the absence of animal and vegetable remains in the so-called primitive strata, had been appealed to, as proving that there had been a period when the planet was uninhabited by living beings, and when, as was also inferred, it was uninhabitable, and, therefore, probably in a nascent state.

The opposite doctrine, that the oldest visible strata might be the monuments of an antecedent period, when the animate world was already in existence, was declared to be equivalent to the assumption, that there never was a beginning to the present order of things. The unfairness of this charge was clearly pointed out by Playfair, who observed, ' that it was one thing to declare that we had not yet discovered the traces of a beginning, and another to deny that the earth ever had a beginning.'

We regret, however, to find that the bearing of our arguments in the first volume has been misunderstood in a similar manner, for we have been charged with endeavouring to establish the proposition, that ' the existing causes of change have operated with absolute uniformity from all eternity *.'

It is the more necessary to notice this misrepresentation of our views, as it has proceeded from a friendly critic whose theoretical opinions coincide in general with our own, but who has, in this instance, strangely misconceived the scope of our argument. With equal justice might an astronomer be accused of asserting, that the works of creation extend throughout *infinite* space, because he refuses to take for granted that the remotest stars now seen in the heavens are on the utmost verge of the material universe. Every improvement of the telescope has brought thousands of new worlds into view, and it would, therefore, be rash and unphilosophical to imagine that we already survey the whole extent of the vast scheme, or that it will ever be brought within the sphere of human observation.

But no argument can be drawn from such premises in favour of the infinity of the space that has been filled with worlds ; and if the material universe has any limits, it then follows that it must occupy a minute and infinitessimal point in infinite space. So, if in tracing back the earth's history, we arrive at the monuments of events which may have happened millions of ages

* Quarterly Review, No. 86, Oct. 1830, p. 464.

before our times, and if we still find no decided evidence of a commencement, yet the arguments from analogy in support of the probability of a beginning remain unshaken; and if the past duration of the earth be finite, then the aggregate of geological epochs, however numerous, must constitute a mere moment of the past, a mere infinitessimal portion of eternity.

It has been argued, that as the different states of the earth's surface, and the different species by which it has been inhabited, have had each their origin, and many of them their termination, so the entire series may have commenced at a certain period. It has also been urged, that as we admit the creation of man to have occurred at a comparatively modern epoch— as we concede the astonishing fact of the first introduction of a moral and intellectual being, so also we may conceive the first creation of the planet itself.

We are far from denying the weight of this reasoning from analogy; but although it may strengthen our conviction, that the present system of change has not gone on from eternity, it cannot warrant us in presuming that we shall be permitted to behold the signs of the earth's origin, or the evidences of the first introduction into it of organic beings.

In vain do we aspire to assign limits to the works of creation in *space*, whether we examine the starry heavens, or that world of minute animalcules which is revealed to us by the microscope. We are prepared, therefore, to find that in *time* also, the confines of the universe lie beyond the reach of mortal ken. But in whatever direction we pursue our researches, whether in time or space, we discover everywhere the clear proofs of a Creative Intelligence, and of His foresight, wisdom, and power.

As geologists, we learn that it is not only the present condition of the globe that has been suited to the accommodation of myriads of living creatures, but that many former states also have been equally adapted to the organization and habits of prior races of beings. The disposition of the seas, continents, and islands, and the climates have varied; so it appears that the species have been changed, and yet they have all been so modelled, on types analogous to those of existing plants and animals, as to indicate throughout a perfect harmony of design and unity of purpose. To assume that the evidence of the beginning or end of so vast a scheme lies within the reach of our philosophical inquiries, or even of our speculations, appears to us inconsistent with a just estimate of the relations which subsist between the finite powers of man and the attributes of an Infinite and Eternal Being.

TABLE I.

Showing the Relations of the Alluvial, Aqueous, Volcanic, and Hypogene Formations of different ages.

Periods.	Formations.		Some of the Localities where the Formations occu
I. RECENT. A.	Alluvial.	. . . . .	Beds of existing rivers, &c., vol. i ch. xiv.
	Aqueous.	*a.* Marine.	Coral reefs of the Pacific, vol. i ch. xviii.
		b. Freshwater.	Bed of Lake Superior, &c., vol. ch. xiii.
	Volcanic.	. . . . .	Etna, Vesuvius, vol. i. ch. xix. xx. xx
	Hypogene.	*a.* Plutonic.	*Concealed ;* foci of active volcano vol. iii. ch. xxv.
		b. Metamorphic.	*Concealed ;* around the foci of activ volcanos, vol. iii. ch. xxvi.
II. TERTIARY. 1. Newer Pliocene. B.	Alluvial.	. . . . .	Loess of the Rhine—gravel coverin the Newer Pliocene strata (Sicily.
	Aqueous.	*a.* Marine.	Val di Noto, Sicily.
		b. Freshwater.	Colle, in Tuscany.
	Volcanic.	. . . . .	Val di Noto, Sicily.
	Hypogene.	*a.* Plutonic.	*Concealed ;* foci of Newer Pliocene vo canos—underneath the Val Noto, vol. iii. p. 107, and ch. xx
		b. Metamorphic.	*Concealed ;* near the foci of Newer Pl ocene volcanos—underneath tł Val di Noto, vol. iii. p. 109, an ch. xxvi.
2. Older Pliocene. C.	Alluvial.	. . . . .	Norfolk ? vol. iii. p. 173.
	Aqueous.	*a.* Marine.	Subapennine formations.
		b. Freshwater.	Near Sienna, vol. iii. p. 160.
	Volcanic.	. . . . .	Tuscany, vol. iii. p. 159.
	Hypogene.	*a.* Plutonic.	*Concealed ;* foci of Older Plioce volcanos—beneath Tuscany.
		b. Metamorphic.	*Concealed ;* probably near the sa foci.
3. Miocene. D.	Alluvial.	. . . . .	Mont Perrier, Auvergne—Orleana vol. iii. p. 217.
	Aqueous.	*a.* Marine.	Bordeaux. Dax.
		b. Freshwater.	Saucats, near Bordeaux, vol. iii. p. 20
	Volcanic.	. . . . .	Hungary, vol. iii. ch. xvi.
	Hypogene.	*a.* Plutonic.	*Concealed ;* foci of Miocene volcan —beneath Hungary.
		b. Metamorphic.	*Concealed ;* probably around the sa foci.
4. Eocene. E.	Alluvial.	. . . . .	Summit of North and South Down vol. iii. p. 311.
	Aqueous.	*a.* Marine.	Paris and London basins.
		b. Freshwater.	Isle of Wight—Auvergne.
	Volcanic.	. . . . .	Oldest volcanic rocks of the Limag d'Auvergne, vol. iii. ch. xix.
	Hypogene.	*a.* Plutonic.	*Concealed ;* foci of Eocene volcano beneath the Limagne d'Auverg
		b. Metamorphic.	*Concealed ;* probably near the sa foci.

TABLE I. *continued.*

Periods.	Formations.		Some of the Localities where the Formations occur.
1. Cretaceous group. F. Table II.	Alluvial.		
	Aqueous.	*a.* Marine.	Wiltshire. North Downs. Flamborough Head.
		b. Freshwater.	
	Volcanic.	. . . . .	Northern flanks of the Pyrenees? near Dax?
	Hypogene.	*a.* Plutonic.	
		b. Metamorphic.	
2. Wealden group. G. Table II.	Alluvial.	. . . . .	Portland ' Dirt-bed.'
	Aqueous.	*a.* Marine.	
		b. Freshwater.	Weald of Surrey, Kent, and Sussex, vol. iii. ch. xxi.
	Volcanic.		
	Hypogene.	*a.* Plutonic.	
		b. Metamorphic.	
3. Oolite group. H. Table II.	Alluvial.		
	Aqueous.	*a.* Marine.	Oxford. Bath. Jura chain.
		b. Freshwater.	
	Volcanic.	. . . . .	Hebrides?
	Hypogene.	*a.* Plutonic.	*Concealed ;* beneath the Hebrides.
		b. Metamorphic.	
4. Lias group. I. Table II.	Alluvial.		
	Aqueous.	*a.* Marine.	Lyme Regis. Whitby. Aberthaw.
		b. Freshwater.	
	Volcanic.	. . . . .	Hebrides?
	Hypogene.	*a.* Plutonic.	
		b. Metamorphic.	Alps? ch. xxvi. p. 371. Valorsine in Savoy?
5. New Red Sandstone group. K. Table II.	Alluvial.		
	Aqueous.	*a.* Marine.	Cheshire. Staffordshire. Vosges. Westphalia (Muschelkalk).
		b. Freshwater.	
	Volcanic.	. . . . .	Near Exeter, Devon.
	Hypogene.	*a.* Plutonic.	*Concealed ;* beneath Devonshire?
		b. Metamorphic.	
6. Carboniferous group. L. Table II.	Alluvial.		
	Aqueous.	*a.* Marine.	Clifton. Dudley. Mendip. [Fife.
		b. Freshwater.	Coal measures of Somersetshire and
	Volcanic.	. . . . .	Forfarshire. Edinburgh. Durham. High Teesdale.
	Hypogene.	*a.* Plutonic.	*Concealed ;* beneath Edinburgh, Northumberland, Durham.
		b. Metamorphic.	Near the Plutonic rocks of the same period.

III. SECONDARY.

DIAGRAM

*Shewing the relative position which the Plutonic and Sedi-
mentary Formations of different ages may occupy ;
(in illustration of* TABLE I.)

No. 91.

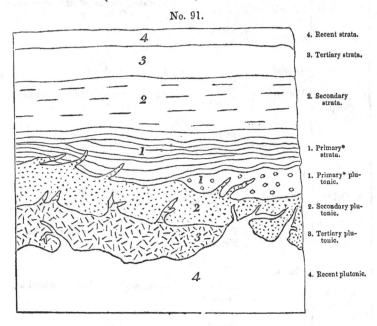

In the above diagram an attempt is made to shew the inverted order in which
the sedimentary and plutonic formations may occur in the earth's crust; subter-
position in the plutonic, like superposition in the sedimentary rocks, being for the
most part characteristic of a newer age. By aid of this illustration, and what we
have said in Chap. 25 and 26, the reader will comprehend why so large a portion
of the plutonic rocks of later periods are concealed, and why the more ancient of
this class have risen nearest to the surface, so as to have been denuded in some
regions and exposed to view.

* The primary formations here mentioned are those, whether stratified or unstratified, which are older than the carboniferous deposits.

TABLE II.

Showing the Order of Superposition, or Chronological Succession, of the principal Sedimentary Deposits or Groups of Strata in Europe.

This Table is referred to in the Glossary, and includes the Secondary Formations alluded to in this Work, but not described in detail.

Periods and Groups.		Names of the principal Members and general Mineral nature of the Formations		Some of the Localities where the Formation occurs.
I. Recent Period.	A	The deposits of this period are for the most part concealed under existing lakes and seas.		
		Consolidated sandy and gravelly beds (*a*), travertin limestones (*b*), calcareous sandstones with broken shells (*c*), coral limestone, consisting of corals, shells, &c. (*d*)		*a.* Delta of the Rhone. *b.* Tivoli, and other parts of Italy. *c.* Shore of island of Guadaloupe. *d.* Coral reefs in Pacific, &c.
II. Tertiary Period.	B Newer Pliocene.	MARINE. *Limestone,* sands, clays, sandstones, conglomerates, marls with gypsum; containing *marine* fossils (*a*).	FRESHWATER. Sands, clays, sandstones, lignites, &c.; containing *land* and *freshwater* fossils (*b*).	*a.* Sicily, Ischia, Morea? *b.* Colle in Tuscany.
	C Older Pliocene.	*Subapennine marl, Subapennine yellow sand,* English 'crag,' and other deposits, as in B, containing *marine* fossils (*a*).	Similar deposits to B; containing *land* and *freshwater* fossils (*b*).	*a.* Subapennine formations, Perpignan, Nice, Norfolk and Suffolk. *b.* Near Sienna, &c.
	D Miocene.	*Faluns of the Loire,* and other deposits of similar mineral composition with B and C, containing *marine* fossils (*a*).	Similar deposits to B and C; containing *land* and *freshwater* fossils (*b*).	*a.* Touraine, Bordeaux, Valley of Bormida, Superga near Turin, Basin of Vienna. *b.* Saucats, twelve miles south of Bordeaux.

TABLE II. *continued.*

Periods and Groups.		Names of the principal Members and general Mineral nature of the Formation.		Some of the Localities where the Formation occurs.
II. TERTIARY PERIOD, *continued.*	E — Eocene.	*Calcaire Grossier* (*a*), plastic clay, sands, sandstones, &c., with *marine* fossils (*b*).	*Calcaire siliceux*—sandstones and conglomerates, red marl, green and white marls, limestone, gypseous marls,—with land and freshwater fossils (*c*).	*a.* Paris basin. *b.* Paris, London, and Hampshire basins, Isle of Wight. *c.* Paris Basin, Isle of Wight, Auvergne, Velay, Cantal.
III. SECONDARY PERIOD.	F — Cretaceous Group.	1. *Maestricht Beds.*—Earthy white limestone with siliceous masses, resembling chalk (marine).		St. Peter's Mount, Maestricht.
		2. *Chalk with flints* (marine).		
		3. Chalk without flints (marine).		North and South Downs, and parts of the intervening Weald of Kent, Surrey, and Sussex. Isle of Wight, coasts of Hampshire and Dorsetshire, Yorkshire, North of Ireland.
		4. *Upper green sand* (marine).—Marly stone, and sand with green particles; layers of calcareous sandstone.		
		5. *Gault* (marine).—Blue clay, with numerous fossils, passing into calcareous marl in the lower parts.		
		6. *Lower green sand* (marine).—Grey, yellowish, and greenish sands, ferruginous sands and sandstones, clays, cherts, and siliceous limestones.		
	G — Wealden Group.	1. *Weald clay* (freshwater).—Clay, for the most part without intermixture of calcareous matter, sometimes including thin beds of sand and shelly limestone.		1, 2. Extensively developed in the central parts of Kent, Surrey, and Sussex. 3. Isle of Purbeck, in Dorsetshire.
		2. *Hastings sands* (freshwater).—Grey, yellow, and reddish-brown sands, sandstones, clays, calcareous grits passing into limestone.		
		3. *Purbeck beds* (freshwater).—Various kinds of limestones and marls.		

TABLE II. *continued.*

Periods and Groups.	Names of the principal Members and general Mineral nature of the Formation.	Some of the Localities where the Formation occurs.
H	1. *Portland beds* (marine).—Coarse shelly limestone, fine-grained white limestone, compact limestone — all more or less of an oolitic structure; beds of cherts.	Isle of Portland, Tisbury in Wiltshire, Aylesbury.
	2. *Kimmeridge clay* (marine).— Blue and greyish-yellow slaty clay, containing gypsum, bituminous slate (Kimmeridge coal).	Near Kimmeridge on coast of Dorsetshire—Sunning Well, near Oxford.
	3. *Coral rag* (marine).—Calcareous shelly freestones, largely oolitic; coarse limestone, full of corals; yellow sands; calcareous siliceous grits.	Headington, near Oxford—Farringdon, in Berkshire—Calne and Steeple Ashton in Wiltshire—Somersetshire.
	4. *Oxford clay* (marine).—Dark blue tenacious clay with septaria, bituminous shale, sandy limestone (Kelloway rock), iron pyrites, gypsum.	New Malton, in Yorkshire—Lincolnshire — Cambridgeshire — Huntingdonshire, and midland counties—abundantly near Oxford—Somersetshire—Dorsetshire.
	5. *Cornbrash* (marine).—Grey or bluish rubbly limestone, separated by layers of clay.	Malmsbury, Atford, Wraxall, Chippenham.
	6. *Forest marble* (marine).—Calcareo-siliceous sand and gritstone; thin fissile beds of limestone, with clay partings; coarse shelly limestone.	Whichwood Forest, Oxfordshire—Frome, south-east of Bath.
	7. *Great oolite* (marine).—White and yellow oolitic calcareous freestone, coarse shelly limestone, layers of clay. Oolitic limestone, with remains of land animals, birds, amphibia, plants, sea-shells (*a*).	Kettering, in Northamptonshire—Bath—Burford, in Oxfordshire—Bradford, in Wiltshire. (*a*) Stonesfield, near Woodstock, Oxfordshire.
	8. *Inferior oolite* (marine).—Fuller's earth, soft freestone, sand with calcareous concretions.	Cotteswold Hills — Dundry Hill, near Bristol.

III. SECONDARY PERIOD, continued. — *Oolite, or Jura Limestone Group.*

Limestones of various qualities, clays, sands, and sandstone. containing the same fossils as those occurring in the series of the oolitic group of England, constitute the main body of the Jura chain of mountains, and cover vast tracts of country in Germany.

TABLE II. *continued.*

Periods and Groups.	Names of the principal Members and general Mineral nature of the Formation.	Some of the Localities where the Formation occurs.
I — Lias Group.	*Lias* (marine).—Blue, white, and yellow earthy limestone, usually in thin beds, interstratified with clay, often slaty and bituminous. Dark blue marl, with a few irregular rubbly limestone beds—sandy marlstone.	Lyme Regis, in Dorsetshire, and in many parts of Somersetshire, Gloucestershire, Warwickshire, Nottinghamshire, and Yorkshire—in Sutherlandshire, the Hebrides, and North of Ireland. In France, and, to a considerable extent, in Germany.
K — New Red Sandstone Group.	1. *Keuper, or variegated marls.*—Red, grey, green, blue, and white marls, sandstones, conglomerates, and shells, containing gypsum and rock-salt.	Neighbourhood of Vosges mountains, and many parts of Wurtemberg and Westphalia, and other parts of Germany.
	2. *Muschelkalk* (marine).—Grey, blue, and blackish limestone, with many fossils, particularly encrinites; siliceous layers and nodules; magnesian limestone, marls of different colours, gypsum, and rock-salt.	Extensively developed in Germany and France. Hitherto no beds in England have been identified with the formation.
	3. *Variegated sandstone.* — Red, white, blue, and green silico-argillaceous sandstone, often micaceous, and containing gypsum and rock-salt.	Nottinghamshire—Yorkshire. It is uncertain whether the variegated sandstone of England belongs to the Keuper formation of Germany, or to the variegated sandstone which lies under the Muschelkalk in Westphalia, Wurtemberg, the Vosges, &c.
	4. *Magnesian limestone* (marine).—Compact shelly limestone, yellow magnesian limestone, marl slate, red marl, and gypsum.	Nottinghamshire, Derbyshire, Yorkshire, Durham, Northumberland. Departments of Saone and Loire. Hartz mountains, Thuringia, Westphalia.
	5. *Red conglomerate.*—Sandstones, conglomerates, sands, and marls.	Neighbourhood of Exeter—Yorkshire—Durham—Westphalia—Wurtemberg—Vosges mountains.

III. SECONDARY PERIOD, *continued.*

TABLE II. *continued.*

Periods and Groups.		Names of the principal Members and general Mineral nature of the Formation.	Some of the Localities where the Formation occurs.
III. SECONDARY PERIOD, *continued.*	Carboniferous Group.	1. *Coal measures* (freshwater ?).—Sandstones, grits, conglomerates, clays with ironstone, shales, and limestone, interstratified with beds of coal.	Northumberland, Durham, Yorkshire, Lancashire, Derbyshire, Staffordshire, Gloucestershire, Somersetshire, South Wales, Valleys of the Forth and Clyde. District of Liege, Westphalia, Silesia, Bohemia, &c.
		2. *Mountain limestone* (marine).—Grey, compact, and crystalline limestone, abounding in lead ore in North of England, and alternating with coal measures in Scotland	Mendip Hills, Somersetshire, Derbyshire, Yorkshire, Lancashire, Westmoreland, Durham, Northumberland, Lanarkshire, Linlithgowshire, many parts of Ireland. North-west of Germany, Belgium, North of France.
		3. *Old red sandstone.*—Coarse and fine siliceous sandstones and conglomerates of various colours, red predominating.	Extensively developed in Shropshire and Herefordshire, Brecknockshire, Dumfriesshire, Forfarshire. Silesia, Bohemia.
		4. *Grauwacke* and *transition limestone* (marine).—Coarse and fine slates, sandstones, and conglomerates—crystalline limestones.	Westmoreland, Cumberland, Wales, Somersetshire, Devonshire, South of Scotland, South of Ireland. North of France, North-west of Germany, &c.

The letter "L" appears at the top of the Carboniferous Group column.

INSTRUCTIONS

FOR USING

M. DESHAYES'S TABLES OF SHELLS,

APPENDIX I.

THE object of these Tables is to give a list, *not* of the *characteristic* shells of the different tertiary formations, of which some figures are given in plates 1, 2, and 3, but to show the connexion of different periods by indicating the shells *common* to two or more periods, or common to some tertiary period and to the *recent* epoch.

The names also of a considerable number of species are given, as being found common to two or more formations of the *same* tertiary period. The localities where the fossil species are met with, and the known habitations of the living species, are also given.

No allusion is made to any *secondary* fossil shells; the word *fossil*, therefore, must always be understood to refer to tertiary formations.

The number of species of recent and fossil shells which were examined and compared in constructing these tables amounted to 7,816, as follows :—

	Living Species.				Fossil Species.
Univalves	3,616	.	.	.	2,098
Bivalves	1,164	.	.	.	938
	4,780				3,036

Of these 3,036 fossil species, 426 were identified with individuals found among the 4,780 living species; 123 of them are only known in a fossil state, but are mentioned as being common to more than one tertiary *period;* and 233 are enumerated by name, although not common to two tertiary periods, or to some tertiary period and the recent epoch, merely because they have been found in two or more formations of the *same period.* Thus the number of fossil species named in the tables amounts to 782, consisting of—

Species found both living and fossil	426
Species fossil only, but common to more than one tertiary period	123
Species fossil only, and named merely as found in two or more formations of the same period	233
	782

A few will be found without specific names, because they have not yet been described or named by any authors.

The tables are continuous from p. 2 to p. 45, and the description of each species extends across two pages.

The following examples will best illustrate the object of the tables. If we take the first genus, Aspergillum (p. 2), we find that—

Column 1 gives the name of the genus.

„ 2 shows that four living species of the genus are known to M. Deshayes.

„ 3 that he has seen one fossil species.

„ 4 is left blank, because the single fossil species has not yet been identified with any living species.

„ 5 is also blank, because the fossil species is only known in one period or formation.

„ 6 is also blank, because the fossil species not having been identified with a living species, it was unnecessary to mention the habitation of any of the four living species.

The columns of the three periods are left blank, because the fossil species has not been found in more than one period. In the column of localities on the right of the right-hand page, in the subdivision headed Bordeaux, the figure 1 denotes that one species of fossil Aspergillum has been found in that locality.

To select another example: if we take the genus Solen (p. 2), we find that—

Column 2 shows that twenty-six living species of the genus are known to M. Deshayes.

„ 3 that he has seen nineteen fossil species of the genus.

Column 4 gives the name of the species Solen *vagina,* because
that species is found both living and fossil.

,, 5 is left blank, because the names of those species only are
placed in this column which have no living analogues,
but are found in more than one period, or in more than
one *formation* of the *same period.* [Thus, in the next
line, Solen siliquarius has no living analogue, but it
occurs in two formations of the Miocene period, viz. at
Bordeaux and in Touraine.]

,, 6 shows that the *living* species of Solen vagina inhabits the
European Ocean and Mediterranean.

The two asterisks in the column of the Pliocene period show that the
species is found in two formations of that period, viz. in the Subapennine
hills and the English crag.

The asterisk in the column of the Miocene period shows that this species is found in the basin of Vienna.

The word Baden in the next column indicates that the species is also
found fossil in that locality.

The column of the Eocene period is blank, because the shell has not
been found in any formation belonging to that period.

The figures in the column of localities will be understood by what we
said above. In summing up these figures it will be found that they
amount to thirty-one, whereas it is stated, in the third column of the left-
hand page, that only nineteen fossil species have been found. The disagreement arises from this—that the same species occur in more than
one locality, and thus come to be counted more than once in the column
of localities.

N. B. In some cases, before the totals of the species in the columns of
localities can tally with the figures in the third column, the species enumerated in the supplementary table of localities, p. 46, must be taken into
account.

A note of interrogation added to the asterisk (*?) indicates a doubt
as to the correct identification of the shell, either because the shell is a
variety which has a somewhat distant analogy to the recognized type of

the species, or because the specimen examined was in rather an imperfect state.

The specific names of the tertiary fossil shells which have been found by M. Deshayes to belong to one period only, and for which he has not yet discovered any living analogues, are *not* given, as their enumeration would have required more space than could be allotted to such a subject in a treatise on Geology; but their aggregate number is included in the subdivisions of the column in the right-hand page headed 'No. of species in each genus in the following localities,' and in the supplementary table, p. 46.

APPENDIX I.

TABLES OF FOSSIL SHELLS,

BY

MONSᴿ. G. P. DESHAYES,

Member of the Geological Society of Paris, &c.

N.B. For a full explanation of the object of these Tables, and instructions
as to the manner of using them, see the preceding four pages.

GENERA.	No. of Species Living.	Fossil (tert.)	Species found both living and fossil (tertiary).	Species found fossil in more than one tertiary formation.	Habitations of living Species.	PLIOCENE PERIOD. Sicily.	Italy, Subap.	English Crag.	Various Localities.
Aspergillum	4	1	. . . . .	. . . . .	. . . . .	..	..	..	. . . .
Clavagella..	2	7	aperta . . .	. . . . .	{ Mediterr. Indian Ocean	*	..	..	. . . .
———	..	..	. . . . .	coronata . .	. . . . . .	..	..	..	. . . .
———	..	..	. . . . .	bacillaris . .	. . . . . .	*	..	..	. . . .
Septaria . .	1								
Teredina . .	..	2	. . . . .	personata . .	. . . . . .	..	..	..	. . . .
Teredo . . .	5	5	. . . . .	new species .	. . . . . .	..	..	..	. . . .
Pholas . . .	15	9	candida . .	. . . . .	Temp. Europe	..	..	*?	. . . .
Fistulana. .	5	7	gigantea . .	. . . . .	Indian Ocean	..	..	..	. . . .
———	..	..	hians . . .	. . . . .	Mediterr. . . .	..	*	..	. . . .
———	..	..	. . . . .	ampullaria	European Ocean	..			
Solen . . .	26	19	Vagina. . .	. . . . .	{ European Ocean Mediterr.	..	*	*	. . . .
———	..	..	. . . . .	siliquarius .	. . . . .	..	..	..	. . . .
———	..	..	Legumen .	. . . . .	Mediterr. . . .	..	*	..	. . . .
———	..	..	coarctatus .	. . . . .	ibid.	*	*	..	. . . .
———	..	..	strigilatus .	. . . . .	ibid.	..	*	..	. . . .
———	..	..	candidus . .	. . . . .	ibid.	*	*	..	. . . .
———	..	..	Siliqua. . .	. . . . .	ibid.	*	..	..	. . . .
Pholadomya	1	1	. . . . .	. . . . .	. . . . . .	..	..	..	. . . .
Glycimeris .	1								
Panopœa . .	2	3	Aldrovandi .	. . . . .	Mediterr. . . .	*	*	..	Morea . . .
Mya. . . .	4	5	truncata . .	. . . . .	{ Temp. Europ. and	..	..	..	Uddevalla .
———	..	..	arenaria . .	. . . . .	Northern Ocean	..	..	*	. . . .
———	..	..	Tugon . . .	. . . . .	Senegal . . .	..	..	..	Morea . . .
Thracia . .	2	4	corbuloides .	. . . . .	Mediterr. . . .	*	..	..	. . . .
———	..	..	pubescens	. . . . .	ibid.	*	..	..	. . . .
Hemicyclostera	1	2	new species .	. . . . .	{ Europ. Ocean & Mediterr.	*	..	..	. . . .
Lutraria . .	11	6	elliptica . .	. . . . .	ibid. ibid. .	*	*	..	. . . .
———	..	..	rugosa . . .	. . . . .	Indian Ocean .	..	*	..	. . . .
———	..	..	new species .	. . . . .	Senegal . . .	..	..	..	. . . .
Mactra. . .	32	14	albina . . .	. . . . .	ibid.	..	..	..	. . . .
———	..	..	solida . . .	. . . . .	European Ocean	..	..	*	. . . .
———	..	..	crassatella ?	. . . . .	ibid.	..	..	*	. . . .
———	..	..	triangula . .	. . . . .	Mediterr. . . .	..	*	..	{ Morea Perpignan
Mesodesma .	7								
Erycina . .	3	23	. . . . .	. . . . .	. . . . . .	..	..	..	. . . .
Crassatella	9	24	. . . . .	lamellosa . .	. . . . . .	..	..	..	. . . .
———			. . . . .	tumida . .	. . . . . .	..	..	..	. . . .
Ungulina . .	1								
Solemya . .	2								
Amphidesma	3	1	. . . . .	. . . . .	. . . . . .	..	..	..	. . . .

Bordeaux, Dax.	Touraine.	Turin.	Vienna.	Various Localities.	Paris.	London.	Various Localities.	Sicily.	Italy, Subap.	English Crag.	Baden.	Bordeaux.	Dax.	Touraine.	Turin.	Ronca.	Vienna.	Angers.	Paris.	London.	Valognes.	Belgium.
												1										
					*	*		2	1			1							4	1		
							Pauliac															
					*	*				1									1	1		
					*		Belgium			1									1			4
									1	1			2	2					1	3		
*					*																	
*									1				2						1	6	1	
					*	*																
			*	Baden																		
*	*																					
*								4	5	1		4	2	2				2	9		2	
*	*		*																			
																						1
								1	1			1							1			
										2		2	1									
*																						
								2				2										
								1					1									
								1	2			2	1	1								
*																						
*																						
									1	2		3	3				1	1	2		1	1
		(0)																				
								1				2	3							14		
					*	*	Valognes						1	1			1	2	2	14	4	5
				Ronca	*																	
								1														

GENERA.	Living.	Fossil (cert.)	Species found both living and fossil (tertiary).	Species found fossil in more than one tertiary formation.	Habitations of living Species.	PLIOCENE PERIOD.			
						Sicily.	Italy, Subap.	English Crag.	Various Localities.
Corbula . .	10	35	nucleus . .	. . . .	Mediterr. European Seas	*	*	. .	. . . .
——. .	. .	. .	. . . .	complanata	. . . . .	. .	. .	*	. . . .
——. .	. .	. .	. . . .	striata . . .	. . . . .	. .	. .	. .	. . . .
——. .	. .	. .	. . . .	new species .	. . . . .	. .	. .	. .	. . . .
Pandora . .	7	3	new species .	. . . .	Mediterr. . .	*	. .	. .	. . . .
——. .	. .	. .	rostrata . .	. . . .	Mediterr. European Ocean	. .	*	. .	. . . .
Saxicava . .	5	11	minuta . .	. . . .	ibid. . .	*	. .	. .	. . . .
——.	. .	. .	pholadis . .	. . . .	ibid.	*	. .	. .	. . . .
Petricola . .	13	10	ochroleuca .	. . . .	Mediterr.	. .	. .	. .	. . . .
——.	. .	. .	lamellosa . .	. . . .	ibid.	*	. .	. .	. . . .
——.	. .	. .	striata . . .	. . . .	European Ocean	. .	. .	. .	. . . .
Venerupis .	8	6	Irus	. . . .	ibid. Mediterr.	*	. .	. .	. . . .
Sanguinolaria	2	2	. . . . .	. . . .	. . . . .	. .	. .	. .	. . . .
Psammobia .	18	4	vespertina .	. . . .	Mediterr. European Ocean	*	. .	. .	. . . .
——.	. .	. .	muricata . .	. . . .	ibid.	. .	*	. .	. . . .
Psammotæa	8	1	. . . . .	. . . .	. . . . .	. .	. .	. .	. . . .
Tellina . .	68	54	strigosa . .	. . . .	Senegal . . .	. .	. .	. .	. . . .
——. .	. .	. .	planata . .	. . . .	Mediterr. . .	. .	*	. .	. . . .
——. .	. .	. .	lacunosa . .	. . . .	African Ocean	. .	*	. .	. . . .
——. .	. .	. .	Oudardii . .	. . . .	Mediterr. . .	*	. .	. .	. . . .
——. .	. .	. .	tenuis . .	. . . .	Mediterr. European Ocean	*	. .	. .	. . . .
——. .	. .	. .	pulchella . .	. . . .	ibid. . .	*	. .	. .	. . . .
——. .	. .	. .	Lantivii . .	. . . .	ibid.	*	. .	. .	. . . .
——. .	. .	. .	serrata . .	. . . .	ibid.	*	*	. .	. . . .
——. .	. .	. .	exilis . . .	. . . .	Unknown . . .	. .	*	. .	. . . .
——. .	. .	. .	new species .	. . . .	Senegal . . .	. .	. .	. .	. . . .
——. .	. .	. .	carnaria . .	. . . .	European Ocean Mediterr.	. .	. .	. .	. . . .
——. .	. .	. .	. . . .	subrotunda .	. . . . .	. .	. .	. .	. . . .
——. .	. .	. .	. . . .	new species .	. . . . .	. .	. .	. .	. . . .
——. .	. .	. .	. . . .	elliptica . .	. . . . .	. .	*	. .	. . . .
——. .	. .	. .	. . . .	bipartita . .	. . . . .	. .	. .	. .	. . . .
——. .	. .	. .	donacina .	. . . .	Mediterr. . .	*	. .	. .	. . . .
Tellinides. .	1								
Corbis . .	2	2	. . . . .	pectunculus		. .	. .	. .	. . . .
Lucina . .	20	59	tigrina . . .	. . . .	Indian Ocean, Senegal	. .	. .	. .	. . . .
——. .	. .	. .	punctata .	. . . .	ibid.	. .	. .	. .	. . . .
——. .	. .	. .	columbella .	. . . .	Senegal . . .	. .	. .	. .	. . . .
——. .	. .	. .	divaricata .	. . . .	Almost all Seas	*	*	. .	Morea, Baden, Perpignan
——. .	. .	. .	lactea . . .	. . . .	Mediterr. . .	*	*	. .	. . . .
——. .	. .	. .	gibbosula . .	. . . .	Unknown . .	*	*	. .	. . . .
——. .	. .	. .	squamosa . .	. . . .	Mediterr. . .	*	. .	. .	. . . .
——. .	. .	. .	radula . . .	. . . .	ibid.	*	. .	. .	. . . .

MIOCENE PERIOD.					EOCENE PERIOD.			No. of Species in each Genus in the following Localities.														
Bordeaux, Dax.	Touraine.	Turin.	Vienna.	Various Localities.	Paris.	London.	Various Localities.	Sicily.	Italy, Subap.	English Crag.	Baden.	Bordeaux.	Dax.	Touraine.	Turin.	Ronca.	Vienna.	Angers.	Paris.	London.	Valognes.	Belgium.
··	··	··	··		··	··																
*	*	··	*		*	··		1	1	2	2	4	4	4	··	··	2	··	21	5	··	3
					*	*																
*	*	··	*	Transylvania	··	··																
··	··	··	··		··	··	··	1	1	··	··	··	··	··	··	··	··	··	1			
··	··	··	··		··	··																
··	··	··	··		··	··		5	··	··	··	1	··	1	··	··	··	··	5			
··	··	··	··		··	··																
*	*	··	··		··	··		··	4	··	··	2	2	1	··	··	··	··	2			
*?	··	··	··		··	··																
··	··	··	··		··	··		1	3	··	··	1	··	··	··	··	··	··	2			
··	··	··	··		··	··		··	··	··	··	··	1	··	··	··	··	··	1			
··	··	··	··		··	··		1	1	··	··	1	1	1	··	··	··	··	2			
*	*	··	*		··	··																
··	··	··	··		··	··		··	··	··	··	··	··	··	··	··	··	··	1			
*	··	··	··	Volhynia	··	··																
··	··	··	··		··	··																
*	··	··	··		··	··																
··	··	··	··		··	··																
*	··	··	··		··	··																
··	··	··	··		··	··																
··	··	··	··		··	··		6	9	4	··	8	6	2	··	··	··	··	18	4	··	6
··	··	··	··		··	··																
*	··	··	··		··	··																
*	··	··	··		··	··																
··	··	··	··		*	*																
*	*	··	··		··	··																
*	··	··	··		··	··																
*	··	··	··	Volhynia	··	··																
··	··	··	··	Ronca	*	··		··	··	··	··	··	··	··	··	··	··	··	2			
*	··	··	··		··	··																
*	··	··	··		··	··																
*	*	··	*	{ Angers / Volhynia	··	··																
*	*	·?	*	Angers	*	*	Valognes	8	6	··	1	12	9	5	··	··	1	1	31	1	··	7
··	··	··	··		··	··																
*	··	··	··		*	··																
··	··	··	··		··	··																

GENERA.	No. of Species Living.	No. of Species Fossil (tert.)	Species found both living and fossil (tertiary).	Species found fossil in more than one tertiary formation.	Habitations of living Species.	PLIOCENE PERIOD. Sicily.	Italy, Subap.	English Crag.	Various Localities.
Lucina, *cont.*	..	..	amphidesmoides	.....	Mediterr.	*	*	..	.....
———	..	..	.....	scopulorum	.....	..	..	..	.....
———	..	..	.....	hiatelloides	.....	..	*	*	.....
———	..	..	.....	lupinus	.....	*	*	..	Ischia .....
Donax	29	15	elongata	.....	Senegal	..	..	..	.....
———	..	..	.....	transversa	.....	..	..	..	.....
———	..	..	.....	new species	.....	..	..	..	.....
Capsa	2								
Astarte	3	19	incrassata	.....	Mediterr.	..	*	..	.....
———	..	..	new species	.....	ibid.	*	*	..	.....
———	..	..	.....	new species	.....	*	*	..	.....
———	..	..	Danmoniensis	.....	Temp. Europe	..	..	*?	.....
———	..	..	.....	scalaris	.....	..	..	..	.....
Cyclas	13	2	.....	.....	.....	..	..	..	.....
Cyrena	14	25	.....	Gravesii	.....	..	..	..	.....
———	..	..	.....	cuneiformis	.....	..	..	..	.....
———	..	..	.....	antiqua	.....	..	..	..	.....
———	..	..	.....	Brongniarti	.....	..	..	..	.....
———	..	..	.....	semistriata	.....	..	..	..	.....
Galathea	1								
Cyprina	2	7	Islandica	.....	North Seas	..	..	*	.....
———	..	..	.....	scutellaria	.....	..	..	..	.....
———	..	..	.....	Pedemontana	.....	..	*	..	.....
———	..	..	.....	gigas	.....	..	*	..	Perpignan
Cytheræa	85	59	Erycina	.....	Indian Ocean	..	*	..	.....
———	..	..	Chione	.....	Senegal, Medit.	*	*	..	Morea, Perpignan
———	..	..	nitidula	.....	Unknown	..	..	..	.....
———	..	..	citrina	.....	New Holland	..	..	..	.....
———	..	..	exoleta	.....	{ Mediterr. European Ocean	*	*	*	.....
———	..	..	concentrica	.....	Indian Ocean	..	*	..	.....
———	..	..	lincta	.....	Medit. Senegal	*	*	..	.....
———	..	..	rufescens	.....	Mediterr.	..	*	..	.....
———	..	..	multilamella	.....	Indian Ocean	*	*	..	.....
———	..	..	.....	elegans	.....	..	..	..	.....
———	..	..	.....	deltoidea	.....	..	..	..	.....
———	..	..	.....	suberycinoides	.....	..	..	..	.....
———	..	..	Venetiana	.....	Mediterr.	*	..	..	.....
Venus	101	43	verrucosa	.....	{ Mediterr. European Ocean	*	*	ab p	Ischia .....
———	..	..	plicata	.....	Indian Ocean	..	*	..	{ Morea, Perpignan
———	..	..	gallina	.....	South Europ. Oc.	..	*	..	
———	..	..	decussata	.....	{ Europ. Oc. Aust. Seas, Lam.	..	..	..	.....

(7)

MIOCENE PERIOD.					EOCENE PERIOD.			No. of Species in each Genus in the following Localities.														
Bordeaux, Dax.	Touraine.	Turin.	Vienna.	Various Localities.	Paris.	London.	Various Localities.	Sicily.	Italy, Subap.	English Crag.	Baden.	Bordeaux.	Dax.	Touraine.	Turin.	Ronca.	Vienna.	Angers.	Paris.	London.	Valognes.	Belgium.
..	..	..	..		..	..																
*	*	..	...	Volhynia	..	..		8	6	..	1	12	9	5	..	..	1	1	31	1	7	
*	*	..	..		..	..																
..	..	..	..		..	..																
*?	..	..	..		..	..																
*	*	..	*		..	..		..	..	..	..	4	3	3	..	..	2	2	7			
..	*	..	...	Angers	..	..																
..	..	..	..		..	..																
..	..	..	..		..	..																
..	..	..	..		..	..		2	2	5	..	1	1	2	..	..	..	3	..	2?	..	4
..	..	..	..		..	..																
..	*	..	..	Angers	..	..																
..	..	..	..		..	..		..	..	..	..	..	..	..	..	..	..	..	1			
..	..	..	..		*	*																
..	..	..	..		*	*																
..	..	..	..		*	*		..	1	..	..	2	2	..	1	1	1	..	13	4	..	2
*	..	..	*		..	..																
..	..	..	..		..	..	{ Mayence, Belgium															
..	..	..	..		..	..																
..	..	..	..		*	..	Belgium	1	3	1	..	1	1	..	1	..	..	..	1	..	..	1
*	..	..	*		..	..																
..	..	..	..		..	..																
*	..	..	*	..	Volhynia	..	..															
*?	*?	..	..	*?		..	..															
..	..	..	..		*	..																
..	..	..	..		*?	*?																
..	..	..	..		..	..																
..	..	..	..		..	..		5	8	2	..	11	18	7	2	1	3	1	22	5	11	
*	*	..	..		..	..																
..	..	*?	..		..	..																
..	..	..	*		..	..																
..	..	..	..		*	*	Valognes															
..	..	..	..		*	*																
*	..	..	..		*	..																
..	..	..	..		..	..																
..	..	..	..	Salles	..	..		6	8	3	1	11	5	8	..	..	3	3	9	4	2	
..	..	..	*		..	..																
..	..	..	..		..	..																
..	..	..	..		*	..																

GENERA.	No. of Species — Living.	No. of Species — Fossil (tert.)	Species found both living and fossil (tertiary).	Species found fossil in more than one tertiary formation.	Habitations of living Species.	PLIOCENE PERIOD. — Sicily.	Italy, Subap.	English Crag.	Various Localities.
Venus, *cont.*			radiata . . .	. . . . .	Mediterr. . .	*	✻		. . . . .
——— . . .			Brongniarti .	. . . . .	ibid. . . .	*			. . . . .
——— . . .			dysera . . .	. . . . . {	Indian Ocean, Mediterr.	*	✻		. . . . .
——— . . .			. . . . .	Casinoides .	. . . . .				. . . . .
——— . . .			. . . . .	vetula . . .	. . . . .				. . . . .
——— . . .			. . . . .	new species .	. . . . .				. . . . .
——— . . .			. . . . .	rotundata . .	. . . . .		✻		. . . . .
——— . . .			geographica	. . . .	Mediterr. . . .	*			. . . . .
——— . . .			Paphia . . .	. . . .	Indian Ocean . .				. . . . .
Venericardia et Cardita }	25	50	sulcata . .	. . . .	Mediterr. . . .	*	✻	{	Salles, Perpignan
——— . .			Ajar	. . . .	Senegal				. . . . .
——— . .			trapezia . . .	. . . . {	Mediterr. African Ocean	✻			. . . . .
——— . .			squamosa . .	. . . .	Mediterr. . . .	✻			. . . . .
——— . .			crassa . . .	. . . . {	Indian, Afric. & Australian Ocean				. . . . .
——— . .			. . . . .	Jouanetti .	. . . . .				. . . . .
——— . .			. . . . .	hippopea . .	. . . . .				. . . . .
——— . .			. . . . .	planicosta .	. . . . .				. . . . .
——— . .			. . . . .	Coravium .	. . . . .				. . . . .
——— . .			. . . . .	acuticosta .	. . . . .				. . . . .
——— . .			intermedia .	. . . . .	New Holland . .		*		. . . . .
——— . .			. . . . .	angusticosta	. . . . .				. . . . .
——— . .			. . . . .	imbricata . .	. . . . .				. . . . .
——— . .			. . . . .	new species .	. . . . .		*		. . . . .
Cardium . .	53	39	ringens . .	. . . . .	Senegal				. . . . .
——— . .			ciliare . . .	. . . . .	European Ocean	*	*		. . . . .
——— . .			echinatum .	. . . . .	ibid. Senegal	✻	*		. . . . .
——— . .			sulcatum . .	. . . . .	European Ocean	*	✻	{	Salles, Morea, Perpignan
——— . .			edule . . .	. . . . .	ibid. Medit. .	*	*	*	Morea . . .
——— . .			tuberculatum	. . . . .	ibid.		*		. . . . .
——— . .			. . . . .	multicostatum	. . . . .		*		. . . . .
——— . .			planatum. .	. . . . .	Mediterr. . . .		*		. . . . .
——— . .			. . . . .	discrepans .	. . . . .				. . . . .
——— . .			. . . . .	hians . . .	. . . . .		✻	{	Perpignan, Morea . .
——— . .			. . . . .	semigranulatum	. . . . .				. . . . .
——— . .			. . . . .	porulosum .	. . . . .				. . . . .
——— . .			. . . . .	obliquum . .	. . . . .				. . . . .
——— . .			. . . . .	Lima . . .	. . . . .				. . . . .
——— . .			. . . . .	verrucosum .	. . . . .				. . . . .
Cypricardia .	4	7	coralliophaga	. . . . .	Mediterr. . .	✻	*		. . . . .
——— . .			new species .	. . . . .	ibid.	*	✻		. . . . .
——— . .			. . . . .	affinis . . .	. . . . .				. . . . .

MIOCENE PERIOD.					EOCENE PERIOD.			No. of Species in each Genus in the following Localities.														
Bordeaux, Dax.	Touraine.	Turin.	Vienna.	Various Localities.	Paris.	London.	Various Localities.	Sicily.	Italy, Subap.	English Crag.	Baden.	Bordeaux.	Dax.	Touraine.	Turin.	Ronca.	Vienna.	Angers.	Paris.	London.	Valognes.	Belgium.
..	..	..	..		..	..																
*	*	..	..	{ Angers Volhynia	..	..																
*	*	..	*		..	..																
*	*	..	..		..	..		6	8	0	1	11	5	8	..	..	2	3	9	4	..	2
*	*	..	..		..	..																
..	..	..	..		..	..																
..	..	..	*		..	..																
..	..	..	..		..	..																
*	*	..	*	Moravia	..	..																
*	*	..	*		..	..																
..	..	..	..		..	..																
*	*	..	..		*	..																
*	*	..	*	Moravia	..	..		3	7	5	..	6	6	5	3	1	4	5	16	5	8	2
*	*	..	..		..	..																
..	..	..	..		*	*	{ Valognes, Belgium															
..	..	..	..		*	*	Valognes															
..	..	..	..		*	*	ibid.															
..	..	..	..		..	..																
..	..	..	..		*	*	{ Valognes, Belgium															
..	..	..	..		*	..	Castelgomberto															
..	..	..	*		..	..																
*?	..	..	*?		..	..																
..	..	..	..		..	..																
*	*?	..	*	Volhynia	..	..																
..	..	..	..		..	..																
..	..	..	..		..	..																
..	..	..	..		..	..																
*	*	..	..		..	..		4	11	3	..	6	7	3	..	..	5	..	17	6	7	1
..	..	..	..		..	..																
*	*	..	..		..	..																
..	..	..	..		..	..																
..	..	..	..		*	*																
..	..	..	..		*	*	Valognes															
..	..	..	..		*	*																
..	..	..	..		*	*?																
..	..	..	..		*	..	Castelgomberto															
..	..	..	..		..	..																
..	..	..	..		..	..		2	2	..	..	1	2	1	..	1	..	..	2			
*	*	..	..		..	..																

GENERA.	No. of Species		Species found both living and fossil (tertiary).	Species found fossil in more than one tertiary formation.	Habitations of living Species.	PLIOCENE PERIOD.			
	Living.	Fossil (tert.)				Sicily.	Italy, Subap.	English Crag.	Various Localities.
Hiatella . .	1								
Isocardia . .	2	3	cor	. . . . .	European Ocean	*	*	*?	Maryland ? .
—. . . .	. .	. .	. . . . .	new species .	. . . . .	. .	. .	. .	. . . .
Cucullæa . .	1	2	. . . . .	. . . . .	. . . . .	. .	. .	. .	. . . .
Arca . . .	43	54	semitorta .	. . . . .	New Holland, L. Mediterr.	. .	. .	. .	Salles ?
—. . . .			Noæ . . .		Indian Ocean	*	*	. .	. . . .
—. . . .			tetragona . .		ibid. ibid.	*	*	. .	. . . .
—. . . .			umbonata .		Jamaica, Senegal	. .	. .	. .	. . . .
—. . . .			barbata . .		European Ocean, Senegal	*	*	. .	Perpignan
—. . . .			Magellanica		Magellan . .	. .	. .	. .	. . . .
—. . . .			Helbingii . .		Afr. Oc. Senegal	. .	. .	. .	Perpignan, Morea
—. . . .			antiquata . .		Ind. & Afr. Oc. Mediterr.	*	*	. .	
—. . . .			rhombea . .		Indian Ocean .	. .	. .	. .	. . . .
—. . . .			clathrata .		Unknown . .	. .	. .	. .	. . . .
—. . . .			new species .		Mediterr. . .	*			
—. . . .			Gaymardi .		ibid.	*			
—. . . .			Quoyi . . .		ibid.	*			
—. . . .				new species .	. . . . .	. .	. .	. .	
—. . . .				biangula . .					
—. . . .				punctifera .					
—. . . .				quadrilatera .					
—. . . .				barbatula .					
Pectunculus .	19	27		new species .					
—. .			glycimeris .	. . . . .	Mediter. Europ. Ocean	*	*	*	Perpignan, Morea
—. .			pilosus . .		ibid. ibid. Northern Europe	*	*	*	
—. .			violacescens.		Mediterr. . . .	*	*	. .	
—. .				pulvinatus					
—. .				terebratularis					
—. .				angusticostatus					
—. .				cor					
—. .				new species .					
—. .			nummarius .	. . . . .	Mediterr. . .	. .	*	. .	
—. .				dispar . . .					
Nucula . . .	7	23	margaritacea	. . . . .	European Ocean	*	*	. .	. . . .
—. . .			Pella . . .	. . . . .	Mediterr. . .	*	*		
—. . .			emarginata .	. . . . .	ibid.	*	*		
—. . .			new species .	. . . . .	ibid.	*	*		
—. .				ovata . .					
Trigonia . .	1								
Castalia . .	1								
Unio . . .	65	2	. . . . .	. . . . .	. . . . .	. .	. .	. .	. . . .

MIOCENE PERIOD.					EOCENE PERIOD.			No. of Species in each Genus in the following Localities.														
Bordeaux, Dax.	Touraine.	Turin.	Vienna.	Various Localities.	Paris.	London.	Various Localities.	Sicily.	Italy, Subap.	English Crag.	Baden.	Bordeaux.	Dax.	Touraine.	Turin.	Ronca.	Vienna.	Angers.	Paris.	London.	Volognes.	Belgium.
..	..	..	*		..	*		1	1	1?	..	1	1	1	..	..	1	..	1	1		
*	*	..	..		..	..																
..	..	..	..		..	..		..	..	..	..	..	..	..	..	..	..	..	2	..		
..	..	..	..		..	..																
..	..	..	..		..	..																
..	..	..	*		..	..																
*	*	..	..		..	..																
*	*	..	..		..	..																
..	..	..	..		*??	..																
*	*	..	..	Angers	*	..	Valognes															
*	*	..	*	Angers, Nantes	..	..																
*	*	..	..		..	..		9	6	..	..	9	9	0	..	..	3	7	23	7	6	
*	*	..	..	Angers	..	..																
..	..	..	..		..	..																
..	..	..	..		..	..																
..	..	..	..		..	..																
*	*	..	*	Podolia	..	..																
..	..	..	..		*	*	ibid.															
..	..	..	..		*	*																
..	..	..	..		*	*	ibid.															
..	..	..	..		*	*	ibid.															
*	..	..	*	Angers	..	..																
*	*	*	*	Volhynia	..	..																
..	..	..	..		..	..																
..	..	..	..		..	..																
..	..	..	..		*	*	Valognes	3	8	2	..	4	3	2	1	1	4	3	9	5	3	2
..	..	..	..		*	..	Belgium															
..	..	..	..		*	*																
*	*	..	*	Angers	..	..																
*	..	..	*	ibid.	..	..																
..	..	..	..		..	..																
..	..	..	..		*	*	Valognes															
*	..	..	*	Volhynia	*	*	Valognes															
..	..	..	*		..	..																
*	*	..	..		..	..		4	4	2	..	3	3	1	..	..	..	..	6	5	3	2
..	..	..	..		..	..																
..	..	..	..		*	*	Valognes															
..	..	..	..		..	..		..	1	..	..	..	..	..	..	..	..	..	..	1		

GENERA.	No. of Species Living.	No. of Species Fossil (tert.)	Species found both living and fossil (tertiary).	Species found fossil in more than one tertiary formation.	Habitations of living Species.	PLIOCENE PERIOD. Sicily.	Italy, Subap.	English Crag.	Various Localities.
Anodonta	19	1	. . . .	. . . .	. . . . . .	..	..	..	. . . .
Hyria . . .	3	..							
Iridina. . .	3	..							
Chama. . .	15	20	gryphoides .	. . . .	Mediterr. . .	*	*	..	
——— . .			crenulata? .	. . . .	Senegal . .	..	..	..	
——— . .			sinistrorsa .	. . . .	Mediterr. . .	*	*	..	Morea . . .
——— . .			new species .	. . . .	ibid.	*	*	..	
——— . .				echinulata .	. . . . . .	..	*	..	
——— . .				rustica . .	. . . . . .	..	..	..	
——— . .				lamellosa .	. . . . . .	..	..	..	
Etheria . .	4								
Tridacna . .	7	2							
Hippopus. .	1								
Modiola . .	29	21	barbata . .	. . . .	Mediterr. . .	..	*	..	
——— . .			discrepans .	. . . .	ib. New Holland	..	..	..	
——— . .			lithophaga .	. . . .	{ Mediterr. Indian & African Ocean	..	*	..	
——— . .			. . . .	argentina .	. . . . . .	..	..	..	
Mytilus . .	42	15	Chemnitzii .	. . . .	Riv. of Germany	..	..	..	{ Baden, Central Austria
——— . .			edulis . . .	. . . .	European Ocean	..	..	*	. . . .
——— . .			. . . .	new species .	. . . . . .	..	*	..	. . . .
——— . .			. . . .	Brardii . .	. . . . . .	..	..	..	
Pinna . . .	16	3	nobilis . .	. . . .	Mediterr. . .	*?	*?	..	Morea . .
——— . .			. . . .	margaritacea	. . . . . .	..	..	..	
Crenatula .	7								
Perna . . .	10	4	. . . .	. . . .	. . . . . .	..	..	..	
Malleus . .	6								
Avicula et Meleagrina }	30	5	. . . .	. . . .	. . . . . .	..	..	..	
Lima . . .	8	13	inflata . .	. . . .	Mediterr. . .	*	*	..	
——— . .			squamosa .	. . . .	ibid. Afr. Ocean	*	*?	..	
——— . .			linguatula .	. . . .	{ Australian Seas Mediterr.	*	..	..	
——— . .			nivea . . .	. . . .	Mediterr. . .	*	*	..	
——— . .			. . . .	plicata . .	. . . . . .	..	..	..	
Pecten . . .	67	60	Jacobæus .	. . . .	European Ocean	*	*	..	Perpignan .
——— . .			Laurentii .	. . . .	{ American Seas Mediterr.	..	*?	..	Corsica ? .
——— . .			pleuronectes .	. . . .	Indian Ocean .	*	*	..	
——— . .			opercularis .	. . . .	European Ocean	*	*	*	{ Salles, Perpignan, Morea
——— . .			inflexus . .	. . . .	Mediterr. . .	*	..	..	
——— . .			varius . .	. . . .	European Ocean	*	*	..	Morea . .
——— . .			ornatus . .	. . . .	Mediterr. . .	*	..	..	. . . .

MIOCENE PERIOD.					EOCENE PERIOD.			No. of Species in each Genus in the following Localities.															
Bordeaux, Dax.	Touraine.	Turin.	Vienna.	Various Localities.	Paris.	London.	Various Localities.	Sicily.	Italy, Subap.	English Crag.	Baden.	Bordeaux.	Dax.	Touraine.	Turin.	Ronca.	Vienna.	Angers.	Paris.	London.	Valognes.	Belgium.
..	..	..	..		..	..		..	1													
..	..	..	..		..	..																
*?	..	..	*		..	..																
..	..	..	..		..	..		4	4	..	..	1	3	3	..	..	2	2	9	1		2
*	*	..	*	Angers	..	..																
..	..	..	..		*	*																
..	..	..	..		*	..	Castelgomberto															
..	..	..	..		..	..																
..	..	..	..		*?	..		1	3	..	1	1	2	..	..	..	..	..	1	16		
*	..	..	..		*	..																
*	..	..	..		*	..																
..	..	..	..		..	..																
..	..	..	..		..	..		..	2	1	2	2	2	1	3	..	3	..	2	..		1
*	..	..	..		..	..																
*	*	*	*		..	..																
..	..	..	..		..	..		1	1	..	..	1	..	..	..	..	..	..	1	..	..	1
..	..	..	..		*	..	Belgium															
..	..	..	..		..	..		..	1	..	..	..	..	..	..	..	..	..	1	..	1	
..	..	..	..		..	..		..	..	..	..	..	1	..	..	..	..	..	4			
*	..	..	..		..	..																
*	*	..	..		..	..																
..	..	..	..		..	..		4	5	..	..	1	2	1	..	..	..	2	6	..	1	
..	..	..	..		..	..																
*	*	..	..	Angers	*	..																
..	..	..	..		..	..																
..	..	..	..		..	..																
..	..	..	*		..	..		12	14	3	2	7	6	4	..	..	6	3	10	..	4	3
..	..	..	..		..	..																
..	..	..	..		..	..																

GENERA.	No. of Species Living.	No. of Species Fossil (tert.)	Species found both living and fossil (tertiary).	Species found fossil in more than one tertiary formation.	Habitations of living Species.	PLIOCENE PERIOD. Sicily.	Italy, Subap.	English Crag.	Various Localities.
Pecten, *cont.*			coarctatus		Mediterr.	*	*		
——			Bruei		ibid.	*			
——			Dumasii		ibid.	*			
——			distans		ibid.	*	*		
——			pusio		ibid.	*			
——				flabelliformis			*		
——				scabrellus			*	{	Perpignan, Morea
——				Burdigalensis		*			
——			nodosus		Indian Ocean	*?			
——				benedictus					Perpignan
——				striatus				*	
——				inæquicostalis			*	{	Perpignan, Salles
——				plebeius					
——				laticostatus			*		Perpignan
Plicatula	5	7		new species					
Spondylus	25	9	gæderopus		Mediterr.	*	*		
——				new species		*			
——				radula					
Gryphæa	1	3							
Ostrea	54	72	cornucopiæ		Indian Ocean	*	*		
——			edulis		European Ocean	*	*	*?	
——			Virginica		American Ocean	*			
——			hippopus		European Ocean	*			Corsica
——				flabellula					
——				gigantea					
——				edulina					
——				Bellovacina					
——				dorsata					
——			navicularis		Mediterr.		*		
——			Forskali		{ African Ocean, Red Sea		*		
——				Virginiana					
——				undata					
——				new species					
——			new species		{ Mediterr. Indian Ocean	*	*	{	Perpignan, Salles, Morea
Hinnites	2	4		Cortesi			*		
Vulsella	5	1							
Placuna	3	1	papyracea		Red Sea				Egypt
Anomia	10	8	ephippium		{ Mediterr. European Ocean	*	*	*	Perpignan
——			electrica		ibid.	*	*		
——				costata			*	*?	
——				dubia					

MIOCENE PERIOD.					EOCENE PERIOD.			No. of Species in each Genus in the following Localities.														
Bordeaux, Dax.	Touraine.	Turin.	Vienna.	Various Localities.	Paris.	London.	Various Localities.	Sicily.	Italy, Subap.	English Crag.	Baden.	Bordeaux.	Dax.	Touraine.	Turin.	Ronca.	Vienna.	Angers.	Paris.	London.	Valognes.	Belgium.
..	..	..	..		..	..																
..	..	..	..		..	..																
..	..	..	..		..	..																
..	..	..	..		..	..																
..	..	..	..		..	..																
*	..	..	..		..	..																
*	*	..	*	Angers	..	..		12	14	3	2	7	6	4	..	..	6	3	10	..	4	3
*	..	..	*	Transylvania	..	..																
..	..	..	..		..	..																
..	*	..	..	Angers, Doué	..	..																
..	*	..	..	ibid.	..	..																
..	..	..	..		*	..	Belgium, Valognes															
..	..	..	..		..	..																
*	*	..	..	Angers	..	..		..	..	..	..	1	2	1	..	..	..	2	3	..	..	2
..	..	..	..		..	..																
*	*	..	..	Angers	*	..	Castelgomberto? Valognes	1	2	..	..	1	1	1	1	..	..	1	2	4	..	1
..	..	..	..		..	..		..	1	..	..	..	..	..	..	..	..	..	2	..	..	..
..	..	..	..		..	..																
..	..	..	*?		..	..																
..	..	..	..		..	..																
..	..	..	..		*	*	Valognes, Belgium															
..	..	..	..		*	*																
..	..	..	..		*	*	Belgium	5	13	1	..	7	4	6	..	..	4	2	42	10	9	3
..	..	..	..		*	*																
..	..	..	..		..	..																
*	*	..	..		..	..																
*	*	..	..		..	..																
*	*	..	..	Montpellier	..	..																
*	*	..	*	Angers	..	..																
..	..	..	..	Doué	..	..		..	1	..	..	..	..	..	..	..	..	1	..	..	..	..
..	..	..	..		..	..		..	..	..	..	..	..	..	..	..	..	..	1	..	..	..
*	*	*?	*	Angers	..	..		3	3	1	1	3	3	2	2	..	..	1	1	2	2	1
..	..	..	..		..	..																
*	*	*	..		..	..																
..	..	..	..		*	*	Valognes, Belgium															

GENERA.	No. of Species Living.	No. of Species Fossil (tert.)	Species found both living and fossil (tertiary).	Species found fossil in more than one tertiary formation.	Habitations of living Species.	PLIOCENE PERIOD. Sicily.	Italy, Subap.	English Crag.	Various Localities.
Anomia *cont.*	..	..	. . . . .	striata . . .	. . . . . .	..	..	..	. . . .
Crania . . .	3	3	personata . .	. . . . .	Mediterr.. . .	*	..	..	. . . .
Orbicula . .	2								
Terebratula .	15	18	vitrea . . .	. . . . .	Mediterr.. . .	*	*?	..	. . . .
—————.	.	..	caput serpentis	. . . . .	ibid.	*	..	..	. . . .
—————.	.	..	truncata . .	. . . . .	ibid.	*	..	..	. . . .
—————.	..	..	. . . . .	ampulla . .	. . . . . .	*	*	..	Morea . . .
Thecidea . .	1	1	Mediterranea	. . . . .	Mediterr.. . .	*	..	..	. . . .
Lingula . .	1								
Hyalæa . .	10	2	. . . . .	. . . . .	. . . . . .	..	..	..	. . . .
Cleodora . .	14	3	lanceolata .	. . . . .	African Ocean	..	*	..	Asti . . . .
—————.	..	..	. . . .	strangulata .	. . . . . .	..	..	..	. . . .
—————.	..	..	. . . .	new species .	. . . . . .	..	*	..	. . . .
Limacina . .	1								
Cymbulia .	1								
Chitonellus .	1								
Chiton . . .	35	1	. . . . .	. . . . .	. . . . . .	..	..	..	Morea . . .
Dentalium .	23	34	elephantinum	. . . . .	Indies, Mediterr.	..	*	..	Morea . . .
—————.	..	..	. . . .	sexangulare .	. . . . . .	*	*	..	ibid. . . .
—————.	..	..	dentalis . .	. . . . .	Mediterr.. . .	*	*	..	Maryland. .
—————.	..	..	. . . .	fossile . . .	. . . . . .	*	*	..	. . . .
—————.	..	..	novem costatum	. . . .	European Ocean	*	..	..	. . . .
—————.	..	..	. . . . .	pseudo-entalis	. . . . . .	..	..	..	. . . .
—————.	..	..	entalis . . .	. . . . .	European Ocean	*	*	*?	Morea . . .
—————.	..	..	. . . . .	incertum . .	. . . . . .	..	..	..	. . . .
—————.	..	..	eburneum . .	. . . . .	Indian Seas	..	..	..	. . . .
—————.	..	..	fissura . . .	. . . . .	ibid.	..	..	..	. . . .
—————.	..	..	. . . . .	coarctatum .	. . . . . .	..	*	..	. . . .
—————.	..	..	strangulatum	. . . . .	Mediterr.. . .	*	*	..	Morea . . .
—————.	..	..	. . . . .	Bouei . . .	. . . . . .	..	*	..	. . . .
Patella . . .	104	10	equalis . . .	. . . . .	European Ocean	..	..	*	. . . .
Siphonaria .	21	3	. . . . .	. . . . .	. . . . . .				
Umbrella . .	2	1	Mediterranea	. . . . .	Mediterr.. . .	*	..	..	. . . .
Parmophorus	2	2	. . . . .	. . . . .	. . . . . .	..	..	..	. . . .
Emarginula .	7	11	fissura . . .	. . . . .	Europ. Oc. Medt.	..	..	*	. . . .
Fissurella . .	33	8	Græca . . .	. . . . .	Europ. Oc. Ind.	*	*	..	. . . .
—————.	..	..	costaria . .	. . . . .	ibid. ibid.	*	..	*	. . . .
—————.	..	..	neglecta . .	. . . . .	Mediterr.. . .	*	*	..	. . . .
—————.	..	..	. . . . .	mitis . . .	. . . . . .	..	*	..	. . . .
Pileopsis . .	7	6	Ungarica . .	. . . . .	European Ocean	*	*	*	. . . .

MIOCENE PERIOD					EOCENE PERIOD			No. of Species in each Genus in the following Localities.														
Bordeaux, Dax.	Touraine.	Turin.	Vienna.	Various Localities.	Paris.	London.	Various Localities.	Sicily.	Italy, Subap.	English Crag.	Baden.	Bordeaux.	Dax.	Touraine.	Turin.	Ronca.	Vienna.	Angers.	Paris.	London.	Valognes.	Belgium.
					*	*		3	3	1	1	3	3	2	2		1	1	2	2	1	1
								1					1					1				
								4	7	1								1	4			1
								1														
														1			1					
*				Baden					2		1		1									
*																						
																					1	
				Baden																		
				ibid.																		
*	*			Baden	*		Valognes	7	10	2	4	3	7	3				1	14	3	2	
*				Moravia	*	*																
				Baden																		
										1							1		5	1		2
													2									1
								1														
																			2			
	*			Angers						2				3				3	5			2
*					*			3	3	1			2	2				2	4			
*	*			Baden				1	2	1	1		2	1					3			

GENERA.	No. of Species — Living.	No. of Species — Fossil (tert.)	Species found both living and fossil (tertiary).	Species found fossil in more than one tertiary formation.	Habitations of living Species.	PLIOCENE PERIOD. Sicily.	Italy, Subap.	English Crag.	Various Localities.
Hipponyx .	6	12	new species .	. . . . .	European Ocean	*	*	..	. . . . .
	..	..	. . . . .	cornucopiæ .	. . . . .	..	..	..	. . . . .
Crepidula .	14	3	sandalina . .	. . . . .	{ European Ocean New Zealand	*	*	..	. . . . .
	..	..	gibbosa . .	. . . . .	Medit. Indies	..	..	..	. . . . .
Calyptræa .	19	15	Sinensis . .	. . . . .	European Ocean	*	*	..	Morea . .
	..	..	muricata . .	. . . . .	ibid.	*	*	..	ibid. . . .
	..	..	squamula . .	. . . . .	Mediterr. . . .	*	..	..	. . . . .
	..	..	. . . . .	trochiformis .	. . . . .	..	..	..	. . . . .
	..	..	. . . . .	deformis . .	. . . . .	..	..	..	. . . . .
Bullæa . .	2	2	. . . . .	. . . . .	. . . . .	..	..	..	. . . . .
Bulla . .	26	23	lignaria . .	. . . . .	European Ocean	*	*	*	. . . . .
	..	..	ampulla . .	. . . . .	{ ibid. and Indies	*	*	..	. . . . .
	..	..	. . . . .	cylindrica .	. . . . .	..	..	..	. . . . .
	..	..	. . . . .	ovulata . . .	. . . . .	..	..	..	. . . . .
	..	..	. . . . .	angistoma .	. . . . .	..	..	..	. . . . .
	..	..	. . . . .	striatella .	. . . . .	..	..	..	. . . . .
	..	..	. . . . .	clathrata . .	. . . . .	..	..	..	. . . . .
	..	..	. . . . .	cylindroides .	. . . . .	..	..	..	. . . . .
. .	..	..	. . . . .	Lajonkairiana	. . . . .	..	..	..	. . . . .
Dolabella .	10								
Testacella .	2								
Vitrina . .	5								
Helix . .	325	35	aspersa ? var.	. . . . .	{ Temperate and Southern Europe	*	..	..	Teneriffe .
	..	..	algira . . .	. . . . .	ibid.	..	..	..	Cette, Nice
	..	..	cespitum . .	. . . . .	Europe	..	*	..	ibid. ibid.
	..	..	nemoralis . .	. . . . .	ibid.	..	..	..	Quercy . .
	..	..	. . . . .	Ramondi .	. . . . .	..	..	..	. . . . .
	..	..	cornea . . .	. . . . .	Southern Europe	..	..	..	Nice . } osseous breccias
	..	..	pisana . . .	. . . . .	ibid.	..	..	..	ibid. . .
	..	..	lapicida . .	. . . . .	Europe	..	..	..	ibid.
	..	..	vermiculata .	. . . . .	Southern Europe	..	..	..	ibid. . .
	..	..	cælatura . .	. . . . .	Bourbon	..	*	..	. . . . .
Anostoma .	2								
Helicina .	17	1							
Pupa . . .	95	3	cinerea . . .	. . . . .	Temp. Europe	..	..	..	Nice . . . .
	..	..	muscorum .	. . . . .	ibid.	..	..	..	Puy-de-Dôme
Clausilia .	52	2	. . . . .	. . . . .	. . . . .	..	..	..	. . . . .
Bulimus .	83	3	. . . . .	. . . . .	. . . . .	..	..	..	. . . . .
Achatina .	53	3	bulloides . .	. . . . .	Indies	..	*	..	. . . . .
Succinea .	10		. . . . .	. . . . .	. . . . .				
Auricula .	17	18	. . . . .	. . . . .	. . . . .	..	..	..	. . . . .

MIOCENE PERIOD.					EOCENE PERIOD.			No. of Species in each Genus in the following Localities.														
Bordeaux, Dax.	Touraine.	Turin.	Vienna.	Various Localities.	Paris.	London.	Various Localities.	Sicily.	Italy, Subap.	English Crag.	Baden.	Bordeaux.	Dax.	Touraine.	Turin.	Ronca.	Vienna.	Angers.	Paris.	London.	Valognes.	Belgium.
*	*	*	··		··	··	} Belgium	1	1	1	··	1	1	··	··	··	··	··	8	··	4	1
··	··	··	··		*	*?																
*	*	··	*	Moravia	··	··	}	1	1	··	1	2	1	··	··	··	1	··	1			
··	··	*	··		··	··	}															
*	··	··	··		··	··																
··	*	··	··		··	··																
··	··	··	··		··	··	} Valognes	3	2	1	··	5	4	2	··	··	··	··	4	2	3	
··	··	··	··		*	*																
*	*	··	*		··	··	}															
··	··	··	··		··	··		··	1	··	··	··	··	··	··	··	··	··	1			
*	*	*	··	Angers	*	*	Valognes .															
··	··	··	··		··	··																
··	··	··	··		*	*	Valognes .															
*	··	··	··		*	*		3	2	1	1	3	7	2	2	··	··	1	14	5	6	
*	··	··	··		*	··	{ Castel- gomberto															
··	··	··	··		··	··																
*	··	*	··		··	··																
··	··	··	··		*	*																
*	··	··	*		··	··																
··	··	··	··		··	··																
··	··	··	··		··	··																
··	··	··	··		··	··																
··	··	··	··		··	··																
··	··	··	··		*	*	Auvergne.	1	1	··	··	··	3	5	··	1	··	··	6	··	··	1
··	··	··	··		··	··																
··	··	··	··		··	··																
··	··	··	··		··	··																
··	··	··	··		··	··																
··	··	··	··		··	··														1			
··	··	··	··		··	··	}	··	··	··	··	1	··	··	··	··	··	··	1			
··	··	*	··		··	··		··	1	··	··	··	··	··	··	··	··	··	1			
··	··	··	··		··	··		··	1	··	··	··	··	··	··	··	··	··	1			
··	··	··	··		··	··		··	1	··	··	··	··	··	··	··	··	··	1			
··	··	··	··		··	··		··	1	1	··	2	··	6	··	··	··	··	7			

GENERA.	Living.	Fossil (tert.)	Species found both living and fossil (tertiary).	Species found fossil in more than one tertiary formation.	Habitations of living Species.	Sicily.	Italy, Subap.	English Crag.	Various Localities.
						PLIOCENE PERIOD.			
Pedipes . .	3	7	. . . . .	ringens . .	. . . . .	..	..	..	. . . . .
—— . .	..	..	buccinea . .	. . . . .	Mediterr. . .	*	*	*	. . . . .
Scarabæus .	3	..	imbrium . .	. . . . .	Amboyna . .	..	*	..	
Cyclostoma .	49	6	elegans . .	. . . . .	Europe . . .	..	..	..	Nice . . . .
Ancylus . .	4	3	. . . . .	elegans . .	. . . . .	..	..	..	
Planorbis . .	23	26	corneus . .	. . . . .	Temp. Europe	..	..	..	
—— . .	..	..	. . . . .	rotundatus .	. . . . .	..	..	..	
—— . .	..	..	marginatus .	. . . . .	Europe . . .	..	..	*	Tols, Bavaria
—— . .	..	..	carinatus . .	. . . . .	ibid.	..	..	*	ibid. ibid. Cette
—— . .	..	..	. . . . .	cornu . .	. . . . .	..	..	..	
—— . .	..	..	. . . . .	lens	. . . . .	..	..	..	
—— . .	..	..	spirorbis . .	. . . . .	Europe . . .	..	..	..	Nice . . . .
—— . .	..	..	nitidus . . .	. . . . .	ibid.	..	..	..	Quercy . . .
Physa . . .	9	1	. . . . .	. . . . .	. . . . .	..	..	..	
Limnea . .	15	27	peregra . .	. . . . .	Europe . . .	..	*	..	Lauzerte . .
—— . .	..	..	auricularis .	. . . . .	ibid.	..	*	..	ibid. . . .
—— . .	..	..	rivalis . . .	. . . . .	ibid.	..	..	..	Agen . . .
—— . .	..	..	. . . . .	longiscata .	. . . . .	..	..	..	
—— . .	..	..	. . . . .	inflata . .	. . . . .	..	..	..	
—— . .	..	..	. . . . .	cornea . . .	. . . . .	..	..	..	
—— . .	..	..	palustris . .	. . . . .	Europe . . .	..	..	..	
Melania . .	34	25	inquinata . .	. . . . .	Philippine Isles	..	..	..	
—— . .	..	..	. . . . .	lactea . . .	. . . . .	..	..	..	
—— . .	..	..	. . . . .	nitida . . .	. . . . .	..	..	..	
—— . .	..	..	inflexa . . .	. . . . .	Mediterr. . .	..	*	..	
—— . .	..	..	. . . . .	costellata .	. . . . .	..	..	..	
—— . .	..	..	Cambessedesii	. . . . .	Mediterr. . .	*	*	..	
—— . .	..	..	new species .	. . . . . {	Lakes of Como and Geneva	..	..	..	
Rissoa . . .	23	22	lactea . . .	. . . . .	Mediterr. . .	*	..	..	
—— . .	..	..	cochlearella .	. . . . .	Indian Ocean	..	*	..	
Pirena . . .	3	2	. . . . .	. . . . .	. . . . .	..	..	..	
Melanopsis .	10	11	buccinoidea .	. . . . . {	Asia, Spain, Greece	..	*	..	Greece . . .
—— . .	..	..	Dufourei . .	. . . . .	Alicant . . .	..	..	..	
—— . .	..	..	costata . . .	. . . . .	Asia	..	*	..	Greece . . .
—— . .	..	..	nodosa . . .	. . . . .	Spain, &c. .	..	*	..	
—— . .	..	..	acicularis . .	. . . . .	Laybach . . .	..	..	..	
—— . .	..	..	incerta . . .	. . . . .	Asia, Greece	..	..	..	Abydos . . .
Valvata . .	4	1	piscinalis . .	. . . . .	Europe . . .	..	..	..	
Paludina . .	25	41	achatina . .	. . . . .	ibid.	..	*	*	
—— . .	..	..	. . . . .	lenta . . .	. . . . .	..	..	..	
—— . .	..	..	unicolor . .	. . . . .	Asia	..	..	..	

MIOCENE PERIOD.					EOCENE PERIOD.			No. of Species in each Genus in the following Localities.														
Bordeaux, Dax.	Touraine.	Turin.	Vienna.	Various Localities.	Paris.	London.	Various Localities.	Sicily.	Italy, Subap.	English Crag.	Baden.	Bordeaux.	Dax.	Touraine.	Turin.	Ronca.	Vienna.	Angers.	Paris.	London.	Valognes.	Belgium.
·	·	·	·		*	·	Valognes.	1	3	1	1	1	2	1	1	·	1	1	1	·	1	·
·	*	·	*	Angers, Baden	·	·		·	·	·	·	·	·	·	·	·	·	·	·	·	·	·
·	·	·	·		*	·	Auvergne, Cantal	·	·	·	·	·	·	·	·	·	·	·	6	·	·	·
·	·	·	·		·	*	Auvergne, Cantal	·	·	·	·	·	·	·	·	·	·	·	2	1	·	·
·	·	·	·		*?	·	Auvergne.	·	·	·	·	·	·	·	·	·	·	·	·	·	·	·
*	*	·	·		*	·		·	·	·	·	·	·	·	·	·	·	·	·	·	·	·
·	·	·	·		·	·		·	·	·	·	·	·	·	·	·	·	·	·	·	·	·
·	·	·	·		*	·	Auvergne.	·	2	·	2	2	1	1	·	·	·	·	13	7	·	·
·	·	·	·		*	*		·	·	·	·	·	·	·	·	·	·	·	·	·	·	·
·	·	·	·		·	·		·	·	·	·	·	·	·	·	·	·	·	·	·	·	·
·	·	·	·		·	·		·	·	·	·	·	·	·	·	·	·	·	·	·	·	·
·	·	·	·		·	·		·	·	·	·	·	·	·	·	·	·	·	1	·	·	·
·	·	·	·		·	·		·	·	·	·	·	·	·	·	·	·	·	·	·	·	·
·	·	·	·		·	·		·	·	·	·	·	·	·	·	·	·	·	·	·	·	·
*	*	·	·		*	*	Auvergne	·	2	·	1	1	1	·	·	·	·	1	15	3	·	·
·	·	·	·		*	·	ibid.	·	·	·	·	·	·	·	·	·	·	·	·	·	·	·
·	·	·	·		*	·	ibid.	·	·	·	·	·	·	·	·	·	·	·	·	·	·	·
·	·	·	·		*?	·		·	·	·	·	·	·	·	·	·	·	·	·	·	·	·
·	·	·	·		*	*	Abbeville, Tours, Valognes.	·	·	·	·	·	·	·	·	·	·	·	·	·	·	·
·	·	·	·	Ronca	*	·		1	2	·	·	2	3	·	1	1	·	2	17	4	·	7
*	·	·	·		*	*		·	·	·	·	·	·	·	·	·	·	·	·	·	·	·
·	·	·	·	Angers	·	·		·	·	·	·	·	·	·	·	·	·	·	·	·	·	·
*	·	*	·		*	·	Valognes.	·	·	·	·	·	·	·	·	·	·	·	·	·	·	·
*	·	·	·		·	·		·	·	·	·	·	·	·	·	·	·	·	·	·	·	·
·	·	·	·	Arapatack.	·	·		·	·	·	·	·	·	·	·	·	·	·	·	·	·	·
·	·	·	·	Angers	*	·	Valognes.	1	1	·	·	3	9	·	·	·	·	3	4	5	·	2
*	·	·	·		·	·		·	·	·	·	·	·	·	·	·	·	·	1	1	·	·
·	·	·	·		*	*		·	·	·	·	·	·	·	·	·	·	·	·	·	·	·
*	·	·	·	Hungary	·	·		·	3	·	2	1	1	·	·	·	·	1	5	3	·	·
·	·	·	·		*	·		·	·	·	·	·	·	·	·	·	·	·	·	·	·	·
·	·	·	·		·	·		·	·	·	·	·	·	·	·	·	·	·	·	·	·	·
·	·	·	·		·	*		·	·	·	·	·	·	·	·	·	·	·	·	·	·	·
·	·	·	·		·	·		·	·	·	·	·	·	·	·	·	·	·	·	·	·	·
·	*	·	·		*?	·		·	·	·	·	·	·	·	·	·	1	·	·	·	·	·
·	·	·	·		·	·		·	6	2	2	2	4	·	·	·	·	1	17	3	·	1
·	·	·	·		*	*		·	·	·	·	·	·	·	·	·	·	·	·	·	·	·
·	·	·	·		*	·	Alsace.	·	·	·	·	·	·	·	·	·	·	·	·	·	·	·

GENERA.	No. of Species		Species found both living and fossil (tertiary).	Species found fossil in more than one tertiary formation.	Habitations of living Species.	PLIOCENE PERIOD.			
	Living.	Fossil (tert.)				Sicily.	Italy, Subap.	English Crag.	Various Localities.
Paludina,*cont.*	....	....	impura	....	Europe	..	*		
Ampullaria	24	14	....	Willemetii			..	..	....
Pileolus	....	2		./			..	..	....
Navicella	.	2							
Neritina	60	17	fluviatilis	....	Europe	..	*	..	....
———				conoidea					....
———				globulus					....
———				new species					....
Nerita	18	16		Caronis					....
———				tricarinata					....
———				mammaria					....
———				new species					....
Natica	56	41	millepunctata	....	European Ocean	*	*	*	Morea,Perpignan
———			Guillemini		Mediterr.	*	..		
———			canrena		European Ocean	*	*	*	Morea,Perpignan
———			Valenciennesii		{ Mediterr. English Channel	*	..	*	
———			Dilwynii		ibid. and Senegal	*	*	..	
———			glaucina		Ind.& Europ. Oc.	*	*	*	Morea,Perpignan
———			monilifera		English Channel	..	..	*	
———				new species					....
———				sigaretina					....
———			mamilla		Indian Ocean				....
———				new species					....
———			zebra		{ Indian Ocean, Senegal	..	*	..	
———				epiglottina					....
———				mutabilis					....
———				glaucinoides					....
———				intermedia					....
———				new species					....
Sigaretus	11	4	....	canaliculatus					....
———			....	laevigatus		..	..	..	....
———			depressus	....	{ Seas of the Molucca Islands	..	*	..	....
Stomatella	6	1	{ locality unknown of fossil species						
Stomatia	.	2							
Haliotis	29	1	tuberculata	....	{ European and African Ocean	*	..	..	....
Tornatella	7	11	fasciata	....	European Ocean	..	*	*	....
———				inflata		..	*	..	....
———				semisulcata		..	*	..	....
Pyramidella	11	8	....	terebellata					....
———				acicula		..	..	..	....

MIOCENE PERIOD.					EOCENE PERIOD.			No. of Species in each Genus in the following Localities.															
Bordeaux, Dax.	Touraine.	Turin.	Vienna.	Various Localities.	Paris.	London.	Various Localities.	Sicily.	Italy, Subap.	English Crag.	Baden.	Bordeaux.	Dax.	Touraine.	Turin.	Ronca.	Vienna.	Angers.	Paris.	London.	Valognes.	Belgium.
..	..	..	..	. . . .	*	..	Castelgomberto	..	1	..	..	..	2	..	..	1	..	..	7	1	1	
..	..	..	..	. . . .	..	..	. . . .	..	..	..	..	..	..	..	..	..	..	..	..	1	..	1
..	..	..	..	. . . .	..	..	. . . .															
..	..	*	..	Ronca. . .	*	..	. . . .		3			1	4		1		1		9	3		
..	..	..	..	. . . .	*	*	. . . .															
*	..	..	*	Podolia . .	..	..	. . . .															
*	*	*	..	. . . .	..	..	. . . .															
..	..	..	..	. . . .	*	*	Valognes .		..		1	1	4	5	1	1		1	4	3	3
..	..	..	..	. . . .	*	*	. . . .															
*	*	..	..	. . . .	..	..	. . . .															
*	*	*	*	. . . .	..	..	. . . .															
..	..	..	..	Transylvania	..	..	. . . .															
*	*	..	*	Volhynia, Baden, Moravia	..	..	. . . .															
..	..	..	..	. . . .	..	..	. . . .															
..	..	..	..	. . . .	..	..	. . . .															
*	*	*	*	Ronca, Angers, Transylvania	..	*?	Valognes .															
..	..	..	..	. . . .	*?	*?	. . . .															
*	*	..	..	. . . .	..	..	. . . .															
*?	..	..	..	. . . .	*	..	Castelgomberto	6	8	6	4	8	9	7	5	5	..	2	16	6	9	1
..	..	*	..	. . . .	..	..	. . . .															
*	..	*	..	. . . .	..	..	. . . .															
..	..	..	..	. . . .	..	..	Valognes															
..	..	..	..	. . . .	*	*	Belgium															
..	..	..	..	Ronca . .	*	*	ibid. . .															
..	..	..	..	. . . .	*	..	. . . .															
..	..	..	..	. . . .	*	..	Castelgomberto															
*	*	..	..	. . . .	..	..	. . . .															
*	..	..	..	. . . .	*	*	. . . .															
*	..	..	..	. . . .	*	..	. . . .	..	1	..	..	1	2	1	..	..	..	..	2	1	1
*	*	..	..	. . . .	..	..	. . . .															
..	..	..	..	. . . .	..	..	. . . .	1														
..	..	..	..	. . . .	..	..	. . . .															
*	..	..	..	. . . .	*	..	. . . .	..	3	1	..	5	8	..	..	..	..	..	3	2	
*	..	..	..	. . . .	..	..	. . . .															
*	*	..	..	Angers . .	*	..	Valognes .		..			4	3	1	..	..	..	1	5	..	3
*	..	..	..	. . . .	*	..	ibid. . .															

GENERA.	No. of Species — Living.	No. of Species — Fossil (tert.)	Species found both living and fossil (tertiary).	Species found fossil in more than one tertiary formation.	Habitations of living Species.	PLIOCENE PERIOD. Sicily.	PLIOCENE PERIOD. Italy, Subap.	PLIOCENE PERIOD. English Crag.	Various Localities.
New Genus. Bulinus Terebellatus, Lam.	..	1	. . . .	terebellatus	. . . . . .	..	*	..	
Vermetus.	8	1	. . . .			..	..	..	
Siliquaria	4	6	anguina	. . . .	Mediterr. Indian Ocean	..	*	..	
Magilus	2								
Scalaria	14	22	communis	. . . .	European Ocean	*	..	*?	
——			pseudoscalaris		Mediterr.	*	*	..	
——			tenuicostata		ibid.	*	..	..	
——			lamellosa		Mediterr. Indian Ocean	..	*	..	
——			varicosa		Unknown	..	*	..	
——				crassicostata		..	*	..	
——				multilamella					
——				cancellata		..	*	..	
——				subulata		..	..	*	
Delphinula	7	12		marginata					
Solarium	12	16	variegatum		Mediter. Indies	*	*		
——			carocollatum		Mediterr.	..	*	..	
——			{ pseudo perspectivum		ibid.	..	*	..	
——				umbrosum		..	*	..	
——				new species					
——				patulum					
——				plicatulum					
——				plicatum					
Omalaxon	..	5							
Rotella	4								
Trochus	103	70	magus		European Ocean, Senegal	*	*	..	Ischia
——			fagus		Mediterr.	*			
——			cingulatus		Adriatic	*	*	..	
——			agglutinans		Mediterr. East Ind. America, &c.	*	*	..	
——			Adansoni		Mediterr.	*	..	..	Ischia
——			conulus		European Ocean	*	*	*?	
——			cinerarius		ibid.	*			
——			conuloides		ibid.	*			
——			Matoni		ibid.	*			
——			zizyphinus		ibid.	*		*	
——				infundibulum					
——			strigosus		Unknown	..	*	..	
——				patulus		..	*	..	Morea
——				crenulatus		*			
——			obliquatus		Mediterr.	*			
——				monilifer					

Bordeaux, Dax.	Touraine.	Turin.	Vienna.	Various Localities.	Paris.	London.	Various Localities.	Sicily.	Italy, Subap.	English Crag.	Baden.	Bordeaux.	Dax.	Touraine.	Turin.	Ronca.	Vienna.	Angers.	Paris.	London.	Valognes.	Belgium.
		MIOCENE PERIOD.				EOCENE PERIOD.					No. of Species in each Genus in the following Localities.											
*	*	··	*	Angers, Baden	*	··	Valognes .	··	1	··	1	1	1	1	··	··	··	1	1	1	··	1
··	··	··	··		··	··		··	··	··	··	··	··	··	··	··	··	··	1	··	··	··
*	*	··	··	Angers ..	··	··		··	1	··	··	··	1	1	··	··	··	··	1	5	··	3
*?	*?	··	··		··	··																
··	··	··	··		··	··																
··	··	··	··		··	··																
*	··	··	··		··	··		2	8	3	1	2	8	1	··	··	··	··	7	4	··	··
*	··	··	··		··	··																
*	··	··	··		*	··																
*	··	··	··		··	··																
*	··	··	··		··	··																
*	··	··	··		*	··	Valognes .	··	1	··	··	1	2	1	1	··	··	··	8	5	··	··
··	··	··	··		··	··																
*	··	··	··		··	··																
*	··	··	··		··	··																
··	··	*	··		··	··		1	6	··	··	2	3	1	1	··	··	1	7	3	3	··
*	*	··	··	Angers ..	··	··																
··	··	··	··		*	*	Valognes															
··	··	··	··		*	*	ibid. .															
··	··	··	··		*	*	ibid. .															
··	··	··	··		··	··		··	··	··	··	··	··	··	··	··	··	··	5	··	··	··
··	··	··	··		··	··																
··	··	··	··	Volhynia ?	··	··																
··	··	··	··		*	*	Valognes, Castel-gomberto															
··	··	··	··		··	··																
··	··	··	··		··	··																
··	··	··	··		··	··		11	13	2	1	6	13	11	3	2	1	9	18	4	5	··
··	··	··	··		··	··																
*	*	*	··		··	··																
*	*	*	*	Angers, Volhynia, Transylvania	··	··																
··	*	··	··	Angers ..	··	··																
··	··	··	··		··	··																
··	··	··	··		*	*																

GENERA.	No. of Species Living.	Fossil (tert.)	Species found both living and fossil (tertiary).	Species found fossil in more than one tertiary formation.	Habitations of living Species.	PLIOCENE PERIOD. Sicily.	Italy, Subap.	English Crag.	Various Localities.
Trochus, *cont.*	..	..	. . . .	boscianus . .	. . . . .	..	..	..	. . . .
———	..	..	. . . .	carinatus .	. . . . .	..	*	..	. . . .
———	..	..	. . . .	new species .	. . . . .	..	..	..	. . . .
Pleurotomaria	..	1	. . . .	. . . .	. . . . .	..	..	..	. . . .
Monodonta .	42	8	. . . .	Pharaonis .	. . . . .	..	*	..	. . . .
Turbo . . .	56	34	rugosus . .	. . . .	Mediterr. . .	*	*	..	. . . .
———	..	..	new species .	. . . .	Indian Ocean .	*	*	..	. . . .
———	..	..	costatus . .	. . . .	Mediterr. . .	*	*	..	. . . .
———	..	..	. . . .	sculptus . .	. . . . .	..	..	..	. . . .
Littorina .	24	10	littorea . .	. . . .	{ Seas of Northern Europe	..	..	*	. . . .
———	..	..	. . . .	new species .	. . . . .	..	..	..	. . . .
———	..	..	striata . . .	. . . .	European Ocean	..	..	*	. . . .
Planaxis . .	4	5	. . . .	. . . .	. . . . .	..	..	..	. . . .
Phasianella .	9	4	pullus . .	. . . .	European Ocean	..	..	..	. . . .
Turritella .	24	45	terebra . . .	. . . .	ibid.	*	*	..	. . . .
———	..	..	new species .	. . . .	Senegal . . .	..	..	..	. . . .
———	..	..	Ligar. . . .	. . . .	ibid.	..	..	..	. . . .
———	..	..	. . . .	terebellata .	. . . . .	..	..	..	. . . .
———	..	..	. . . .	imbricataria.	. . . . .	..	..	..	. . . .
———	..	..	new species .	. . . .	Mediterr. . .	..	..	*	. . . .
———	..	..	. . . .	Desmarestina	. . . . .	..	*	..	. . . .
———	..	..	. . . .	spirata . .	. . . . .	..	*	..	. . . .
———	..	..	. . . .	subangulata .	. . . . .	*	*	..	. . . .
———	..	..	. . . .	vermicularis.	. . . . .	..	*	..	. . . .
———	..	..	. . . .	tornata . .	. . . . .	*	*	..	. . . .
———	..	..	. . . .	turris . . .	. . . . .	..	..	..	. . . .
———	..	..	. . . .	granulosa . .	. . . . .	..	..	..	. . . .
Proto . . .	2	4	. . . .	cathedralis	. . . . .	..	..	..	. . . .
Cerithium .	87	220	vulgatum . .	. . . .	{ European Ocean, Senegal	*	*	..	Morea, Ischia
———	..	..	Latreillei . .	. . . .	Mediterr. . .	*	*	..	Ischia . . .
———	..	..	doliolum . .	. . . .	ibid.	..	*	..	. . . .
———	..	..	giganteum .	. . . .	New Holland ??	..	..	..	. . . .
———	..	..	. . . .	plicatum . .	. . . . .	..	..	..	. . . .
———	..	..	. . . .	hexagonum .	. . . . .	..	..	..	. . . .
———	..	..	. . . .	pleurotomoides.	. . . . .	..	..	..	. . . .
———	..	..	. . . .	cinctum . .	. . . . .	..	..	..	. . . .
———	..	..	. . . .	Cornucopiæ .	. . . . .	..	..	..	. . . .
———	..	..	. . . .	geminatum .	. . . . .	..	..	..	. . . .
———	..	..	. . . .	ventricosum.	. . . . .	..	..	..	. . . .
———	..	..	. . . .	Lamarckii . .	. . . . .	..	..	..	. . . .
———	..	..	. . . .	Cordieri . .	. . . . .	..	..	..	. . . .

MIOCENE PERIOD					EOCENE PERIOD			No. of Species in each Genus in the following Localities.														
Bordeaux, Dex.	Touraine.	Turin.	Vienna.	Various Localities.	Paris.	London.	Various Localities.	Sicily.	Italy, Subap.	English Crag.	Baden.	Bordeaux.	Dax.	Touraine.	Turin.	Ronca.	Vienna.	Angers.	Paris.	London.	Valognes.	Belgium.
*	*																					
		*						11	13	2	1	6	13	11	3	2	1	9	18	4		5
*	*																					
																			1			
*	*			Angers					2			1	1	3	1				1	1		1
			*	Moravia																		
					*	*																
								3	5			5	3	3	4	2	1	2	16	2		5
*	*			Angers																		
*	*									2	1	1	3	1					1	3		1
													1						4			
*					*								1	1					2			2
	*																					
*																						
*																						
					*	*	Valognes															
		*??			*	*	id. Belgium															
*?																						
*								3	19	3	5	8	6	7	5	2	6	1	17	4	6	1
*	*																					
	*			Angers																		
*	*			Transylvania																		
*			*	Cracovia, Volhynia																		
				Ronca	*																	
*	*	*											2	3	2	1	1					
					*	*	{ Klein-spaunen, Mayence															
*		*	*	{ Montpelier, Podolia	*																	
		*		Ronca.	*																	
					*	*	Valognes	4	14	2	3	27	19	11	8	7	6	9	137	11	53	4
					*	*	ibid.															
						*	{ ibid. Castel-gomberto															
					*	*																
					*																	
					*		Auvergne.															
					*	*																

GENERA.	No. of Species Living.	No. of Species Fossil (tert.)	Species found both living and fossil (tertiary).	Species found fossil in more than one tertiary formation.	Habitations of living Species.	PLIOCENE PERIOD. Sicily.	Italy, Subap.	English Crag.	Various Localities.
Cerithium, *cont*	..	..	. . . . .	tricinctum	. . . . .	..	*	*?	. . . . .
—— .	..	..	. . . . .	margaritaceum	. . . . .	..	*	..	. . . . .
—— .	..	..	. . . . .	corrugatum .	. . . . .	..	*	..	. . . . .
—— .	..	..	. . . . .	inconstans .	. . . . .	..	..	..	. . . . .
—— .	..	..	. . . . .	papaveraceum	. . . . .	..	..	..	. . . . .
—— .	..	..	. . . . .	crenatum .	. . . . .	..	*	..	. . . . .
—— .	..	..	. . . . .	pictum . .	. . . . .	..	..	..	. . . . .
—— .	..	..	alucaster . .	. . . . .	Mediterr. .	..	*	..	. . . . .
—— .	..	..	. . . . .	multisulcatum	. . . . .	..	..	..	. . . . .
—— .	..	..	. . . . .	new species .	. . . . .	..	..	..	. . . . .
—— .	..	..	granulosum .	. . . . .	Mediterr.. . .	*	*	..	. . . . .
—— .	..	..	. . . . .	inversum . .	. . . . .	..	..	..	. . . . .
—— .	..	..	bicinctum. .	. . . . .	Mediterr.. . .	*	*	..	. . . . .
—— .	..	..	. . . . .	new species .	. . . . .	..	..	..	. . . . .
—— .	..	..	. . . . .	new species .	. . . . .	..	..	..	. . . . .
—— .	..	..	. . . . .	salmo . . .	. . . . .	..	..	..	. . . . .
—— .	..	..	. . . . .	pupæforme .	. . . . .	..	..	..	. . . . .
Triforis . .	3	2	. . . . .	. . . . .	. . . . .	..	..	..	. . . . .
Pleurotoma .	71	156	. . . . .	intorta . . .	. . . . .	..	✲	..	. . . . .
—— .	..	..	. . . . .	cataphracta .	. . . . .	..	*	..	. . . . .
—— .	..	..	. . . . .	rustica . . .	. . . . .	..	✲	..	. . . . .
—— .	..	..	. . . . .	oblonga . .	. . . . .	..	✲	..	. . . . .
—— .	..	..	. . . . .	new species .	. . . . .	..	✲	..	. . . . .
—— .	..	..	. . . . .	new species .	. . . . .	..	*	..	. . . . .
—— .	..	..	. . . . .	interrupta .	. . . . .	..	*	..	. . . . .
—— .	..	..	. . . . .	rotata . . .	. . . . .	..	*	..	. . . . .
—— .	..	..	. . . . .	reticulata .	. . . . .	..	*	..	. . . . .
—— .	..	..	Cordieri . .	. . . . .	Mediterr.. . .	*	..	..	. . . . .
—— .	..	..	caumarmondi	. . . . .	ibid.	*	..	..	. . . . .
—— .	..	..	vulpecula. .	. . . . .	ibid.. . . .	*	*	..	. . . . .
—— .	..	..	craticulata .	. . . . .	ibid.. . . .	*	..	..	. . . . .
—— .	..	..	new species .	. . . . .	ibid.. . . .	*	..	..	. . . . .
—— .	..	..	. . . . .	tuberculosa .	. . . . .	..	..	..	. . . . .
—— .	..	..	. . . . .	denticula . .	. . . . .	..	..	..	. . . . .
—— .	..	..	. . . . .	Borsoni .	. . . . .	..	..	..	. . . . .
—— .	..	..	. . . . .	new species .	. . . . .	..	*	..	. . . . .
—— .	..	..	. . . . .	turella . . .	. . . . .	..	*	..	. . . . .
—— .	..	..	. . . . .	pustulosa .	. . . . .	..	*	..	. . . . .
—— .	..	..	. . . . .	plicata . . .	. . . . .	..	..	..	. . . . .
—— .	..	..	. . . . .	new species .	. . . . .	..	..	..	. . . . .
—— .	..	..	. . . . .	new species .	. . . . .	..	..	..	. . . . .

MIOCENE PERIOD.					EOCENE PERIOD.			No. of Species in each Genus in the following Localities.														
Bordeaux, Dax.	Touraine.	Turin.	Vienna.	Various Localities.	Paris.	London.	Various Localities.	Sicily.	Italy. Subap.	English Crag.	Baden.	Bordeaux.	Dax.	Touraine.	Turin.	Ronca.	Vienna.	Angers.	Paris.	London.	Valognes.	Belgium.
*	*	*	..		..	..																
*	..	*	*	Weissenau.	..	..																
*	*	*	..		..	..																
*	..	..	*	Austria ..	..	..																
*	*	..	..	Montpelier.	..	..																
*	*	..	..	{ Volhynia,Transylvania	..	..																
*	*	..	*	Podolia, Hungary	..	..																
*	..	..	..		..	..																
*	*	*	..	Montpelier.	..	..		4	14	2	3	27	19	11	8	7	6	9	137	11	53	4
..	*	..	..	Baden ..	..	..																
*	..	..	..		..	..																
..	..	..	..	Angers ..	*	..	Valognes.															
..	..	..	..		..	..																
*	..	..	..	Angers ..	..	..																
*	*	..	..		..	..																
*	..	..	..	Moravia ..	..	..																
*	..	..	*	Volhynia ..	..	..																
..	..	..	..		..	..		..	..	..	..	..	..	..	..	..	..	..	2			
*	..	..	..		..	..																
*	..	..	..		..	* ?																
*?	*?	..	..		..	..																
*	*	..	*		..	..																
*	*	..	..		..	..																
..	..	..	*	Angers ..	..	..																
..	..	..	*		..	..																
*	*	*	..		..	..																
..	..	..	..		..	..																
..	..	..	..		..	..																
..	..	..	..		..	..		5	33	..	12	30	22	15	1	..	5	17	41	10	12	3
..	..	..	..		..	..																
..	..	..	..		..	..																
*	*	..	*	Cracovia .	..	..																
*	..	..	..	Baden ..	..	..																
*	*	..	*	ibid. ..	..	..																
*	..	..	..		..	..																
*	*	..	*	ibid. ..	..	..																
*	..	..	..		..	..																
*?	..	..	..	Angers ..	*	..																
..	*	..	..	ibid. ..	..	..																
..	..	..	..	ibid. ..	*	..																

GENERA.	No. of Species Living.	No. of Species Fossil (tert.)	Species found both living and fossil (tertiary).	Species found fossil in more than one tertiary formation.	Habitations of living Species.	PLIOCENE PERIOD. Sicily.	PLIOCENE PERIOD. Italy, Subap.	PLIOCENE PERIOD. English Crag.	Various Localities.
Pleurotoma *continued*	..	..	. . . . .	dentata . .	. . . . .	..	..	..	. . . .
——	..	..	. . . . .	clavicularis .	. . . . .	..	..	..	. . . .
——	..	..	. . . . .	multinoda . .	. . . . .	..	..	..	. . . .
——	..	..	. . . . .	lineolata . .	. . . . .	..	..	..	. . . .
Turbinella .	32	3	. . . . .	. . . . .	. . . . .	..	..	..	. . . .
Cancellaria .	13	42	cancellata .	. . . . {	Mediterr. Senegal	☆	☆	..	. . . .
——	..	..	. . . . .	varicosa . .	. . . . .	..	☆	..	. . . .
——	..	..	. . . . .	contorta . .	. . . . .	..	*	..	. . . .
——	..	..	. . . . .	hirta. . . .	. . . . .	..	☆	..	. . . .
——	..	..	. . . . .	evulsa . . .	. . . . .	..	..	..	. . . .
——	..	..	. . . . .	Lyra . . .	. . . . .	..	☆	..	. . . .
Fasciolaria .	7	5	. . . . .	new species .	. . . . .	..	..	..	. . . .
Fusus . .	67	111	craticulatus .	. . . . .	Mediterr.	*	..	..	. . . .
——	..	..	rostratus . .	. . . . .	ibid. . .	☆	..	..	. . . .
——	..	..	strigosus . .	. . . . .	ibid. . . .	☆	..	..	. . . .
——	..	..	lignarius . .	. . . . .	ibid. . . .	☆	☆	..	. . . .
——	..	..	sinistrorsus .	. . . . .	Unknown. .	*	..	..	. . . .
——	..	..	Tarentinus .	. . . . .	Mediterr. . .	☆	..	..	. . . .
——	..	..	antiquus . .	. . . . {	Seas of Northern Europe	..	..	*	. . . .
——	..	..	brevicauda .	. . . . .	ibid. . . .	..	..	*	. . . .
——	..	..	carinatus .	. . . . .	ibid. . . .	..	..	☆	. . . .
——	..	..	despectus .	. . . . .	ibid. . . .	..	..	*?	. . . .
——	..	..	Peruvianus .	. . . . .	Unknown, Peru??	..	..	*	. . . .
——	..	..	. . . . .	crispus . . .	. . . . .	☆	☆	..	. . . .
——	..	..	. . . . .	mitræformis .	. . . . .	..	☆	..	. . . .
——	..	..	. . . . .	subulatus . .	. . . . .	..	*	..	. . . .
——	..	..	. . . . .	abreviatus .	. . . . .	..	..	..	. . . .
——	..	..	. . . . .	Burdigalensis	. . . . .	..	..	..	. . . .
——	..	..	. . . . .	new species .	. . . . .	..	..	..	. . . .
——	..	..	. . . . .	bulbiformis .	. . . . .	..	..	..	. . . .
——	..	..	. . . . .	Noæ. . . .	. . . . .	..	..	..	. . . .
——	..	..	. . . . .	scalaris .	. . . . .	..	..	..	. . . .
——	..	..	. . . . .	acicularis .	. . . . .	..	..	..	. . . .
——	..	..	. . . . .	ficulneus .	. . . . .	..	..	..	. . . .
——	..	..	. . . . .	excisus . .	. . . . .	..	..	..	. . . .
——	..	..	. . . . .	polygonus .	. . . . .	..	..	..	. . . .
——	..	..	. . . . .	minax . . .	. . . . .	..	..	..	. . . .
——	..	..	. . . . .	costulatus. .	. . . . .	..	..	..	. . . .
——	..	..	. . . . .	subcarinatus	. . . . .	..	..	..	. . . .
——	..	..	. . . . .	longævus . .	. . . . .	..	..	..	. . . .
——	..	..	. . . . .	new species .	. . . . .	..	..	..	. . . .

Miocene Period: Bordeaux, Dax.	Touraine.	Turin.	Vienna.	Various Localities.	Eocene Period: Paris.	London.	Various Localities.	Sicily.	Italy, Subap.	English Crag.	Baden.	Bordeaux.	Dax.	Touraine.	Turin.	Ronca.	Vienna.	Angers.	Paris.	London.	Valognes.	Belgium.
..	..	..	..	Angers, Ronca	*	*	Valognes .															
..	..	..	..		*	*	ibid.	5	33	..	12	30	22	15	1	..	5	17	41	10	12	3
..	..	..	..		*	*	ibid.															
..	..	..	..		*	*	ibid.															
..	..	..	..		..	..		..	1	..	..	1	2									
*	*?	..	*		..	..																
..	..	..	*		..	..																
*	..	..	..	Moravia	..	..		1	16	..	3	12	5	5	..	..	2	2	5	3	3
*	..	..	..		..	..																
..	..	..	..		*	*																
..	..	..	..	Baden	..	..																
*	*	..	*		..	..		..	1	..	..	2	3	1	..	..	1	..	1			
..	..	..	..		..	..																
..	*	..	..		..	..																
*	*	..	..		..	..																
..	..	..	..		..	..																
..	..	..	..		..	..																
..	..	..	..		..	..																
..	..	..	..		..	..																
..	..	..	..		..	..																
..	..	..	..		..	..																
*	..	..	..		..	..																
*	..	..	..		..	..		10	13	8	3	11	11	10	4	3	2	4	42	14		18
*	*?	*	..		..	..																
*	*	..	..		..	..																
*	*	..	..		..	..																
..	..	..	..		*	*																
..	..	..	..		*	*																
..	..	..	..		*	*																
..	..	..	..		*	*																
=	..	..	..		*	*																
..	..	..	..		*	*																
..	..	..	..	Ronca	*	*?																
..	..	..	..		*	*																
..	..	..	..	Ronca	*	..																
..	..	..	..	ibid.	*	..																
..	..	..	..		*	*																
..	..	..	..		*	*																

GENERA.	No. of Species Living.	No. of Species Fossil (tert.)	Species found both living and fossil (tertiary).	Species found fossil in more than one tertiary formation.	Habitations of living Species.	PLIOCENE PERIOD. Sicily.	Italy, Subap.	English Crag.	Various Localities.
Fusus, cont..				new species .					
————				new species .					
————				new species .					
————				sublavatus .					
————									
————									
Pyrula	31	21	reticulata .		Indian Ocean		*		
————			ficus .		ibid.		*		
————			melongena .		ibid. and Senegal				
————			spirillus .		ibid. ibid.				
————				clava .					
————				ficoides .			*		
————				lævigata .					
————				subcarinata .					
————				elegans .					
Struthiolaria	2	1?							
Ranella	19	8	gigantea .		Mediterr.	*	*		
————			granulata .		Indian Ocean? Senegal?				
————			pygmæa .		European Ocean				
————			tuberosa .		Isle of France		*		
————				lævigata .			*		Perpignan, Morea
Murex	75	89	cornutus .		Senegal		*		
————			Brandaris .		Mediterr.	*	*		Toulon, Perpignan, Morea
————			trunculus .		ibid. & Indian Oc. Senegal	*	*		Morea
————			erinaceus .		Europ. Oc. Medit.	*	*		Perpignan, Morea
————			tripterus .		Indian Ocean		*?		
————			cristatus .		Mediterr.	*	*		
————			fistulosus .		ibid.	*	*		
————			tubifer .		Unknown		*		
————				rectispina .			*		
————				suberinaceus					
————			new species .		Senegal				
————			elongatus .		ibid. and Indian Ocean				
————			angularis .		Senegal				
————			saxatilis, var.		ibid.				
————			new species .		Sicily				
————				polymorphus			*		
————				contabulatus					
————				tricarinatus					
————				tripterus .					
————			Lasseignei .		Mediterr.	*	*		

MIOCENE PERIOD.					EOCENE PERIOD.			No. of Species in each Genus in the following Localities.														
Bordeaux, Dax.	Touraine.	Turin.	Vienna.	Various Localities.	Paris.	London.	Various Localities.	Sicily.	Italy, Subap.	English Crag.	Baden.	Bordeaux.	Dax.	Touraine.	Turin.	Ronca.	Vienna.	Angers.	Paris.	London.	Valognes.	Belgium.
*	..	*	..		..	..																
*	..	..	*		..	..																
*	..	..	*		..	..		10	13	8	3	11	11	10	4	3	2	4	42	14	18	
..	*	..	..	Moravia	..	..																
*	..	..	..	ibid.	..	..																
..	..	..	..		*	*																
*	*	*	..	Angers	..	..																
*	..	*	..		..	..																
*	*	..	..		..	..																
*	*	..	..		..	..																
*	..	..	..		*?	..		..	3	..	1	7	6	2	1	..	..	..	..	10	3	3
*	..	..	..		..	..																
..	..	..	..		*	*	Valognes															
..	..	..	..		*	*	ibid.															
..	..	..	..		*	*																
..	..	..	..		..	..		..	..	..	..	..	..	..	..	..	..	..	..	1		
*?	..	..	..		..	..																
*?	..	..	..		..	..		1	4	..	1	5	6	..	2	..	1					
*?	..	..	..		..	..																
*	..	*	..		..	..																
*	..	*	*	Baden	..	..																
..	..	..	..		..	..																
..	..	..	..	Angers	..	..																
*?	*	..	..		..	..																
*	*	..	..		..	..																
*?	..	..	..		..	..																
..	..	..	..		..	..																
*	..	..	..	Baden	..	*																
*	*	..	..		*	*																
*	..	..	..	Baden	..	..		7	20	5	6	15	20	18	3	..	2	9	24	8	9	
*	..	*	..		..	..																
*	*	..	..		..	..																
*	*	..	..		..	..																
*	..	..	..		..	..																
*	..	..	..		..	..																
*	..	..	..		..	..																
..	..	..	..	Angers	*	..	Valognes															
..	..	..	..		*	*																
..	..	..	..		*	*																
*	..	..	..		..	..																

GENERA.	No. of Species Living.	Fossil (tert.)	Species found both living and fossil (tertiary).	Species found fossil in more than one tertiary formation.	Habitations of living Species.	Sicily.	Italy, Subap.	English Crag.	Various Localities.
Murex, *cont.*	..	..	new species	.....	Mediterr.	*	..	..	.....
———	..	..	.....	new species	.....	..	*	..	.....
———	..	..	.....	sublavatus	.....	..	*	..	.....
———	..	..	.....	new species	.....	..	*	..	.....
Triton	43	25	nodiferum	.....	Mediterr. and Indian Ocean	*	※	..	.....
———	..	..	lampas	.....	Indian Ocean	..	*?	..	.....
———	..	..	scrobiculator	.....	Mediterr.	..	*?	..	.....
———	..	..	succinctum	.....	ibid., New Holland? Lam.	..	※	..	.....
———	..	..	clathratum	.....	Indian Ocean	..	..	..	.....
———	..	..	unifilosum	.....	Mediterr.	..	※	..	.....
———	..	..	.....	intermedium	.....	..	※	..	.....
———	..	..	.....	cancellinum	.....	..	※	..	.....
———	..	..	.....	gibbosum	.....	..	..	..	.....
———	..	..	.....	new species	.....	..	*	..	.....
———	..	..	.....	nodularium	.....	..	..	..	.....
———	..	..	.....	viperinum	.....	..	..	..	.....
Rostellaria	7	8	.....	macroptera	.....	..	..	..	.....
———	..	..	.....	columbella	.....	..	..	..	.....
———	..	..	.....	fissurella	.....	..	..	..	.....
———	..	..	pes pelicani	.....	European Ocean	*	*	..	Morea, Perpignan
———	..	..	pes carbonis	.....	Mediterr.	..	..	..	.....
———	..	..	new species	.....	ibid.	..	※	..	.....
Pterocera	..	7	.....	.....	.....	..	..	..	.....
Strombus	45	9	gigas	.....	Indian Ocean & West Indies	..	*?	..	.....
———	..	..	.....	new species	.....	..	※	..	.....
———	..	..	.....	Bonelli	.....	..	..	..	.....
———	..	..	.....	ornatus	.....	..	..	..	.....
Cassidaria	7	8	echinophora	.....	Mediterr.	※	*	..	Morea, Perpignan
———	..	..	Thyrrena	.....	ibid.	*	※	..	.....
———	..	..	.....	new species	.....	..	※	..	.....
———	..	..	.....	carinata	.....	..	..	..	.....
———	..	..	.....	cithara	.....	..	..	..	.....
Cassis	30	15	flammea	.....	Indian and African Oceans	..	..	..	.....
———	..	..	granulosa	.....	Mediterr.	..	※	..	.....
———	..	..	crumena	.....	Unknown	..	※	..	.....
———	..	..	Saburon	.....	Medit. Senegal	*	*	..	Morea
———	..	..	new species	.....	Unknown	..	..	..	.....
———	..	..	bisulcata	.....	Senegal?	..	*	..	.....
———	..	..	.....	cypræformis	.....	..	※	..	.....
Ricinula	14	1	.....	new species	.....	..	..	..	.....

MIOCENE PERIOD.					EOCENE PERIOD.			No. of Species in each Genus in the following Localities.														
Bordeaux, Dax.	Touraine.	Turin.	Vienna.	Various Localities.	Paris.	London.	Various Localities.	Sicily.	Italy, Subap.	English Crag.	Baden.	Bordeaux.	Dax.	Touraine.	Turin.	Ronca.	Vienna.	Angers.	Paris.	London.	Valognes.	Belgium.
..	..	..	..	. . . . .	..	..	. . . . .															
*	..	..	..	Cracovia	..	..	. . . . .	7	20	5	6	15	20	18	3	..	..	2	9	24	8	9
*	..	..	..	ibid.	..	..	. . . . .															
*	..	..	..	Baden	..	..	. . . . .															
*	..	..	..	. . . . .	..	..	. . . . .															
..	..	..	..	. . . . .	..	..	. . . . .															
..	..	..	..	. . . . .	..	..	. . . . .															
..	..	*	..	. . . . .	..	..	. . . . .															
..	..	..	..	. . . . .	*	..	. . . . .	1	8	..	..	..	3	7	1	2	..	1	1	6	2	5
*	..	..	..	. . . . .	..	..	. . . . .															
*	..	*	..	. . . . .	..	..	. . . . .															
*	*	*	*	. . . . .	..	..	. . . . .															
*	..	*	..	. . . . .	..	..	. . . . .															
..	*	..	*	Cracovia	..	..	. . . . .															
..	..	..	..	. . . . .	*	*	Valognes															
..	..	..	..	. . . . .	*	*	ibid.															
..	..	..	..	. . . . .	*	*	Belgium	1	2	..	..	3	3	2	1	..	..	1	..	4	3	2
..	..	..	..	. . . . .	*	..	ibid.															
..	..	..	..	. . . . .	*	*	Valognes															
..	..	..	*	Moravia	..	..	. . . . .															
*	*	*	..	. . . . .	..	..	. . . . .															
*	*	..	..	. . . . .	..	..	. . . . .															
*?	..	..	..	. . . . .	..	..	. . . . .															
..	*	..	..	. . . . .	..	..	. . . . .	..	2	..	..	1	4	1	1	1	..	..	..	2	1	1
*	..	*	..	. . . . .	*	*	. . . . .															
..	..	..	..	. . . . .	..	..	. . . . .															
..	..	..	..	. . . . .	..	..	. . . . .															
..	..	..	..	. . . . .	..	..	. . . . .	2	2	..	..	..	..	..	2	..	1	..	..	3	1	1
*	..	*	..	. . . . .	*	*	{ Valognes, { Belgium															
*	..	*	..	. . . . .	..	..	. . . . .															
*?	..	..	..	. . . . .	..	..	. . . . .															
..	..	..	..	. . . . .	..	..	. . . . .															
..	..	..	..	. . . . .	..	..	. . . . .															
*	..	*	..	. . . . .	..	..	. . . . .	1	5	1	2	4	4	..	2	..	1	..	2	1	1	
*	..	..	..	. . . . .	..	..	. . . . .															
*	..	..	*	Baden	..	..	. . . . .															
*	..	..	..	. . . . .	..	..	. . . . .															
*	..	*	..	. . . . .	..	..	. . . . .	..	..	..	..	1	1	..	1							

GENERA.	Living.	Fossil (tert.)	Species found both living and fossil (tertiary).	Species found fossil in more than one tertiary formation.	Habitations of living Species.	Sicily.	Italy, Subap.	English Crag.	Various Localities.
Purpura	76	4	hæmastoma	. . . .	Medit., Senegal	..	*	..	. . . . .
————	..	..	. . .	new species	. . . . .	..	..	..	. . .
————	..	..	. . .	new species	. . . . .	..	..	..	. . .
Monoceros	6	1	. . .	. . .	.	.	.	.	.
Harpa	8	2	. .	. .					
Dolium	7	1	pomum?	. .	Indian Ocean	..	*?	..	. . .
Buccinum	143	95	undatum	. .	European Seas, Senegal	..	..	*	. . .
————			reticulatum	. .	ibid. ibid.	..	*	..	. . .
————			maculosum	. .	Mediterr.	*	..	..	Morea
————			mutabile	.	ibid.	*	*	..	Perpignan, Morea
————			clathratum	.	Indian Ocean	..	*	..	ibid.
————			Neriteum	.	ibid. Medit.	..	*	..	Morea
————			Desnoyersi	.	Senegal	..	..	..	. . .
————			prismaticum	.	Mediterr.	*	*	..	Ischia
————			asperulum	.	ibid.	*	*	..	. . .
————			musivum	.	ibid.	*	*	..	. . .
————			inflatum	.	ibid.	*	*	..	Perpignan, Morea
————			polygonum	.	Unknown	..	*	..	. . .
————			D'Orbignii	.	Mediterr.	*	..	..	. . .
————			Linnæi	.	ibid.	*	..	..	. . .
————			new species	.	Indian Ocean	..	..	..	. . .
————			politum	.	ibid.	..	..	..	. . .
————			new species		Unknown	..	..	..	. . .
————			ditto		Mediterr.	*	..	..	. . .
————			ditto		ibid.	*	..	..	. . .
————			ditto		Unknown	..	..	..	. . .
————				serratum		..	*	..	Perpignan
————				baccatum		..	..	..	. . .
————				new species		..	..	..	. . .
————				tessellatum		..	*	..	. . .
————				new species		..	..	..	. . .
————				semistriatum		*	*	..	Perpignan, Morea
————				callosum		..	*	..	. . .
————				new species		..	..	..	. . .
————				Andræi		..	..	..	. . .
————				new species		..	..	..	. . .
————				angulatum		..	*	..	. . .
————				{new species, reticulatum affinis}		..	..	..	. . .
————				turritum		..	..	..	. . .
Eburna	5	1		new species		..	..	..	. . .

MIOCENE PERIOD.					EOCENE PERIOD.			No. of Species in each Genus in the following Localities.														
Bordeaux, Dax.	Touraine.	Turin.	Vienna.	Various Localities.	Paris.	London.	Various Localities.	Sicily.	Italy, Subap.	English Crag.	Baden.	Bordeaux.	Dax.	Touraine.	Turin.	Ronca.	Vienna.	Angers.	Paris.	London.	Valognes.	Belgium.
*		*							1			1	2		1		1		1			
	*		*																			
									1													
																			2			
									1													
*	*		*	Podolia																		
*?			*?	Volhynia																		
*	*		*																			
		*																				
*			*																			
	*																					
				Angers																		
*																						
	*							10	27	9	7	21	14	18	3	1	8	6	9	4	5	2
	*?																					
			*																			
*			*	Podolia																		
*	*																					
*																						
*	*																					
				Baden, Moravia																		
*	*		*																			
*			*																			
*					*																	
*			*																			
*	*																					
*	*		*																			
*	*																					
*		*		Ronca								1	1		1	1						

GENERA.	No. of Species Living.	No. of Species Fossil (tert.)	Species found both living and fossil (tertiary).	Species found fossil in more than one tertiary formation.	Habitations of living Species.	Sicily.	Italy, Subap.	English Crag.	Various Localities.
Terebra . .	44	16	Faval . . .	. . . .	{ Senegal, Indian Ocean	..	*	..	{ Perpignan, Morea
——— . .	..	..	new species .	. . . .	Unknown . . .	..	..	..	. . . .
——— .	..	..	strigilata . .	. . . .	Ind. Oc. Senegal	..	*	..	. . . .
——— .	..	..	pertusa . . .	. . . .	Unknown . . .	..	..	..	. . . .
——— .	..	..	new species .	. . . .	ibid.	..	*	..	. . . .
——— .	..	..	. . . .	striata . . .		..	..	..	. . . .
——— .	..	..	. . . .	plicatula . .		..	..	..	. . . .
Columbella .	33	4	. . . .	new species .		..	..	..	. . . .
Mitra . . .	112	66	. . . .	scrobiculata .		..	*	..	{ Morea, Perpignan
——— .	..	..	. . . .	fusiformis . .		..	*	..	. . . .
——— .	..	..	lutescens	. . . .	Mediterr. . .	*	..	..	. . . .
——— .	..	..	cornea . . .	. . . .	ibid.	..	..	..	. . . .
——— .	..	..	. . . .	incognita . .		..	..	..	. . . .
——— .	..	..	. . . .	columbellata		..	..	..	. . . .
——— .	..	..	. . . .	graniformis .		..	..	..	. . . .
——— .	..	..	. . . .	cupressina .		..	*	..	. . . .
Voluta . . .	66	32	Lamberti . .	. . . .	{ Unknown South Seas ?	..	..	*	. . . .
——— .	..	..	. . . .	papillaris . .		..	..	..	. . . .
——— .	..	..	. . . .	magorum . .		..	*	..	. . . .
——— .	..	..	. . . .	rarispina . .		..	..	..	. . . .
——— .	..	..	. . . .	spinosa . . .		..	..	..	. . . .
——— .	..	..	. . . .	costaria . .		..	..	..	. . . .
——— .	..	..	. . . .	crassicosta .		..	..	..	. . . .
——— .	..	..	. . . .	athleta . . .		..	..	..	. . . .
——— .	..	..	. . . .	ambigua . .		..	..	..	. . . .
——— .	..	..	. . . .	digitalina . .		..	..	..	. . . .
——— .	..	..	. . . .	crenulata . .		..	..	..	. . . .
Marginella .	48	17	cypræola . .	. . . .	Mediterr. . .	*	*	..	. . . .
——— .	..	..	miliacea . .	. . . .	Senegal . . .	..	..	..	. . . .
——— .	..	..	. . . .	eburnea . .		..	..	..	. . . .
——— .	..	..	monilis . .	. . . .	Senegal . . .	..	*?	..	. . . .
Volvaria . .	1	2	. . . .	bulloides . .		..	..	..	. . . .
——— .	..	..	. . . .	acutiuscula .		..	..	..	. . . .
Ovula . . .	18	6	spelta . . .	. . . .	Mediterr. . .	..	..	..	. . . .
——— .	..	..	birostris .	. . . .	Indian Ocean	..	*	..	. . . .
——— .	..	..	new species .	. . . .	Mediterr. . .	*	..	..	. . . .
Cypræa . .	135	19	lurida . . .	. . . .	ibid.	*	*	..	. . . .
——— .	..	..	rufa . . .	. . . .	ibid.	*	*	..	. . . .
——— .	..	..	annulus . .	. . . .	African Ocean	..	..	..	. . . .
——— .	..	..	coccinella . .	. . . .	European Ocean	*	*	*	. . . .
——— .	..	..	new species .	. . . .	Sicily	..	*	..	. . . .

MIOCENE PERIOD					EOCENE PERIOD			No. of Species in each Genus in the following Localities.														
Bordeaux, Dax.	Touraine.	Turin.	Vienna.	Various Localities.	Paris.	London.	Various Localities.	Sicily.	Italy, Subap.	English Crag.	Baden.	Bordeaux.	Dax.	Touraine.	Turin.	Ronca.	Vienna.	Angers.	Paris.	London.	Valognes.	Belgium.
*	*		*	Baden																		
*																						
*	*																					
*									5	4	10	7	2	3			3		1	1	1	
*																						
*	*																					
					*	*	Valognes															
*	*		*	Angers					1								2			1	2	
*				Baden																		
	*																					
								1	13		2	4	6	15					8	24	3	15
*	*																					
*	*																					
					*	*?	Valognes															
			*																			
*	*			Angers																		
*		*																				
		*																				
*	*	*	*																			
					*	*																
					*	*	Valognes		2	1		4	4	2	4	1	1		24	8	7	3
*					*	*																
					*	*																
					*	*																
					*	*	Valognes															
		*?			*	*																
*	*			Angers																		
*	*			ibid.				1	5			2	2	2	1	1			2	9		4
		*			*		Valognes															
					*		Valognes												2	1	1	
					*	*																
	*								2						2					2		
*		*		Transylvania				3	6	3		4	8	5	6				3	4	1	2
*?																						

GENERA.	No. of Species Living.	No. of Species Fossil (tert.)	Species found both living and fossil (tertiary).	Species found fossil in more than one tertiary formation.	Habitations of living Species.	Sicily.	Italy, Subap.	English Crag.	Various Localities.
Cypræa, *cont.*			new species: sphæriculata? Lam.		Unknown		*		
——				Duclosiana					
——				new species					
——				inflata					
——				leporina					
——			sanguinolenta		Senegal				
——				lyncoides					
Oliva	78	13	hiatula		Senegal				
——			flammulata		African Ocean, Madagascar				
——				clavula			*?		
——				Branderi					
——				mitreola					
Ancillaria	9	9		glandiformis			*		
——				canalifera					
——				buccinoidea					
——				inflata					
Terebellum	1	2		convolutum					
——				fusiforme					
Conus	181	49	Mediterraneus		Mediterr.	*			Morea
——				antediluvianus					
——				deperditus					
——				scabriusculus					
——				Brongniarti					
——				Brocchii			*		
——				alsiosus					
——				acutangulus					
——				mercati			*		
——				ponderosus					
——				distans					
——				pyrula			*		
——				new species					
Beloptera		1							
Sepiostera		3							
Sepia	4								
Nautilus	2	4		Deshayesii			*		
Nodosaria	00	21	lævigata		Mediterr.				
——			oblonga		ibid.				
——			sulcata		ibid.		*		
Frondicularia	2	5							
Lingulina	1	2	carinata		West Ind. Seas		*		
Rimulina		1							

MIOCENE PERIOD					EOCENE PERIOD			No. of Species in each Genus in the following Localities.														
Bordeaux, Dax	Touraine	Turin	Vienna	Various Localities	Paris	London	Various Localities	Sicily	Italy, Subap.	English Crag	Baden	Bordeaux	Dax	Touraine	Turin	Ronca	Vienna	Angers	Paris	London	Valognes	Belgium
*		*																				
	*			Angers, Nantes																		
					*	*	Valognes	3	6	3		4	8	5	6			3	4	1	2	
*																						
	*	*																				
*	*			Angers																		
*	*																					
*	*	*	*	Angers																		
*	*	*							1			4	4	3	2		1	1	6	3	3	1
					*	*																
					*	*	Valognes, Belgium															
*	*	*	*	Volhynia																		
*	*	*	*	Angers	*	*	Valognes															
					*	*	ibid.		1			2	2	4	2		2		5	4	3	
					*	*	ibid.															
					*	*	Pauliac, Valognes, Belgium												2	2	1	1
					*	*																
					*	*	Valognes															
					*	*																
					*	*																
					*	*																
*		*	*	Moravia																		
*?																						
*		*	*					1	10			2	14	7	9	3	1	7	2	7	4	2
*	*			Angers, Volhynia, Moravia																		
*			*																			
		*																				
*				Angers																		
			*																			
*	*		*																			
																				1		
																				3		2
*					*	*								2						3		
*									18			3	1							2		
*																						
									4						1							
									1													

f

GENERA.	Living.	Fossil (tert.)	Species found both living and fossil (tertiary).	Species found fossil in more than one tertiary formation.	Habitations of living Species.	Sicily.	Italy, Subap.	English Crag.	Various Localities.
Vaginulina .	8								
Marginulina	9	5	raphanus .	. . . . .	Mediterr. .	.	.	*	. .
Planularia .	3	2	auris . . .	. . . . .	ibid. .	.	.	*	. .
Pavonina. .	1								
Bigenerina .	4								
Textularia. .	15	15	gibbosa . .	. . . . .	Adriatic .	.	.	*	. .
—— .	.	. .	angularis .	. . . . .	ibid. .	.	.	*	. .
—— .	.	. .	sagittula .	. . . . .	Mediterr. .	.	.	*	. .
Vulvulina .	3								
Dimorphina.	1								
Polymorphina	23	11	. . . . .	inequalis .	. . . . .	.	.	*	.
——	.	. .	pupa. . .	. . . . .	Mediterr. .	.	.	*	.
——	.	. .	communis .	. . . . .	Mediterr. .	.	.	*	. .
——	.	. .	caudata .	. . . . .	ibid. .	.	.	*	.
——	.	. .	lævigata .	. . . . .	ibid. .	.	.	*	.
——	.	. .	gibba . .	. . . . .	ibid. .	.	.	*	.
——	.	. .	ovata . .	. . . . .	ibid. .	.	.	.	.
Virgulina .	. .	1	. . . . .	. . . . .	. . . . .	.	.	.	.
Sphæroidina .	1	1	bulloides .	. . . . .	Medit. Ind. Oc.	.	.	*	. .
Clavulina. .	2	3	communis .	. . . . .	Mediterr. .	.	.	*	. .
Uvigerina .	2	3	. . . . .	. . . . .	. . . . .	.	.	.	.
Bulimina .	13	4	. . . . .	. . . . .	. . . . .	.	.	.	.
Valvulina .	1	7	. . . . .	pupa. . .	. . . . .	.	.	.	.
—— .	.	. .	. . . . .	globularis	. . . . .	.	.	.	.
Rosalina . .	6	2	. . . . .	. . . . .	. . . . .	.	.	.	.
Rotalia. . .	25	31	pileus . .	. . . . .	Mediterr. .	.	.	.	.
—— .	.	. .	subrotunda .	. . . . .	ibid. .	.	.	*	. .
—— .	.	. .	armata. .	. . . . .	Cayenne .	.	.	.	'
—— .	.	. .	. . . . .	carinata .	. . . . .	.	.	*	. .
—— .	.	. .	communis .	. . . . .	{ Mediterr. Madagascar	.	.	.	Etang de Tau
—— .	.	. .	Italica . .	. . . . .	Mediterr. .	.	.	*	. .
Calcarina . .	5								
Globigerina .	9	4	elongata .	. . . . .	Mediterr. .	.	.	*	. .
Gyroidina .	9	2	lævis. . .	. . . . .	ibid. .	.	.	*	. .
Truncatulina .	5	5	tuberculata	. . . .	{ ibid. Ind. Oc. European Ocean	*	.	*	. .
Planulina .	4								
Planorbulina	4								
Operculina .	2	2	. . . . .	. . . . .	. . . . .	.	.	.	.
Soldania . .	3	3	. . . . .	. . . . .	. . . . .	.	.	.	.
Cassidulina .	1								
Anomalina .	3	2	. . . . .	. . . . .	. . . . .	.	.	.	Etang de Tau

MIOCENE PERIOD.					EOCENE PERIOD.			No. of Species in each Genus in the following Localities.														
Bordeaux, Dax.	Touraine.	Turin.	Vienna.	Various Localities.	Paris.	London.	Various Localities.	Sicily.	Italy, Subap.	English Crag.	Baden.	Bordeaux.	Dax.	Touraine.	Turin.	Ronca.	Vienna.	Angers.	Paris.	London.	Valognes.	Belgium.
									4				1									
									2													
*									6				3						1	1		
				Angers																		
					*?																	
*																						
					*?				8				5	6	1				3	10			
*																						
*				Angers	*																	
*					*?																	
									1													
									1													
*									2					1						1		
									2					1								
									3					1								
					*		Valognes												4			5
					*		ibid.						1						1			
*																						
*				Angers					5				9	3					1	12		2
*																						
									1				1	1						1		
*									1				1							1		
*								1	1				2	1						1		
													1	1								
									3													
												1										

GENERA.	No. of Species Living.	No. of Species Fossil (tert.)	Species found both living and fossil (tertiary).	Species found fossil in more than one tertiary formation.	Habitations of living Species.	PLIOCENE PERIOD. Sicily.	PLIOCENE PERIOD. Italy, Subap.	PLIOCENE PERIOD. English Crag.	PLIOCENE PERIOD. Various Localities.
Vertebralina	1								
Polystomella	8	3	angularis		Senegal, Australian, Indian, Mediterr. and African Oceans.	..	*	..	. . . .
———	..	..	strigilata		Coasts of Africa	...	..	..	Etang de Tau
Dendritina	2	1				..	..	..	
Peneroplis	3	2				..	..	..	
Spirolina	..	6				..	..	..	
Robulina	17	10	cultrata		Adriatic	...	..	..	
———	..	..	calcar		ibid.	..	*	..	
———	..	..	marginata		West Indian Seas	...	..	..	
Cristellaria	10	14	cassis		Mediterr.	..	*	..	
———	..	..	tuberculata		ibid.	..	*	..	
———	..	..	Italica		Mediterr.	..	*	..	
Nonionina	15	11	umbilicata		ibid.	..	*	..	
———	..	..	communis		ibid., Madagascar, West Indies	..	*	..	
Nummulites	1	13				..	..	..	
Biloculina	4	8	bulloides		Mediterr.	..	*	..	
———	..	..		ringens			..	..	
———	..	..	longirostris		Mediterr.	..	*	..	
———	..	..	depressa		ibid.	..	*	..	
———	..	..	lævis		ibid.	*	..	..	
Spiroloculina	6	9	depressa		ibid.	..	..	..	
Triloculina	16	12		trigonula			..	..	
———	..	..	gibba		Mediterr.	..	*	..	
———	..	..	inflata		ibid.	..	*	..	
———	..	..	oblonga		ibid., Indian & European Oceans	..	*	..	
———	..	..	Brongniarti		West Indies	..	*	..	
Articulina	..	1				..	..	..	
Quinqueloculina	30	23	undulata		Mediterr.	..	*	..	
———	..	..	triangularis		ibid. St. Helena	..	*	..	
———	..	..	bicarinata		Mediterr.	..	*	..	
———	..	..	seminulum		ibid.	..	*	..	
Adelosina	2	2				..	..	..	
Amphistegina	6	1				..	..	..	
Heterostegina	2								
Orbiculina	1								
Alveolina	1	5				..	..	..	
Fabularia	..	1				..	..	..	

MIOCENE PERIOD.					EOCENE PERIOD.			No. of Species in each Genus in the following Localities.														
Bordeaux, Dax.	Touraine.	Turin.	Vienna.	Various Localities.	Paris.	London.	Various Localities.	Sicily.	Italy, Subap.	English Crag.	Baden.	Bordeaux.	Dax.	Touraine.	Turin.	Ronca.	Vienna.	Angers.	Paris.	London.	Valognes.	Belgium.
	*			Angers, Nantes					2						1				1			
												1										
													1						1			
																			6			
			*						8			1							1			
*																						
									14													
*									5			4	3						1			
*												3	2						5		2	2
*					*		Valognes															
								1	2			4							3			
									4			1	2						3			
					*		Valognes															
*									4				4						5		2	
*					*																	
																			1			
*									7			2	1						14			
									2													
												1										
														1					3			
																			1		1	

SUPPLEMENTARY TABLE,

Containing Localities for which there was not sufficient space in the preceding tables.

GENERA.	Scotland	Perpignan	Morea	Osseous breccias	Teneriffe	Quercy	Nice	St. Helena	Lauzerte	Agen	Val d'Arno	Etang de Tau	India	Mayence	Egypt	Germany	Maryland	Doué	Montpellier	Auvergne	Alsace	Bavaria, Hungary, &c.	St. Esprit	Klein Spaunen	Localities unknown
Cyclas	1																								
Cyrena													1	2											
Tridacna														1	1										
Perna																		1							
Pecten		6	4																3	3					
Hinnites																			2						
Helix					6	1	1							1							2	2			7
Pupa									1		1														
Bulimus										1															
Achatina																						1			
Planorbis				2		2															3		1		
Limnea											4	2									3				
Paludina													5			2									5
Ampullaria																								2	
Cerithium															1					4					4
Textularia												4													
Anomalina												1													

GENERAL RESULTS

A COMPARISON OF THE SPECIES EXAMINED IN COMPILING THE FOREGOING TABLES.

PLIOCENE PERIOD.

Italy, Sicily, the Morea, Perpignan, and the English Crag. The fossils of Perpignan and the Morea are, with the exception of three or four species, the same as those of Italy.

In Italy . 569, of which 238 are still living, and 331 extinct (or *unknown*)

Sicily . 226 ,, 216 ,, 10 ,,

The Crag 111 ,, 45 ,, 66 ,,

906

No. of species common to Italy and Sicily . 103

Italy and the Crag * 4

Sicily and the Crag 4

Italy, Sicily, and the Crag 18

129

No. of species proper to Sicily . 65

to the Crag . 23

By subtracting from the total number of species enumerated as belonging to the above localities . . . 906

those species which are common to different localities . 129

We find the real number of the species of this epoch to be . 777

The number of living analogues is 350, which is in the proportion of 49 in 100.

MIOCENE PERIOD.

Bordeaux, Dax, Touraine, Turin, Baden, Vienna, Moravia, Hungary, Cracovia, Volhynia, Podolia, Transylvania, Angers, and Ronca †.

The species of Moravia, Hungary, Cracovia, Volhynia, Podolia, and Transylvania, are the same, with a very few exceptions, as those of Vienna and Baden.

* The statement that there are only 4 species common to Italy and the Crag, may seem inconsistent with the fact that 18 are common to those places and to Sicily; but the reader will understand that there are only 4 species which are common to Italy and the Crag, and which are not also common to some *other Pliocene locality*. The same remark is applicable to similar statements in the sequel.

† Ronca may very probably belong to the Eocene epoch; but in this, as in respect to a few other localities mentioned in the tables, the number of analogues is too small to lead to certain conclusions.

No. of species.
Bordeaux and Dax* 594, of which 136 are still living, and 458 extinct.

Touraine	.	.	298	,,	68	,,	230 ,,
Turin	.	.	97	,,	17	,,	80 ,,
Vienna	.	.	124	,,	35	,,	89 ,,
Baden	.	.	99	,,	26	,,	73 ,,
Angers	.	.	166	,,	25	,,	141 ,,
Ronca	.	.	40	,,	3	,,	37 ,,

1418

No. of species.

Common to Bordeaux, Dax, Touraine	62
ib. ib. Turin	18
ib. ib. Vienna	23
ib. ib. Baden	13
ib. ib. Angers	8
ib. ib. Ronca	0
ib. ib. Touraine and Turin . . .	12
ib. ib. Touraine and Vienna . .	17
ib. ib. Touraine and Baden . . .	4
ib. ib. Touraine and Angers . .	14
ib. ib. Touraine and Ronca . . .	0
ib. ib. Touraine, Turin and Vienna .	8
ib. ib. Touraine, Turin and Angers .	2
ib. ib. Touraine, Vienna and Angers .	7
ib. ib. Turin and Vienna . . .	6
ib. ib. Turin and Ronca . . .	1
ib. ib. Baden and Angers . . .	1
Touraine and Angers	10
Touraine and Turin	3
Touraine and Vienna	15
Touraine and Baden	2
Turin and Ronca	2
Vienna and Angers	2
Angers and Ronca	1
Touraine, Vienna and Baden	2
Touraine, Vienna, Angers and Baden . . .	1
Bordeaux, Dax, Touraine, Turin, Vienna and Angers	3
ib. ib. ib. Turin, Vienna and Baden	3
ib. ib. ib. Vienna and Baden .	14
ib. ib. ib. Vienna, Angers and Baden	2

Carried over . . . 256

* There are at Bordeaux . . 446 species
and at Dax . . . 473

making a total of . . 919
but from the great number of species common to the two localities there are, in reality, only 594 species, as above mentioned.

	No. of Species
Brought over	256
Common to Bordeaux, Dax, Touraine, Turin, Vienna, Angers & Baden	1
ib. ib. ib. Angers and Baden . .	2
ib. ib. Vienna and Baden . . .	4
	263
By adding to the above 134 species which are common to the Miocene, and the two other epochs	134
the total number of analogues will be found to be . .	397
By subtracting from the total number of species of the above localities	1418
those species which are common to different localities . .	397
We find the real number of species of this epoch to be .	1021

The number of living analogues is 176, which is in the proportion of rather less than 18 in 100; the number of fossil analogues, after subtracting those which pass from the Miocene into both the Pliocene and Eocene epochs, is 168, which is very nearly in the same proportion.

The species which pass from the Miocene into the Pliocene period are in number 196, of which 114 are living, and 82 fossil, which is very nearly in the proportion of 20 in 100 of the total number of species of the latter epoch. Thus it is remarkable that there are 18 in 100 living analogues, 18 in 100 of analogous fossil species, and that 20 in 100 of these species pass from the Miocene to the Pliocene epoch.

The 114 living species, and the 82 fossil ones, which are common to the Miocene and Pliocene periods, are distributed, in the last-mentioned epoch, in the following manner:—

LIVING.		FOSSIL.	
Crag . . .	4	Crag . .	4
Italy . . .	48	Sicily . .	1
Sicily . .	5	Italy . .	71
Sicily and Italy .	46	Sicily and Italy .	5
Sicily, Italy, and the Crag	11	Sicily and the Crag	1
	114		82

The preceding distribution of species will show that Italy is represented in the Miocene period by 181 species, Sicily by 69, and the Crag by 20.

EOCENE PERIOD.

Paris, London, Valognes, Belgium, Castelgomberto, and Pauliac.

A small number of species only have been examined from Belgium, Pauliac, and Castelgomberto, but which agreed, with few exceptions, with species of the Paris basin. So also in regard to Valognes.

Number of species, Paris   .   1122 of which 38 are still living, and 1084 extinct (or unknown).

London   .   239 of which 12 are still living, and 227 extinct (or unknown).

Valognes   .   332

Belgium   .   49

———

1742

By subtracting from these localities the number of analogous species   .   .   .   504

The real number of species of this epoch is   .   .   .   1238

The number of fossils of this period identified with living species is 42, which is to 1238 in the proportion of $3\frac{1}{4}$ in 100. The number of fossil species which pass from the Eocene into the two other periods is 46, that is to say, in nearly the same proportion as the living analogues. Among the fossil species, four only are common to the three epochs, which are the following :—

1 Dentalium coarctatum.	3 Bulimus terebellatus.
2 Tornatella inflata.	4 Corbula complanata.

The 42 other fossil species, which go no farther than the Miocene epoch, are distributed in the following manner :—

Bordeaux and Dax	17
Turin	3
Angers	2
Ronca	7
Bordeaux, Dax and Touraine . .	4
ib. ib. and Turin . . .	1
ib. ib. Touraine and Angers . .	2
ib. ib. Turin, Vienna and Baden .	1
ib. ib. Touraine, Turin, Vienna and Angers .	1
ib. ib. Touraine, Vienna, Angers and Baden .	1
Turin and Ronca	2
Angers and Ronca	1
	42

Of the 42 living species, the following 13 are common to the three epochs,—

1 Dentalium entalis,	7 Murex tubifer,
2 ——— strangulatum,	8 Polymorphina gibba,
3 Fissurella græca,	9 Triloculina oblonga,
4 Bulla lignaria,	10 Lucina divaricata,
5 Rissoa cochlearella,	11 ——— gibbosula,
6 Murex fistulosus,	12 Isocardia cor,

13 Nucula margaritacea.

Of the other species, 7 go no farther than the Miocene epoch, and are distributed in the following manner,—

Bordeaux and Dax			3
ib.	ib. and Baden		1
ib.	ib. and Touraine		1
ib.	ib. and Angers		1
ib.	ib. Touraine and Angers		1
			7

Total number of species in the three periods,—

In the Pliocene	.	.	777
In the Miocene	.	.	1021
In the Eocene	.	.	1238
			3036

From the above lists it will appear that there are 17 species which are common to the three epochs, and which may therefore be said to characterise the entire tertiary formations of Europe. Thirteen of them are species still living, while four are only known as fossil. There is not a single species common to the Pliocene and Eocene epochs which is not also found in the Miocene.

Geographical Distribution of the living Species which have their fossil Analogues.

Pliocene Epoch, 350 species.

In the Mediterranean	. . . .	242
In the Indian Ocean	. . . .	25
At Senegal	. . . .	5
Common to the Mediterranean and Senegal	.	14
——————————— and the African Ocean		8
————— Indian Ocean and to Senegal	.	7
——————————— and to America	.	5
In the Northern European Ocean	. .	43
—— Pacific Ocean	. . .	1
		350

Fossil in Sicily and Italy. Fossil in the Crag.

Miocene Epoch, 176 species, (100 species common to the preceding epoch.)

Species.

At Senegal, of which 13 are common to the Indian Ocean, and 12 to the Mediterranean 79

In the Mediterranean and Southern European Ocean, of which 10 are common to the Indian Ocean, and 12 to Senegal . 86

In the Indian Ocean, 10 of which are common to the Southern European Ocean 29

Carried over 194

Species.

Brought over 194

In the Equatorial Seas of America, 2 of which are common to the
Indian Ocean 7
In the Pacific Ocean 2
———
203

Number common to different localities   27
———
176

*Eocene Epoch, 42 species, of which 26 are common to the
two preceding epochs.*

In the Mediterranean, 5 of which are common to India and New
Holland 19
In the Indian Ocean 7
In New Holland 3
In Senegal 3
—
32

Of the fluviatile and terrestrial species, 5 are still living in Europe,
1 in the Philippine Islands, and 4 in Asia, Spain and Greece   10
—
42

APPENDIX II.

N. B. Those which are marked with an asterisk in this and the following lists
are unknown as recent.

FOSSIL SHELLS FROM THE FLANKS OF ETNA, IMME-
DIATELY ABOVE THE BAY OF TREZZA.

In clay and volcanic tuff. (See p. 79.)

Mactra triangula, *Broc.*
Corbula nucleus.
Astarte, unnamed.
  „   ditto.
Venus Brongniarti, *Payraudeau.*
  „   radiata.
  „   species doubtful.
Cytherea exoleta.
  „   lincta.
Cardium edule.
  „   sulcatum.
Arca antiquata, *Lamk.*
  „ barbata, *Lamk.*
Pectunculus glycimeris.
  „   pilosus.
Nucula, new species.
  „   margaritacea.
Chama unicornis.
Pecten ornatus.
  „   an, new species ?
  „   Bruei, *Payr.*
  „   opercularis.
  „   unicolor.
  „   Jacobæus.
Ostrea edulis.
Anomia ephippium.
Dentalium entalis.
  „   * new species.
  „   strangulatum, *Desh.*
  „   novem costatum.
Pileopsis Ungarica.
Calyptræa Chinensis.
Natica millepunctata.

Natica Dillwynii, *Payr.*
  „   Guillemini, *Payr.*
  „   glaucina.
Trochus magus.
  „   conulus.
  „   Adansoni, *Payr.*
Turbo rugosus.
Monodonta Viellotii, *Payr.*
Turritella terebra, *Broc.*
Cerithium sulcatum.
Pleurotoma, new species.
Cancellaria cancellata.
Fusus, new species.
  „   new species.
  „   craticulatus, *Blainv.*
  „   strigosus.
  „   lignarius.
Murex trunculus.
  „   Brandaris.
  „   erinaceus.
Triton an pileare ?
Rostellaria pes pelicani.
Cassidaria echinophora.
Buccinum musivum, *Broc.*
  „   mutabile.
  „   maculosum.
  „   unnamed.
  „   Calmelii.
* „   semistriatum.
Colombella rustica.
Mitra lutescens.
Conus Mediterraneus

VILLASMONDE. (See p. 65.)

Mactra lactæa.
Corbula nucleus.
Pecten opercularis.
Dentalium novem costatum.

Natica glaucina.
„ canrena.
Rostellaria pes pelicani.

MILITELLO. (V. DI NOTO.) (See p. 65.)

Cytherea chione?
Cardium edule.
Arca antiquata.
Pecten Jacobæus.
„ varius.

Pecten opercularis.
Ostrea edulis.
* Dentalium, new species.
Buccinum prismaticum?
An Turbo rugosus???

GIRGENTI.—(*In limestone and clay.*—See p. 65.)

Corbula nucleus.
Mactra triangula? *Broc.*
Pectunculus pilosus.
Modiola, species doubtful.
Pecten Jacobæus.

Pecten opercularis.
Dentalium entalis.
Natica millepunctata.
* Turritella tornata.
* Buccinum semistriatum.

SYRACUSE. (See p. 67.)

Thracia pubescens.
* Tellina?
Cardium sulcatum.
„ edule.
„ echinatum.
Isocardia cor.
Arca antiquata.
Pectunculus pilosus.
„ glycimeris.
Pecten, an nodosus?
„ Jacobæus.
„ Audouini, *Payr.*
„ opercularis.

Pecten coarctatus, *Broc.*
„ varius.
Ostrea edulis.
* Dentalium, new species.
„ strangulatum, *Desh.*
„ sexangulare, *Broc.*
Haliotis tuberculata.
Trochus conulus.
Turritella terebra.
Murex trunculus?
* Buccinum semistriatum.
Cypræa rufa.
Conus Mediterraneus??

CALTANISETTA.—(*In clay and yellow sand.*—See p. 67.)

Lucina lactea?
Venus multilamella.
Cardium echinatum.
„ edule.
Arca antiquata.
Pecten Jacobæus.
Ostrea edulis.
* Dentalium, new species.
„ fossile.
„ sexangulare.

Natica Guillemini, *Payr.*
„ millepunctata.
Turritella subangulata, *Broc.*
* „ tornata, *Broc.*
„ terebra.
Cerithium vulgatum.
Fusus lignarius.
Rostellaria pes pelicani.
* Buccinum semistriatum.
Mitra lutescens.

CALTAGIRONE. (See p. 67.)

Mactra triangula, *Broc.*
Amphidesma, new species.
Corbula nucleus.
Psammobia angulata (Tellina, *Lin.*)
Cytherea lincta.
Venus multilamella.
Cardita sulcata, *Brug.*
,, squamosa.
Cardium echinatum.
Arca antiquata.
* Nucula Italica, *Def.*
Pecten opercularis.
,, Bruei, *Payr.*
Dentalium, new species.
,. entalis.
,, novem costatum.
Fissurella costaria, *Desh.*
Calyptræa chinensis.
Bulla lignaria.
Natica canrena.

Natica Dilwynii, *Payr.*
,, Valenciennesii, *Payr.*
,, Guillemini, *Payr.*
Scalaria tenuicostata, *Mich.*
Turritella terebra.
Cerithium Latreillei, *Payr.*
* Pleurotoma, new species.
,, vulpecula, *Broc.*
,, craticulata, *Broc.*
Fusus craticulatus, *Blain.*
,, lignarius.
,, rostratus.
Murex Brandaris.
Rostellaria pes pelicani.
* Buccinum semistriatum.
,, mutabile.
,, prismaticum, *Broc.*
,, turbinellus, *Broc.*
Mitra lutescens.
Cypræa oryza, *Duclos.*

CASTROGIOVANNI. (See p. 67.)

Lucina lactea.
* Nucula Italica, *Def.*
Pecten Jacobæus.

Pecten opercularis.
Natica Guillemini, *Payr.*

VIZZINI.

Murex Brandaris.

PIAZZA.

Ostrea edulis,

VAL DI NOTO.

Pectunculus glycimeris.

Pecten opercularis.

PALERMO.—(*In limestone and clay.*—See p. 65.)

* Clavagella bacillaris, *Desh.*
Solen coarctatus, *Broc.*
Panopea Aldrovandi.
Thracia corbuloides, *Desh.*
,, pubescens, *Desh.*
Lutraria solenoides.
Corbula nucleus.

Tellina Donacina.
,, new species.
Lucina radula.
,, new species, a lupinus? *Broc.*
,, lactea.
Astarte, new species, an incrassata ?
,, new species.

PALERMO—continued.

Cytherea rugosa, *Broc.*
   ,,   exoleta.
   ,,   Chione.
Venus radiata, *Broc.*
   ,,   species doubtful.
Cardita squamosa.
Cyprina Islandicoides.
Cardium sulcatum.
   ,,   edule.
   ,,   echinatum.
   ,,   Deshayesii, *Payr.*
Isocardia cor.
Arca antiquata.
Pectunculus pilosus.
Nucula, new species.
   ,,   margaritacea.
Chama gryphoides?
   ,,   unicornis.
Pecten ornatus.
   ,,   coarctatus, *Broc.*
   ,,   an, new species.
   ,,   opercularis.
   ,,   new species.
   ,,   Jacobæus.
   ,,   varius??
   ,,   pleuronectes.
Ostrea cornucopiæ?
   ,,   edulis.
   ,,   Virginica.
   ,,   hippopus.
Terebratula truncata.
*   ,,   ampulla.
* Dentalium, new species.
*   ,,   fossile, *Lin. Var.*
   ,,   strangulatum, *Desh.*
Patella bonnardii, *Payr.*
Emarginula curvirostris, *Desh.*
Fissurella, species doubtful.
   ,,   Græca.
Pileopsis Ungarica.
Bulla lignaria.
Auricula buccinea.
Natica millepunctata.
   ,,   Guillemini, *Payr.*

Natica Canrena.
   ,,   Valenciennesii, *Payr.*
Scalaria communis.
   ,,   pseudo scalaris.
Solarium stramineum ?
Trochus magus.
   ,,   agglutinans.
   ,,   cingulatus, *Broc.*
   ,,   Adansoni, *Payr.*
   ,,   conulus.
   ,,   cinereus.
Turbo rugosus.
Turritella terebra.
Cerithium tricinctum.
*   ,,   margaritaceum.
*   ,,   new species.
Pleurotoma Cordieri.
   ,,   Caumarmondi, *Mich.*
   ,,   new species.
Fasciolaria tarentina.
Fusus sinistrorsus, *Desh.*
   ,,   strigosus.
   ,,   rostratus.
*   ,,   clavatus.
   ,,   craticulatus.
*   ,,   new species.
   ,,   lignarius.
Ranella gigantea.
*   ,,   lævigata.
Murex Brandaris.
   ,,   trunculus.
Triton unifilosum, *Bon.*
Rostellaria pes pelicani.
Cassidaria echinophora.
Cassis saburon.
Dolium pomum.
Buccinum prismaticum, *Broc.*
   ,,   new species.
   ,,   new species.
*   ,,   semistriatum.
   ,,   new species.
   ,,   mutabile.
Conus Mediterraneus.

2.—FOSSIL SHELLS COLLECTED BY THE AUTHOR IN ISCHIA, AND NAMED BY M. DESHAYES. (See p. 126.)

Solen coarctatus, *Broc.*
Lucina lupinus, *Broc.*
Venus radiata, *Broc.*
„ verrucosa.
Cardium sulcatum.
„ edule.
Pectunculus violacescens
Arca, new species, An arca quoyi?
Payr.
Nucula margaritacea.
Pecten varius.
„ Jacobæus.
„ Dumasii, *Payr.*
„ opercularis.
Dentalium novem costatum.

Melania Cambessedesii, *Payr.*
Natica Guillemini, *Payr.*
„ Valenciennesii? *Payr.*
Trochus magus.
„ conuloides.
„ new species.
Turritella terebra.
Cerithium Latreillei, *Payr.*
„ new species.
„ vulgatum.
„ doliolum, *Broc.*
Rostellaria pes pelicani.
Buccinum prismaticum, *Broc.*
Cypræa lurida.

The four following shells have since been sent to me from Ischia. They are all of recent species :—

Pectunculus pilosus.
Natica glaucina ?

Trochus crenulatus.
Turritella duplicata.

FOSSIL SHELLS FROM THE WESTERN BORDERS OF THE RED SEA, COLLECTED BY MR. JAMES BURTON, COMMUNICATED BY G. B. GREENOUGH, ESQ., P.G.S.

Lutraria solenoides ?
„ plicatella ? ?
Mactra stultorum ?
Corbula.
Psammobia.
Tellina lingua felis.
„ rugosa.
„ virgata.
„ rostrata.
Lucina globosa.
Cyprina.
Cytherea tigerina.
„ picta.
„ castrensis.
„ erycina.
„ scripta ?
Venus geographica.
„ reticulata.
„ ovata ? ?
„ paphia.
Cardium rugosum.
„ Æolicum ?
„ retusum.
Cardita turgida.
„ calyculata.

Cypricardia.
Chama lazarus.
Tridacna squamosa.
Arca scapha.
„ antiquata.
„ Noæ.
Pectunculus pectiniformis.
Pecten maximus.
„ pes felis.
Spondylus gaderopus.
Parmophorus elegans.
Fissurella Græca.
Bulla ampulla.
„ solida.
Helix desertum.
Nerita.
Natica melanostoma.
„ mamillaris.
„ mamilla.
„ Græca.
„ alba.
Haliotis tuberculata ?
„ striata ?
Tornatella.
Pyramidella.

RED SEA SHELLS—continued.

Solarium perspectivum.
Trochus maculatus.
„ virgatus.
„ Mauritianus.
Monodonta tectum.
„ Pharaonis.
„ Ægyptica.
Turbo chrysostomus.
„ petholatus.
Turritella
Cerithium nodulosum.
„ sulcatum.
„ virgatus.
Pleurotoma virgo.
Turbinella lineolata.
Cancellaria contabulata.
Fasciolaria trapezium.
Pyrula abbreviata.
„ rapa.
„ citrina.
„ reticulata.
„ francolinus.
Ranella granifera.
„ crumena.
Murex crassispina.
„ scorpio.
Triton variegatum.
„ lampas.
„ pileare.
„ maculosum.
Strombus gigas (young).
„ bituberculatus.
„ lineolatus.
„ gibberulus.
„ terebellatus.
Cassis vibex.
„ saburon.
„ erinaceus.
Ricinula arachnoides.
Dolium perdix.

Dolium pomum.
Buccinum coronatum.
„ arcularia.
„ senticosum.
Terebra crenulata.
„ subulata.
„ myosurus.
„ maculata.
„ duplicata.
Colombella turturina?
Mitra striatula.
„ coronata?
Cypræa mappa.
„ Arabica.
„ talpa.
„ caurica.
„ vitellus.
„ erosa.
„ carneola.
„ turdus?
„ lurida?
„ flaveola?
„ nucleus.
„ stercus muscarum?
„ caput serpentis?
Ancillaria.
Oliva erythrostoma.
Conus arenatus.
„ generalis.
„ literatus?
„ betulinus.
„ striatus.
„ episcopus.
„ tessellatus.
„ textile.
„ nussatella.
„ clavus.
„ terebra??
„ capitaneus.
Terebellum subulatum.

The above shells were named by Mr. GRAY, F.R.S., and Mr. Frembley.

FOSSIL SHELLS COLLECTED BY THE AUTHOR FROM THE LOESS OF THE VALLEY OF THE RHINE.

(See p. 151.)

(See p. 151.)

Helix obvoluta, *Drap.*
„ ericetorum, *ib.*
„ ourthusianella, *ib.*
„ plebeium, *ib.*
„ pomatia.
„ nemoralis, *ib.*
„ fruticum, *Drap.*
„ arbustorum.

Helix striata, *Drap.*
Succinea elongata.
Cyclas palustris, *Drap.*
„ lacustris, *ib.*
Valvata piscinalis.
Limnea ovata, *Drap.*
Paludina impura.

SIENNA. (See pp. 160 and 163.)

Serpula arenaria.
  ,,   ditto, var.
* ,,   new species.
* ,,   glomerata.
Mactra triangula.
Tellina complanata.
* Cytheræa rugosa, *Broc.*
Cardita intermedia.
* ,,   new species.
Cardium edule.
Arca antiquata.
Pectunculus pilosus.
  ,,   nummarius.
* ,,   auritus.
Nucula margaritacea.
Pecten Jacobæus.
  ,,   opercularis.
⁜ ,,   striatus?
* ,,   laticostatus.
Ostrea edulis, *Lin.*
  ,,   edulis? junior.
* ,,   (nobis incognita).
Dentalium sexangulare.
* ,,   fossile.
Auricula buccinea, *Desh.*
Melanopsis buccinoides.
Natica glaucina.
  ,,   punctata.
  ,,   Marochiensis.
Trochus fermonii, *Payr.*
* ,,   new species, with its colour.
Turbo rugosus.
Turritella terebra.
* ,,   imbricataria, *Broc.*
  ,,   subangulata.
* ,,   tornata.
* ,,   varicosa.
Cerithium vulgatum.
  ,,   doliolum.
* ,,   tricinctum.
* ,,   new species.
* ,,   new species.
* Pleurotoma cataphracta.
* ,,   interrupta, *Broc.*
* ,,   oblonga, *Broc.*

* Pleurotoma rotata, *Broc.*
* ,,   reticulata, *Broc.*
* ,,   textilis, *Broc.*
* ,,   turricula, *Broc.*
* ,,   dimidiata, *Broc.* (dentata, *Lamk.*)
* ,,   dimidiata, var.
Cancellaria cancellata.
* ,,   varicosa.
* ,,   Lyrata.
Fusus lignarius
* ,,   mitræformis.
  ,,   subulatus.
* ,,   longiroster.
* ,,   thiara, *Broc.*
* Ranella bimarginata.
Murex cornutus, var.
  ,,   tubifer.
* ,,   horridus.
* ,,   spirispina.
  ,,   pomum.
* ,,   bracteatus, *Broc.*
* ,,   new species.
Triton unifilosus, *Bonelli.*
* ,,   reticularis, *Broc.*
* ,,   new species.
Rostellaria pes pelicani.
* Strombus Bonelli, *Brong.*
Cassis saburon.
Buccinum clathratum? *Broc.*
* ,,   serratum, *Broc.*
* ,,   costulatum.
* ,,   gibbosulum.
* Terebra plicaria.
* ,,   duplicata.
* Mitra pyramidella, *Broc.*
* ,,   scrobiculata.
Marginella cypræola.
* Conus antediluvianus, *Broc.*
* ,,   ponderosus.
* ,,   mercator.
* ,,   pyrula, *Broc.*
* ,,   betulinoides.
* ,,   virginalis.

60

SECONDARY FOSSIL SHELLS.

FOSSIL SHELLS COMMON TO THE MAESTRICHT BEDS AND THE WHITE CHALK; COMMUNICATED BY M. DESHAYES. (See p. 325.)

Catillus (Inoceramus) Cuvieri?
  (specimens imperfect.)
Plagiostoma spinosa.
  ,,   Hoperi.
Pecten fragilissimus.
Ostrea vesicularis.
  ,, carinata.

Crania Parisiensis.
Terebratula octoplicata.
  ,,   carnea.
  ,,   pumilus (magus, *Sow.*)
  ,,   Defrancii.
Belemnites mucronatus.

FOSSIL SHELLS COMMON TO THE MAESTRICHT BEDS AND THE UPPER GREEN-SAND; COMMUNICATED BY M. DESHAYES. (See p. 325.)

Plagiostoma spinosa.
Ostrea vesicularis.
  ,, carinata.

Belemnites mucronatus.
Baculites Faujasii.

GLOSSARY

Of Geological and other Scientific Terms used in this Work.

Several of the Author's friends, who had read the first and second volumes of the Principles of Geology, having met with difficulties from their previous un-acquaintance with the technical terms used in Geology and Natural History, suggested to him that a Glossary of those words would render his work much more accessible to general readers. The Author willingly complied with this sug-gestion, but finding that his own familiarity with the subject made him not a very competent judge of the terms that required explanation, he applied to the friends above alluded to for their assistance, and from lists of words supplied by them, the following Glossary has been constructed. It will be obvious to men of science, that in order to attain the object in view, it was necessary to employ illustrations and language as familiar as possible to the general reader.

ACEPHALOUS. The Acephala are that division of molluscous animals which, like the oyster and scallop, are without heads. The class Acephala of Cuvier comprehends many genera of bivalve shells, and a few genera of mollusca which are devoid of shells. *Etym.*, *a*, *a*, without, and κεφαλη, *cephale*, the head.

ALGÆ. An order or division of the cryptogamic class of plants. The whole of the sea-weeds are comprehended under this divi-sion, and the application of the term in this work is to marine plants. *Etym.*, *Alga*, sea-weed.

ALUM-STONE, ALUMEN, ALUMINOUS. Alum is the base of pure clay, and strata of clay are often met with containing much iron-pyrites. When the latter substance decomposes, sulphuric acid is pro-duced, which unites with the aluminous earth of the clay to form sulphate of alumine, or common alum. Where manu-factories are established for obtaining the alum, the indurated beds of clay employed are called Alum-stone.

ALLUVIAL. The adjective of alluvium, which see.

ALLUVION. Synonymous with alluvium, which see.

ALLUVIUM. Earth, sand, gravel, stones, and other transported matter which has been washed away and thrown down by rivers, floods, or other causes, upon land not *permanently* submerged beneath the waters of lakes or seas. *Etym.*, *Alluo*, to wash upon. For a further explanation of the term, as used in this work, see vol. ii. chap. xiv., and vol. iii. p. 145.

AMMONITE. An extinct and very numerous genus of the order of molluscous animals, called Cephalopoda, allied to the modern genus Nautilus, which inhabited a chambered shell, curved like a coiled snake. Species of it are found in all geological periods of the secondary strata; but they have not yet been seen in the tertiary beds. They are named from their resemblance to the horns on the statues of Jupiter Ammon.

AMORPHOUS. Bodies devoid of regular form. *Etym.*, a, a, without, and μορφη, *morphe*, form.

AMYGDALOID. One of the forms of the Trap-rocks, in which agates and simple minerals appear to be scattered like almonds in a cake. *Etym.*, αμυγδαλα, *amygdala*, an almond.

ANALCIME. A simple mineral of the Zeolite family, of frequent occurrence in the trap-rocks.

ANALOGUE. A body that resembles or corresponds with another body. A recent shell of the same species as a fossil-shell, is the analogue of the latter.

ANOPLOTHERE, ANOPLOTHERIUM. A fossil extinct quadruped belonging to the order Pachydermata, resembling a pig. It has received its name because the animal must have been singularly wanting in means of defence, from the form of its teeth and the absence of claws, hoofs, and horns. *Etym.*, ανοπλος, *anoplos*, unarmed, and θηριον, *therion*, a wild beast.

ANTAGONIST POWERS. Two powers in nature, the action of the one counteracting that of the other, by which a kind of equilibrium or balance is maintained, and the destructive effect prevented that would be produced by one operating without a check.

ANTENNÆ. The articulated horns with which the heads of insects are invariably furnished.

ANTHRACITE. A shining substance like black-lead; a species of mineral charcoal. *Etym.*, ανθραξ, *anthrax*, coal.

ANTHRACOTHERIUM. A name given to an extinct quadruped, supposed to belong to the Pachydermata, the bones of which were found in lignite and coal of the tertiary strata. *Etym.*, ανθραξ, *anthrax*, coal, and θηριον, *therion*, wild beast.

ANTHROPOMORPHOUS. Having a form resembling the human. *Etym.*, ανθρωπος, *anthropos*, a man, and μορφη, *morphe*, form.

ANTICLINAL AXIS. If a range of hills, or a valley, be composed of strata, which on the two sides dip in opposite directions, the imaginary line that lies between them, towards which the strata on each side rise, is called the anticlinal axis. In a row of houses with steep roofs facing the south, the slates represent inclined strata dipping north and south, and the ridge is an east and west anticlinal axis. For a farther explanation, with a diagram, see vol. iii. p. 293.

ANTISEPTIC. Substances which prevent corruption in animal and vegetable matter, as common salt does, are said to be antiseptic. *Etym.*, αντι, *against*, and σηπω, *sepo*, to putrefy.

ARENACEOUS. Sandy. *Etym.*, *Arena*, sand.

ARGILLACEOUS. Clayey, composed of clay. *Etym.*, *Argilla*, clay.

ARRAGONITE. A simple mineral, a variety of carbonate of lime, so called from having been first found in Arragon, in Spain.

AUGITE. A simple mineral of a dark green, or black colour, which forms a constituent part of many varieties of volcanic rocks.

AVALANCHES. Masses of snow which, being detached from great heights in the Alps, acquire enormous bulk by fresh accumulations as they descend; and when they fall into the valleys below often cause great destruction. They are also called *lavanges*, and *lavanches*, in the dialects of Switzerland.

BASALT. One of the most common varieties of the Trap-rocks. It is a dark green or black stone, composed of augite and felspar, very compact in texture, and of considerable hardness, often found in regular pillars of three or more sides, called basaltic columns. Very remarkable examples of this kind of rock are seen at the Giant's Causeway, in Ireland, and at Fingal's Cave, in the island of Staffa, one of the Hebrides. The term is used by Pliny, and is said to come from *basal*, an Æthiopian word signifying iron, not an improbable derivation, inasmuch as the rock often contains much iron, and as many of the figures of the Egyptian temples are formed of basalt.

' BASIN ' of Paris, ' BASIN ' of London. Deposits lying in a great hollow or trough surrounded by low hills or high land, sometimes used in geology almost synonymously with ' formation.'

BELEMNITE. An extinct genus of the order of molluscous animals called Cephalopoda, that inhabited a long, straight, and chambered conical shell. *Etym.*, βελεμνον, *belemnon*, a dart.

BITUMEN. Mineral pitch, of which the tar-like substance which is often seen to ooze out of the Newcastle coal when on the fire, and which makes it cake, is a good example. *Etym.*, *Bitumen*, pitch.

BITUMINOUS SHALE. An argillaceous shale, much impregnated with bitumen, which is very common in the coal measures.

BLENDE. A metallic ore, a compound of the metal zinc with sulphur. It is often found in brown shining crystals, hence its name among the German miners, from the word *blenden*, to dazzle.

BLUFFS. High banks presenting a precipitous front to the sea or a river. A term used in the United States of North America.

BOTRYOIDAL. Resembling a bunch of grapes. *Etym.*, βοτρυς, *botrys*, a bunch of grapes, and ειδος, *eidos*, form.

BOWLDERS. A provincial term for large rounded blocks of stone lying on the surface of the ground, or sometimes imbedded in loose soil, different in composition from the rocks in their vicinity, and which have been therefore transported from a distance.

BRECCIA. A rock composed of angular fragments connected together by lime or other mineral substance. An Italian term.

CALC SINTER. A German name for the deposits from springs holding carbonate of lime in solution—petrifying springs. *Etym.*, *Kalk*, lime, *sintern*, to drop.

CALCAIRE GROSSIER. An extensive stratum, or rather series of strata, belonging to the Eocene tertiary period, originally found in, and specially belonging to, the Paris Basin. See Table II. E, p. 390. *Etym.*, *Calcaire*, limestone, and *grossier*, coarse.

CALCAREOUS ROCK. Limestone. *Etym.*, *Calx*, lime.

CALCEDONY. A siliceous simple mineral, uncrystallized. Agates are partly composed of calcedony.

CARBON. An undecomposed inflammable substance, one of the simple elementary bodies. Charcoal is almost entirely composed of it. *Etym.*, *Carbo*, coal.

CARBONATE OF LIME. Lime combines with great avidity with carbonic acid, a gaseous acid only obtained fluid when united with water,—and all combinations of it with other substances are called *Carbonates*. All limestones are carbonates of lime, and quick lime is obtained by driving off the carbonic acid by heat.

CARBONATED SPRINGS. Springs of water, containing carbonic acid gas. They are very common, especially in volcanic countries, and sometimes contain so much gas, that if a little sugar be thrown into the water it effervesces like soda-water.

CARBONIC ACID GAS. A natural gas which often issues from the ground, especially in volcanic countries. *Etym.*, *Carbo*, coal, because the gas is obtained by the slow burning of charcoal.

CARBONIFEROUS. A term usually applied, in a technical sense, to the lowest group of strata of the secondary rocks, see Table II. L, p. 393; but any bed containing coal may be said to be carboniferous. *Etym.*, *Carbo*, coal, and *foro*, to bear.

CATACLYSM. A deluge. *Etym.*, κατακλυξω, *catacluso*, to deluge.

CEPHALOPODA. A class of molluscous animals, having their organs of motion arranged round their head. *Etym.*, κεφαλη, *cephale*, head, and ποδα, *poda*, feet.

CETACEA. An order of vertebrated mammiferous animals inhabiting the sea. The whale, dolphin, and narwal, are examples. *Etym.*, *Cete*, whale.

CHALK. A white earthy limestone, the uppermost of the secondary series of strata. See Table II. F, p. 390.

CHERT. A siliceous mineral, approaching in character to flint, but less homogeneous and simple in texture.

CHLORITIC SAND. Sand coloured green by an admixture of the simple mineral chlorite. *Etym.*, κλωρος, *chloros*, green.

COAL FORMATION. This term is generally understood to mean the same as the Coal Measures. See Table II. L, p. 393. There are, however, 'coal formations' in all the geological periods, wherever any of the varieties of coal form a principal constituent part of a group of strata.

COLEOPTERA. An order of insects (Beetles) which have four wings, the upper pair being crustaceous and forming a shield. *Etym.*, κολεος, *coleos*, a shield, and πτερον, *pteron*, a wing.

CONGENERS. Species which belong to the same genus.

CONGLOMERATE. Rounded water-worn fragments of rock, or pebbles, cemented together by another mineral substance, which may be of a siliceous, calcareous, or argillaceous nature. *Etym.*, *Con*, together, *glomero*, to heap.

CONIFERÆ. An order of plants which, like the fir and pine, bear cones or tops in which the seeds are contained. *Etym.*, *Conus*, cone, and *fero*, to bear.

COOMB. A provincial name in different parts of England for a valley on the declivity of a hill, and which is generally without water.

CORNBRASH. A rubbly stone extensively cultivated in Wiltshire for growth of corn. It is a provincial term adopted by Smith. Brash is derived from breçan, Saxon, to break. See Table II. H, p. 391.

CORNSTONE. A provincial name for a red limestone, forming a subordinate bed in the Old Red Sandstone group.

COSMOGONY, COSMOLOGY. Words synonymous in meaning, applied to speculations respecting the first origin or mode of creation of the earth. *Etym.*, κοσμος, *kosmos*, the world, and γονη, *gonee*, generation, or λογος, *logos*, discourse.

CRAG. A provincial name in Norfolk and Suffolk for a deposit, usually of gravel, belonging to the Older Pliocene period. See Table II. C, p. 389.

CRATER. The circular cavity at the summit of a volcano, from which the volcanic matter is ejected. *Etym.*, *Crater*, a great cup or bowl.

CRETACEOUS. Belonging to chalk. *Etym., Creta*, chalk.

CROP OUT. A miner's or mineral surveyor's term, to express the rising up or exposure at the surface of a stratum or series of strata.

CRUST OF THE EARTH. See Earth's crust.

CRUSTACEA. Animals having a shelly coating or crust which they cast periodically. Crabs, shrimps, and lobsters are examples.

CRYPTOGAMIC. A name applied to a class of plants in which the fructification, or organs of reproduction are concealed. *Etym.*, κρυπτος, *kryptos*, concealed, and γαμος, *gamos*, marriage.

CRYSTALS. Simple minerals are frequently found in regular forms, with facets like the drops of cut glass of chandeliers. Quartz being often met with in rocks in such forms, and beautifully transparent like ice, was called *rock-crystal*, κρυσταλλος, *crystallos*, being Greek for ice. Hence the regular *forms* of other minerals are called crystals, whether they be clear or opake.

CRYSTALLIZED. A mineral which is found in regular forms or crystals, is said to be crystallized.

CRYSTALLINE. The internal texture which regular crystals exhibit when broken, or a confused assemblage of ill-defined crystals. Loaf-sugar and statuary-marble have a *crystalline* texture. Sugar-candy and calcareous spar are crystallized.

CYCADEÆ. An order of plants, which are natives of warm climates, mostly tropical, although some are found at the Cape of Good Hope. They have a short stem, surmounted by a peculiar foliage, termed pinnated fronds by botanists, which spreads in a circle. The growth of these plants is by a cluster of fresh fronds shooting from the top of the stem, and pushing the former fronds outwards. These last decay down to their bases, which are broad, and remain covering the sides of the stem. The term is derived from κυκας, *cycas*, a name applied by the ancient Greek naturalist Theophrastus to a palm, said to grow in Ethiopia.

CYPERACEA. A tribe of plants answering to the English sedges; they are distinguished from grasses by their stems being solid and generally triangular, instead of being hollow and round. Together with *gramineæ* they constitute what writers on botanical geography often call *glumaceæ*.

DEBACLE. A great rush of waters, which breaking down all opposing barriers, carries forward the broken fragments of rocks, and spreads them in its course. *Etym., debacler*, French, to unbar, to break up as a river does at the cessation of a long-continued frost.

DELTA. When a great river before it enters the sea divides into separate streams, they often diverge and form two sides of a triangle, the sea being the base. The land included by the three lines, and which is invariably alluvial, is called a delta from its resemblance to the letter of the Greek alphabet which goes by that name Δ. Geologists extend the boundaries of the delta, so as to include all the alluvial land outside the triangle, which has been formed by the river.

DENUDATION. The carrying away of a portion of the solid materials of the land, by which the inferior parts are laid bare. *Etym.*, *denudo*, to lay bare.

DESICCATION. The act of drying up. *Etym.*, *desicco*, to dry up.

DIAGONAL STRATIFICATION. For an explanation of this term, see vol. iii. p. 174.

DICOTYLEDONOUS. A grand division of the vegetable kingdom, founded on the plant having two *cotyledons* or seed-lobes. *Etym.*, δις, *dis*, double, and cotyledon.

DIKES. When a mass of the unstratified or igneous rocks, such as granite, trap, and lava appears as if injected into a great rent in the stratified rocks, cutting across the strata, it forms a dike; and as they are sometimes seen running along the ground, and projecting, like a wall, from the strata on both sides of them being worn away, they are called in the north of England and in Scotland *dikes*, the provincial name for wall. It is not easy to draw the line between dikes and veins. The former are generally of larger dimensions, and have their sides parallel for considerable distances; while veins have generally many ramifications, and these often thin away into slender threads.

DILUVIUM. Those accumulations of gravel and loose materials which, by some geologists, are said to have been produced by the action of a diluvian wave or deluge sweeping over the surface of the earth. *Etym.*, *diluvium,* deluge.

DIP. When a stratum does not lie horizontally, but is inclined, the point of the compass towards which it sinks is called the dip of the stratum, and the angle it makes with the horizon is called the angle of dip or inclination.

DIPTERA. An order of insects, comprising those which have only two wings. *Etym.*, δις, *dis* double, and πτερον, *pteron*, wing.

DOLERITE. One of the varieties of the trap-rocks, composed of augite and felspar.

DOLOMITE. A crystalline limestone, containing magnesia as a constituent part. Named after the French geologist Dolomieu.

DUNES. Low hills of blown sand that skirt the shores of Holland,

Spain, and other countries. *Etym.*, *dun* or *dune* is an Anglo-Saxon word for hill.

EARTH's CRUST.　Such superficial parts of our planet as are accessible to human observation.

ELYTRA.　The wing-sheaths, or upper crustaceous membranes, which form the superior wings in the tribe of beetles, being crustaceous appendages which cover the body and protect the true membranous wing.　*Etym.*, ελυτρον, *elytron*, a sheath.

EOCENE.　See explanation of this word, vol. iii. p. 55.

ESCARPMENT, the abrupt face of a ridge of high land. *Etym.*, *escarper*, French, to cut steep.

ESTUARIES.　Inlets of the land, which are entered both by rivers and the tides of the sea.　Thus we have the estuaries of the Thames, Severn, Tay, &c.　*Etym. Æstus*, the tide.

FALUNS.　A provincial name for some tertiary strata abounding in shells in Touraine, which resemble in lithological characters the ' crag' of Norfolk and Suffolk.

FAULT, in the language of miners, is the sudden interruption of the continuity of strata in the same plane, accompanied by a crack or fissure varying in width from a mere line to several feet, which is generally filled with broken stone, clay, &c., and such a displacement that the separated portions of the once continuous strata occupy different levels.

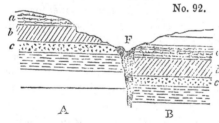

No. 92.

The strata *a*, *b*, *c*, &c., must at one time have been continuous, but a fracture having taken place at the fault F, either by the upheaving of the portion A, or the sinking of the portion B, the strata were so displaced, that the bed *a* in B is many feet lower than the same bed *a* in the portion A.

FAUNA.　The various kinds of animals peculiar to a country constitute its FAUNA, as the various kinds of plants constitute its FLORA.　The term is derived from the FAUNI, or rural deities in Roman Mythology.

FELSPAR　A simple mineral, which constitutes the chief material of many of the unstratified or igneous rocks.　The white angular portions in granite are felspar.　It is originally a German miners' term.　*Etym.*, *feld*, field, and *spath*, a very old mineralogical word in Germany, which seems to have been at first specially applied to a transparent kind of gypsum called selenite.

FELSPATHIC. Of or belonging to felspar.

FERRUGINOUS. Anything containing iron. *Etym., ferrum,* iron.

FLOETZ ROCKS. A German term applied to the secondary strata by the geologists of that country, because these rocks were supposed to occur most frequently in flat horizontal beds. *Etym., flötz,* a layer or stratum ; the word is applied in some parts of Germany to pavements and plastered floors.

FLORA. The various kinds of trees and plants found in any country constitute the Flora of that country in the language of botanists.

FLUVIATILE. Belonging to a river. *Etym., fluvius,* a river.

FORMATION. A group, whether of alluvial deposits, sedimentary strata, or igneous rocks, referred to a common origin or period.

FOSSIL. All minerals used to be called fossils, but geologists now use the word only to express the remains of animals and plants found buried in the earth. *Etym., fossilis,* anything that may be dug out of the earth.

GALENA, a metallic ore, a compound of lead and sulphur. It has often the appearance of highly polished lead. *Etym.,* γαλεω, *galeo* to shine.

GARNET. A simple mineral generally of a deep red colour, crystallized, most commonly met with in mica slate, but also in granite and other igneous rocks.

GAULT. A provincial name in the east of England for a series of beds of clay and marl, the geological position of which is between the upper and the lower greensand. See Table II. F, p. 390.

GEOLOGY, GEOGNOSY. Both mean the same thing, but with an unnecessary degree of refinement in terms, it has been proposed to call our description of the structure of the earth *geognosy.* (*Etym.* γεα, *gea,* earth, and γινωσκω, *ginosco,* to know,) and our theoretical speculations as to its formation *geology.* (*Etym.,* γεα, and λογος, *logos,* a discourse.

GLACIER. The vast accumulations of ice and hardened snow in the Alps and other lofty mountains. *Etym. glace,* French for ice.

GLACIS. A term borrowed from the language of fortification, where it means an easy insensible slope or declivity, less steep than a *talus,* which see.

GNEISS. A stratified primary rock, composed of the same materials as granite, but having usually a larger proportion of mica, and a laminated texture. The word is a German miner's term,

GRAMINEÆ, the order of plants to which grasses belong. *Etym.*, *gramen*, grass.

GRANITE. An unstratified or igneous rock, generally found inferior to or associated with the oldest of the stratified rocks, and sometimes penetrating them in the form of dikes and veins. It is composed of three simple minerals, felspar, quartz, and mica, and derives its name from having a coarse *granular* structure; *granum*, Latin for grain. Westminster, Waterloo, and London bridges, and the paving-stones in the carriage-way of the London streets are good examples of the most common varieties of granite.

GRAUWACKE, a German name, generally adopted by geologists for the lowest members of the secondary strata, consisting of sandstone and slate, and which form the chief part of what are termed by some geologists the *transition* rocks. The rock is very often of a grey colour, hence the name, *grau* being German for grey, and *wacke* being a provincial miner's term.

GREENSAND. Beds of sand, sandstone, limestone, belonging to the Cretaceous Period. See Table II. F, p. 390. The name is given to these beds, because they often, but not always, contain an abundance of green earth or chlorite scattered through the substance of the sandstone, limestone, &c. See vol. iii. p. 324.

GREENSTONE, a variety of trap, composed of hornblende and felspar.

GRIT, a provincial name for a coarse-grained sandstone.

GYPSUM, a mineral composed of lime and sulphuric acid, hence called also *sulphate* of *lime*. Plaster and stucco are obtained by exposing gypsum to a strong heat. It is found so abundantly near Paris, that Paris plaster is a common term in this country for the white powder of which casts are made. The term is used by Pliny for a stone used for the same purposes by the ancients. The derivation of it is unknown.

GYPSEOUS, of, or belonging to, gypsum.

GYROGONITES. Bodies found in fresh-water deposits, originally supposed to be microscopic shells, but subsequently discovered to be the seed-vessel of fresh-water plants of the genus *chara*. See vol. ii. p. 273, and 2d Edit. p. 280. *Etym.* γυρος, *gyros*, curved, and γονος, *gonos*, seed, on account of their external structure.

HEMIPTERA, an order of insects, so called from a peculiarity in their wings, the superior being coriaceous at the base, and membranous at the apex, ἡμισυ, *hemisu*, half, and πτερον, *pleron*, wing.

HORNBLENDE, a simple mineral of a dark green or black colour, which enters largely into the composition of several varieties of the trap rocks.

HYDROPHYTES. Plants which grow in water. *Etym.*, ὑδωρ, *hydor*, water, and φυτον, *phyton*, plant.

HYPOGENE ROCKS. For an explanation of this term, see vol. iii. p. 374.

ICEBERG. The great masses of ice, often the size of hills, which float in the polar and northern seas. *Etym.*, ice, and *berg*, German for hill.

ICHTHYOSAURUS, a gigantic fossil marine reptile, intermediate between a crocodile and a fish. *Etym.*, ιχθυς, *ichthus*, a fish, and σαυρα, *saura*, a lizard.

INDUCTION, a consequence, conclusion or inference, drawn from propositions or principles first laid down, or from the observation and examination of phenomena.

INFUSORY ANIMALCULES. Minute living creatures generated in many *infusions ;* and the term *infusoria* has been given to all such animalcules whether found in infusions or in stagnant water, vinegar, &c.

INSPISSATED. Thickened. *Etym.*, *spissus*, thick.

INVERTEBRATED ANIMALS. Animals which are not furnished with a back-bone. For a further explanation, see " Vertebrated Animals."

ISOTHERMAL. Such zones or divisions of the land, ocean, or atmosphere, which have an equal degree of mean annual warmth, are said to be isothermal, from ισος, *isos*, equal, and θερμη, *therme*, heat.

JURA LIMESTONE. The limestones belonging to the oolite group, see Table II. H, p. 391, constitute the chief part of the mountains of the Jura, between France and Switzerland, and hence the geologists of the Continent have given the name to the group.

KIMMERIDGE CLAY, a thick bed of clay, constituting a member of the Oolite Group. See Table II. H, p. 391. so called because it is found well developed at Kimmeridge in the isle of Purbeck, Dorsetshire.

LACUSTRINE, belonging to a lake. *Etym.*, *Lacus*, a lake.

LAMINÆ. Latin for plates ; used in geology, for the smaller layers of which a stratum is frequently composed.

LAMANTINE. A living species of the herbivorous cetacea or whale tribe, which inhabits the mouths of rivers on the coasts of Africa and South America; the sea-cow.

LAMELLIFEROUS. A stone composed of thin plates or leaves like paper. *Etym., lamella*, the diminutive of *lamina*, plate, and *fero*, to bear.

LANDSLIP. A portion of land that has slid down in consequence of disturbance by an earthquake, or from being undermined, by water washing away the lower beds which supported it.

LAPIDIFICATION—Lapidifying process. Conversion into stone. *Etym., lapis*, stone, and *fio*, to make.

LAPILLI. Small volcanic cinders. *Lapillus*, a little stone.

LAVA. The stone which flows in a melted state from a volcano.

LEUCITE. A simple mineral found in volcanic rocks, crystallized, and of a white colour. *Etym.*, λευκος, *leucos*, white.

LIAS. A provincial name, adopted in scientific language, for a particular kind of limestone, which being characterized, together with its associated beds, by peculiar fossils, is formed in this work into a particular group of the secondary strata. See Table II. I, p. 392.

LIGNIPERDOUS. A term applied to insects which destroy wood. *Etym. lignum*, wood, and *perdo*, to destroy.

LIGNITE. Wood converted into a kind of coal. *Etym., lignum*, wood.

LITHODOMI. Molluscous animals which bore into solid rocks, and lodge themselves in the holes they have formed. *Etym.*, λιθος, *lithos*, stone, and *domus*, house.

LITHOLOGICAL. A term expressing the stony structure or character of a mineral mass. We speak of the lithological character of a stratum as distinguished from its zoological character. *Etym.*, λιθος, *lithos*, stone, and λογος, *logos*, discourse.

LITHOPHAGI. Molluscous animals which bore into solid stones. *Etym.*, λιθος, *lithos*, stone, and φαγειν, *phagein*, to eat.

LITTORAL. Belonging to the sea-shore. *Etym., littus*, the shore.

LOAM. A mixture of sand and clay.

LYCOPODIACEÆ. Plants of an inferior degree of organization to Coniferæ, some of which they very much resemble in foliage, but all recent species are infinitely smaller. Many of the fossil species are as gigantic as recent coniferæ. Their mode of reproduction is analogous to that of ferns. In English they are called club-mosses, generally found in mountainous heaths in the north of England.

MADREPORE. A genus of corals, but generally applied to all the corals distinguished by superficial star-shaped cavities. There are several fossil species.

MAGNESIAN LIMESTONE. An extensive series of beds lying in geological position, immediately above the coal-measures, so called because the limestone, the principal member of the series, contains much of the earth magnesia as a constituent part. See Table II. K, p. 392.

MAMMILLARY. A surface which is studded over with rounded projections. *Etym., mammilla,* a little breast or pap.

MAMMIFEROUS. Animals which give suck to their young. *Etym., mamma,* a breast, and *fero,* to bear.

MAMMOTH. An extinct species of the elephant (E. primigenius), of which the fossil bones are frequently met with in various countries. The name is of Tartar origin, and is used in Siberia for animals that burrow underground.

MARL. A mixture of clay and lime; usually soft, but sometimes hard, in which case it is called indurated marl.

MARSUPIAL ANIMALS. A tribe of quadrupeds having a sack or pouch under the belly, in which they carry their young. The kangaroo is a well-known example. *Etym., marsupium,* a purse.

MASTODON. A genus of fossil extinct quadrupeds allied to the elephant. So called from the form of the hind teeth or grinders, which have their surface covered with conical mammillary crests. *Etym.,* μαστος, *mastos,* mammilla or little pap, and οδων, *odon,* tooth.

MATRIX. If a simple mineral or shell, in place of being detached, be still fixed in a portion of rock, it is said to be in its matrix. *Matrix,* womb.

MECHANICAL ORIGIN, Rocks of. When rocks are composed of sand, pebbles, or fragments, to distinguish them from those of an uniform crystalline texture, which are of chemical origin.

MEDUSÆ. A genus of marine radiated animals, without shells; so called because their organs of motion spread out like the snaky hair of the fabulous Medusa.

MEGALOSAURUS. A fossil gigantic amphibious animal of the saurian or lizard and crocodile tribe. *Etym.,* μεγαλη, *megale,* great, and σαυρα, *saura,* lizard.

MEGATHERIUM. A fossil extinct quadruped, resembling a gigantic sloth. *Etym.,* μεγα, *mega,* great, and θηριον, *therion,* wild-beast.

MELASTOMA. A genus of MELASTOMACEA, an order of plants of the evergreen tree, and shrubby exotic kinds. *Etym.,* μελας, *melas,* black, and στομα, *stoma,* mouth; because the fruit of one of the species stains the lips.

MESOTYPE. A simple mineral, white, and needle-shaped, one of the
Zeolite family, frequently met with in the trap rocks.

METAMORPHIC ROCKS. For an explanation of this term, see vol. iii.
p. 374.

MICA. A simple mineral, having a shining silvery surface, and
capable of being split into very thin elastic leaves or scales.
It is often called *talc* in common life, but mineralogists apply
the term talc to a different mineral. The brilliant scales in
granite are mica. *Etym.*, *mico*, to shine.

MICA-SLATE, MICA-SCHIST, MICACEOUS SCHISTUS. One of the lowest
of the stratified rocks, belonging to the primary class, which is
characterized by being composed of a large proportion of mica,
united with quartz.

MIOCENE. See an explanation of this term, vol. iii. p. 54.

MOLASSE. A provincial name for a soft, green sandstone, associated with marl and conglomerates, belonging to the Miocene
tertiary period, extensively developed in the lower country of
Switzerland. See vol. iii. p. 212.

MOLLUSCÆ, Molluscous Animals. Animals, such as shell-fish, which,
being devoid of bones, have soft bodies. *Etym.*, *mollis*, soft.

MONITOR. An animal of the saurian or lizard tribe, species
of which are found in both the fossil and recent state.

MONOCOTYLEDONOUS. A grand division of the vegetable kingdom,
founded on the plant having only one *cotyledon*, or seed-lobe.
Etym., μονος, *monos*, single.

MOSCHUS. The quadruped resembling the chamois or mountain-
goat, from which the perfume musk is obtained.

MOUNTAIN LIMESTONE. A series of limestone strata, of which the
geological position is immediately below the coal measures, and
with which they also sometimes alternate. See Table II. L, p. 393.

MOYA. A term applied in South America to mud poured out from
volcanos during eruptions.

MURIATE OF SODA. The scientific name for common culinary salt,
because it is composed of muriatic acid and the alkali soda.

MUSACEÆ. A family of tropical monocotyledonous plants, including
the banana and plantains.

MUSCHELKALK. A limestone which, in geological position, belongs to the red sandstone group. This formation has not yet
been found in England, and the German name is adopted by
English geologists. The word means shell-limestone: *muschel*,
shell, and *kalkstein*, limestone. See Table II. K, p. 392.

NAPHTHA. A very thin, volatile, inflammable, and fluid mineral substance, of which there are springs in many countries, particularly in volcanic districts.

NENUPHAR. A yellow water-lily.

NEW RED SANDSTONE. A series of sandy, argillaceous, and often calcareous strata, the predominant colour of which is brick-red, but containing portions which are of a greenish grey. These occur often in spots and stripes, so that the series has sometimes been called the variegated sandstone. The European formation so called lies in a geological position immediately above the coal-measures. See Table II. K, p. 392.

NODULE. A rounded irregular-shaped lump or mass. *Etym.*, diminutive of *nodus*, knot.

NORMAL GROUPS. Groups of certain rocks taken as a rule or standard. *Etym. norma*, rule or pattern.

NUCLEUS. A solid central piece, around which other matter is collected. The word is Latin for kernel

NUMMULITES. An extinct genus of the Order of Molluscous animals, called Cephalopoda, of a thin lenticular shape, internally divided into small chambers. *Etym.*, *nummus*, Latin for money, and λιθος, *lithos*, stone, from its resemblance to a coin.

OBSIDIAN. A volcanic product, or species of lava, very like common green bottle-glass, which is almost black in large masses, but semi-transparent in thin fragments. Pumice-stone is obsidian in a frothy state ; produced most probably by water that was contained in or had access to the melted stone, and converted into steam. There are very often portions in a mass of solid obsidian, which are partially converted into pumice.

OGYGIAN DELUGE. A general inundation of fabulous history, which is supposed to have taken place in the reign of Ogyges in Attica, whose death is fixed in Blair's Chronological Tables in the year 1764 before Christ.

OLD RED SANDSTONE. A stratified rock belonging to the Carboniferous group. See Table L, p. 393.

OLIVINE. An olive-coloured, semi-transparent, simple mineral, very often occurring in the forms of grains and of crystals in basalt and lava.

OOLITE, Oolitic. A limestone, forming a characteristic feature of a group of the secondary strata. See Table II, H, p. 391. It is so named, because it is composed of rounded particles, like the roe or eggs of a fish. *Etym.* ωον, *oon*, egg, and λιθος, *lithos*, stone.

OPALIZED WOOD. Wood petrified by siliceous earth, and acquiring a structure similar to the simple mineral called opal.

OPHIDIOUS REPTILES. Vertebrated animals, such as snakes and serpents. *Etym., οφις, ophis*, a serpent.

ORGANIC REMAINS. The remains of animals and plants; *organized* bodies, found in a fossil state.

ORTHOCERATA. An extinct genus of the order of Molluscous Animals, called Cephalopoda, that inhabited a long chambered, conical shell, like a straight horn. *Etym., ορθος, orthos,* straight, and *κερας, ceras,* horn.

OSSEOUS BRECCIA. The cemented mass of fragments of bones of extinct animals found in caverns and fissures. *Osseus* is a Latin adjective, signifying bony.

OUTLIERS. When a portion of a stratum occurs at some distance, detached from the general mass of the formation to which it belongs, some practical mineral surveyors call it an *outlier*, and the term is adopted in geological language.

OVATE. The shape of an egg. *Etym., ovum*, egg.

OVIPOSITING. The laying of eggs.

OXYGEN. One of the constituent parts of the air of the atmosphere; that part which supports life. For a further explanation of the word, consult elementary works on chemistry.

OXIDE. The combination of a metal with oxygen; rust is oxide of iron.

PACHYDERMATA. An order of quadrupeds, including the elephant, rhinoceros, horse, pig, &c., distinguished by having thick skins. *Etym. παχυς, pachus,* thick, and *δερμα, derma,* skin or hide.

PACHYDERMATOUS. Belonging to pachydermata.

PALÆOTHERIUM, PALEOTHERE. A fossil extinct quadruped, belonging to the order pachydermata, resembling a pig or tapir, but of great size. *Etym. παλαιος, palaios,* ancient, and *θηριον, therion,* wild beast.

PELAGIAN, PELAGIC. Belonging to the *deep* sea. *Etym. pelagus,* sea.

PEPERINO. An Italian name for a particular kind of volcanic rock, formed, like tuff, by the cementing together of volcanic sand, cinders, or scoriæ, &c.

PETROLEUM. A liquid mineral pitch, so called because it is seen to ooze like oil out of the rock. *Etym. petra,* rock, and *oleum,* oil.

PHANEROGAMIC PLANTS. A name given by Linnœus to those plants in which the reproductive organs are apparent. *Etym.* φανερος, *phaneros*, evident, and γαμος, *gamos*, marriage.

PHYSICS. The department of science, which treats of the properties of natural bodies, laws of motion, &c., sometimes called Natural philosophy and mechanical philosophy. *Etym.* φυςις, *physis*, nature.

PHYTOLOGY, PHYTOLOGICAL. The department of science which relates to plants—synonymous with botany and botanical. *Etym.* φυτον, *phyton*, plant, and λογος, *logos*, discourse.

PHYTOPHAGOUS. Plant eating. *Etym.* φυτον, *phyton*, plant, and φαγειν, *phagein*, to eat.

PISLIAR, a misprint for PISTIA, in vol. ii. p. 98, 1st ed., p. 102, 2d ed. The plant mentioned by Malte-Brun is probably the *Pistia stratiotes*, a floating plant, related to English duck-weed, but very much larger.

PISOLITE. A stone possessing a structure like an agglutination of pease. *Etym.* πισον, *pison*, pea, and λιθος, *lithos*, stone.

PIT COAL. Ordinary coal; called so because it is obtained by sinking pits in the ground.

PITCH STONE. A rock of an uniform texture, belonging to the unstratified and volcanic classes, which has an unctuous appearance, like indurated pitch.

PLASTIC CLAY. One of the beds of the Eocene tertiary period (see Table II. E, p. 390.) It is so called because it is used for making pottery. *Etym.* πλασσω, *plasso*, to form or fashion.

PLESIOSAURUS. A fossil extinct amphibious animal, resembling the saurian, or lizard and crocodile tribe. *Etym.* πλησιον, *plesion*, near to, and σαυρα, *saura*, a lizard.

PLIOCENE. See explanation of this term, vol. iii. p. 53.

PLUTONIC ROCKS. For an explanation of this term, see vol. iii. p. 353.

POLYPARIA. CORALS. A numerous class of invertebrated animals, belonging to the great division called Radiata.

PORPHYRY. An unstratified or igneous rock. The term is as old as Pliny, and was applied to a red rock with small angular white bodies diffused through it, which are crystallized felspar, brought from Egypt. The term is hence applied to every species of unstratified rock, in which detached crystals of felspar are diffused through a base of other mineral composition. *Etym.* πορφυρα, *porphyra*, purple.

PORTLAND LIMESTONE, PORTLAND BEDS. A series of limestone strata, belonging to the upper part of the Oolite group (see Table II.

H, p. 391.), found chiefly in England, in the Island of Portland on the coast of Dorsetshire. The great supply of the building stone used in London is from these quarries.

Pozzuolana. Volcanic ashes, largely used as mortar for buildings, similar in nature to what is called in this country Roman cement. It gets its name from Pozzuoli, a town in the bay of Naples, from which it is shipped in large quantities to all parts of the Mediterranean.

Productæ. An extinct genus of fossil bivalve shells, occurring only in the older of the secondary rocks. It is closely allied to the living genus Terebratula.

Pubescence. The soft hairy down on insects. *Etym., pubesco,* the first growth of the beard.

Pumice.—A light spongy lava, of a white colour, produced by gases, or watery vapour getting access to the particular kind of glassy lava called obsidian, when in a state of fusion—it may be called the froth of melted volcanic glass. The word comes from the Latin name of the stone, *pumex.*

Purbeck Limestone, Purbeck Beds. Limestone strata belonging to the Wealden group. See Table II. G, p. 390.

Pyrites (Iron). A compound of sulphur and iron, found usually in yellow shining crystals like brass, and in almost every rock stratified and unstratified. The shining metallic bodies, so often seen in common roofing slate, are a familiar example of the mineral. The word is Greek, and comes from πυρ, *pyr,* fire, because, under particular circumstances, the stone produces spontaneous heat and even inflammation.

Quadrumana. The order of mammiferous animals to which apes belong. *Etym., quadrus,* a derivation of the Latin word for the number four, and *manus,* hand,—the four feet of those animals being in some degree usable as hands.

Qua-qua-versal Dip. The dip of beds to all points of the compass around a centre, as in the case of beds of lava round the crater of a volcano. *Etym., quâ-quâ versum,* on every side.

Quartz. A German provincial term, universally adopted in scientific language, for a simple mineral composed of pure silex, or earth of flints; rock-crystal is an example.

Red Marl. A term often applied to the New Red Sandstone, which is the principal member of the Red Sandstone group. See Table II. K, p. 392.

RETICULATE. A structure of cross lines, like a net, is said to be reticulated, from *rete*, a net.

ROCK SALT. Common culinary salt, or muriate of soda, found in vast solid masses or beds, in different formations, extensively in the New Red Sandstone formation, as in Cheshire, and it is then called *rock*-salt.

RUMINANTIA. Animals which ruminate or chew the cud. *Etym.*, the Latin verb *rumino*, meaning the same thing.

SACCHAROID, SACCHARINE. When a stone has a texture resembling that of loaf-sugar. *Etym.*, σακκαρ, *sacchar*, sugar, and ειδος, *eidos*, form.

SALIENT ANGLE. In a zig-zag line, *a a* are the salient angles, *b b* the re-entering angles.

No. 93.

Etym., *salire*, to leap or bound forward.

SALT SPRINGS. Springs of water containing a large quantity of common salt. They are very abundant in Cheshire and Worcestershire, and culinary salt is obtained from them by mere evaporation.

SANDSTONE. Any stone which is composed of an agglutination of grains of sand, which may be either calcareous or siliceous.

SAURIAN. Any animal belonging to the lizard tribe. *Etym.*, σαυρα, *saura*, a lizard.

SCHIST. Synonymous with slate. *Etym.*, part of the Latin verb *scindo*, to split, from the facility with which slaty rocks may be split into thin plates.

SCHISTOSE ROCKS. Synonymous with *slaty* rocks.

SCORIÆ. Volcanic cinders. The word is Latin for cinders.

SEAMS. Thin layers which separate two strata of greater magnitude.

SECONDARY STRATA. An extensive series of the stratified rocks which compose the crust of the globe, with certain characters in common, which distinguish them from another series below them, called *primary*, and from a third series above them called *tertiary*. See vol. iii. p. 324, and Table II. p. 390.

SECULAR REFRIGERATION. The periodical cooling and consolidation of the globe, from a supposed original state of fluidity from heat. *Sæculum*, age or period.

SEDIMENTARY ROCKS, are those which have been formed by their materials having been thrown down from a state of suspension or solution in water.

SELENITE. Crystallized gypsum, or sulphate of lime—a simple mineral.

SEPTARIA. Flattened balls of stone, generally a kind of iron-stone, which, on being split, are seen to be separated in their interior into irregular masses. *Etym.*, *septa*, inclosures.

SERPENTINE. A rock usually containing much magnesian earth, for the most part unstratified, but sometimes appearing to be an altered or metamorphic stratified rock. Its name is derived from frequently presenting contrasts of colour, like the skin of some serpents.

SHALE. A provincial term, adopted in geological science, to express an indurated slaty clay. *Etym.*, German *schalen*, to peal, to split.

SHELL MARL. A deposit of clay, peat, and other substances mixed with shells, which collects at the bottom of lakes.

SHINGLE. The loose and completely water-worn gravel on the sea-shore.

SILEX. The name of one of the pure earths, being the Latin word for *flint*, which is wholly composed of that earth. French geologists have applied it as a generic name for all minerals composed entirely of that earth, of which there are many of different external forms.

SILICA. One of the pure earths. *Etym.*, *silex*, flint, because found in that mineral.

SILICATE. A chemical compound of silica and another substance, such as silicate of iron. Consult elementary works on chemistry.

SILICEOUS. Of or belonging to the earth of flint. *Etym.*, silex, which see. A siliceous rock is one mainly composed of silex.

SILICIFIED. Any substance that is petrified or mineralized by *siliceous* earth.

SILT. The more comminuted sand, clay, and earth, which is transported by running water. It is often accumulated by currents in banks. Thus we speak of the mouth of a river being silted up when its entrance into the sea is impeded by such accumulation of loose materials.

SIMPLE MINERAL. Individual mineral substances, as distinguished from rocks, which last are usually an aggregation of simple minerals. They are not simple in regard to their nature, for when subjected to chemical analysis, they are found to consist of a variety of different substances. Pyrites is a simple mineral in the sense we use the term, but it is a chemical compound of sulphur and iron.

SOLFATARA. A volcanic vent from which sulphur, sulphureous, watery, and acid vapours and gases are emitted.

SPORULES. The reproductory corpuscula (minute bodies) of cryptogamic plains. *Etym.*, σπορα, *spora*, a seed.

STALACTITE. When water holding lime in solution deposits it as it drops from the roof of a cavern, long rods of stone hang down like icicles, and these are called *stalactites*. *Etym.*, σταλαζω, *stalazo*, to drop.

STALAGMITE. When water holding lime in solution drops on the floor of a cavern, the water evaporating leaves a crust composed of layers of limestone ; such a crust is called *stalagmite*, from σταλαγμα, *stalagma*, a drop, in opposition to *stalactite*, which see.

STILBITE. A white crystallized simple mineral, one of the Zeolite family, frequently included in the mass of the trap rocks.

STRATIFIED. Rocks arranged in the form of *strata*, which see.

STRATIFICATION. An arrangement of rocks in *strata*, which see.

STRATUM, STRATA. When several rocks lie like the leaves of a book, one upon another, each individual forms a *stratum ;*—strata is the plural of the word. *Etym.*, *stratum*, part of a Latin verb signifying to strew or lay out.

STRIKE. The direction or line of bearing of strata, which is always at right angles to their prevailing dip. For a fuller explanation, see vol. iii. p. 346.

SUBAPENNINES. Low hills which skirt or lie at the foot of the great chain of the Apennines in Italy. The term Subapennine is applied geologically to a series of strata of the Older Pliocene period.

SYENITE. A kind of granite, so called because it was brought from Syene in Egypt. For geological acceptation of the term, see vol. iii. p. 358.

SYNCLINAL AXIS. See explanation of this term, vol. iii. p. 293.

TALUS. When fragments are broken off by the action of the weather from the face of a steep rock, as they accumulate at its foot, they form a sloping heap, called a talus. The term is borrowed from the language of fortification, where *talus* means the outside of a wall of which the thickness is diminished by degrees, as it rises in height, to make it the firmer.

TARSI. The feet in insects, which are articulated, and formed of five or a less number of joints.

TERTIARY STRATA. A series of sedimentary rocks, with characters

which distinguish them from two other great series of strata,—
the secondary and primary, which lie *beneath* them. See
Tables, p. 61, &c.

TESTACEA.   Molluscous animals, having a shelly covering.   *Etym.*,
testa, a shell, such as snails, whelks, oysters, &c.

THIN OUT.   When a stratum, in the course of its prolongation in
any direction, becomes gradually less in thickness, the two surfaces approach nearer and nearer; and when at last they meet,
the stratum is said to thin out, or disappear.

TRACHYTE.   A variety of lava essentially composed of glassy felspar, and frequently having detached crystals of felspar in
the base or body of the stone, giving it the structure of por-
·phyry.   It sometimes contains hornblende and augite; and
when these last predominate, the trachyte passes into the
varieties of trap called greenstone, basalt, dolorite, &c.   The
term is derived from τραχυς, *trachus*, rough, because the rock
has a peculiar rough feel.

TRAP and TRAPPEAN ROCKS.   Volcanic rocks composed of felspar,
augite, and hornblende.   The various proportions and state of
aggregation of these simple minerals, and differences in external forms, give rise to varieties, which have received distinct
appellations, such as basalt, amygdaloid, dolorite, greenstone,
and others.   The term is derived from *trappa*, a Swedish
word for stair, because in Sweden the rocks of this class often
occur in large tabular masses, rising one above another, like
the steps of a staircase.   For further explanation, see vol. iii.
p. 359.

TRAVERTIN.   A limestone, usually hard and semi-crystalline, deposited from the water of springs holding lime in solution.   The
word is Italian, and a corruption of the term *Tiburtinus*, the
stone being formed in great quantity by the river Anio, at
Tibur, near Rome, and hence it was called by the ancients
Lapis Tiburtinus.

TROPHI, of Insects.   Organs which form the mouth, consisting of
an upper and under lip, and comprising the parts called man-
dibles, maxillæ, and palpi.

TUFF, or TUFO.   An Italian name for a variety of volcanic rock, of
an earthy texture, seldom very compact, and composed of an
agglutination of fragments of scoriæ and loose matter ejected
from a volcano.

TUFACEOUS.   A rock with the texture of tuff or tufo, which see.

TURBINATED.   Shells which have a spiral or screw-form structure.
Etym., *turbinatus*, made like a top.

VEINS, Mineral. Cracks in rocks filled up by substances different from the rock, which may either be earthy or metallic. Veins are sometimes many yards wide ; and they ramify or branch off into innumerable smaller parts, often as slender as threads, like the veins in an animal, and hence their name.

VERTEBRATED ANIMALS. A great division of the animal kingdom, including all those which are furnished with a back-bone, as the mammalia, birds, reptiles, and fishes. The separate joints of the back-bone are called *vertebræ*, from the Latin verb *verto*, to turn.

VESICLE. A small circular inclosed space, like a little bladder. *Etym.*, diminutive of *vesica*, Latin for a bladder.

VOLCANIC BOMBS. Volcanos throw out sometimes detached masses of melted lava, which, as they fall, assume rounded forms (like bomb-shells), and are often elongated into a pear shape.

VOLCANIC FOCI. The subterranean centres of action in volcanos, where the heat is supposed to be in the highest degree of energy.

ZEOLITE. A family of simple minerals, including stilbite, mesotype, analcime, and some others, usually found in the trap or volcanic rocks. Some of the most common varieties swell or boil up when exposed to the blow-pipe, and hence the name of ξεω, *zeo*, to froth, and λιθος, *lithos*, stone.

ZOOPHYTES. Corals, sponges, and other aquatic animals allied to them, so called because, while they are the habitation of animals, they are fixed to the ground, and have the forms of plants, *Etym.*, ξωον, *zoon*, animal, and φυτον, *phyton*, plant.

INDEX.

Vol. III.

ABERDEENSHIRE, passage from trap into granite in, 361

Abesse, near Dax, section of inland cliff at, —see wood-cut No. 53, 210

Acquapendente, alternations of volcanic tuffs with the Subapennine marls at, 159

Adanson on the age of the baobab tree, 99

Addington hills, 279

Adernò, opposite dip of the strata in two sections near, 78

Adour, section of tertiary strata in the valley of the—see diag. No. 51, 207

Adur, view of the transverse valley of the river—see wood-cut No. 73, 299

Agassiz, M., on fossil fish of the brown coal formation, 200

—— on the fossil fish of the Paris basin, 253

—— on the distinctness of the secondary and tertiary fossil fish, 327

Age of volcanos, mode of computing the, 97

Ages, relative, of rocks how determined, 35

Aidat, Lake, how formed, 269

Aix, in Provence, tertiary strata of, 276

—— fossil insects abundant in the calcareous marl of, 277

Albenga, height of the tertiary strata above the sea at, 165, 166

—— resemblance of the strata at, to the Subapennines, 167

Allan, Mr. T., his discovery of the bones of mammalia in the fresh-water strata of the Isle of Wight, 281

Allier, river, section of volcanic tuff and fresh-water limestone on the banks of the, 258

Alluvium, passage of marine crag strata into, 181

—— ancient, of the valley of the Rhine, 200

—— of the Weald valley, 295

Alluviums formed in all ages, 145

—— of the newer Pliocene period, 139, 145, 151

—— distinction between regular subaqueous strata and, 145

—— marine, 145

—— British, how formed, 147

—— European, in great part tertiary, 150

Alluviums, underlying lavas of Catalonia, 188, 189, 190, 192

—— of the Miocene era, localities of, 217

—— trachytic breccias alternating with, in Auvergne—see wood-cut No. 54, 217

—— of Auvergne, extinct quadrupeds in, 218

—— of different ages covered by lava in Auvergne—see wood-cut No. 61, 266

—— of the Eocene period, 317

Alps, shells drifted into the Mediterranean from the, 48

—— erratic blocks of the, 148

—— Maritime, tertiary strata at the base of the, 164

—— secondary strata penetrated by granite in the, 358

—— strata of oolite altered in the, 371

Altered strata in contact with granite, 370, 371

—— strata, enumeration of the probable conversions of sedimentary strata into well-known metamorphic rocks, 373

Alternations of strata with and without organic remains, how caused, 254

Alum Bay, alternation of the London and plastic clay in, 278

Amer, geological structure of the country near, 185

Anapo, valley of the, 111

Andernach, gorge of, 152

—— loess and volcanic ejections alternating at, 153

Andes, sudden rise of the, said to have caused the historical deluge, 148

Angers, fossil shells found at—see tables Appendix I.

Anglesea, changes caused by a volcanic dike in, 368

Animals, their fossilization partial, 31

—— remains of, in the successive tertiary periods, 59

Anoplotherium found in the fresh-water formation of the Isle of Wight, 281, 317

Anthracite, whence derived, 373

Anticlinal axis of the Weald valley—see wood-cuts Nos. 63 and 64, 288

Anticlinal and synclinal lines described—see wood-cut No. 68, 293

Anticlinal lines, how far those formed at the same time are parallel, 349

Antilles, recent shells imbedded in limestone in the, 133

Antrim, chalk in, converted into marble by trap-dike, 369

———— altered coal and lias in, 369

Apennines, tertiary strata at the foot of the, 155

Apollinaris does not mention the volcanos in his description of Auvergne, 269

Areas of sedimentary deposition, shifting of the, 26

Argillaceous strata, change caused by a dike of lava in, 70

Arno, river, yellow sand like the Subapennines deposited by the, 161

Arun, transverse valley of the, 298, 299

Asia, western, great cavity in, 29, 270

Astroni, crater of, 187

Atlantis of Plato, 330

Atrio del Cavallo, dikes in the, 124

Aurillac, fresh-water formation of, 236

———— silex abundant in the fresh-water strata of, 237

———— resemblance of the fresh-water lime-stone and flints to the chalk, 237

———— proofs of the gradual deposition of the fresh-water marls of, 239

Australian breccias, bones of marsupial animals in, 143

Auvergne, appearance of some of the lavas of, 94

——— position of the Miocene alluviums of —see wood-cut No. 54, 217

——— extinct quadrupeds in the alluviums of, 218

——— age of the volcanic rocks of, 224

——— lacustrine deposits of, 226

——— map of the lacustrine basins and volcanic rocks of—see wood-cut No. 56, 226

——— tertiary red marl and sandstone of, like new red sandstone, 229, 333

——— indusial limestone of, 232

——— dip of the tertiary strata of, 233, 235

——— arrangement and origin of the fresh-water formation of, 233

——— analogy of the tertiary deposits of, to those of the Paris basin, 241

——— geographical connexion of the Paris basin and, 241

——— probably once connected witn the Paris basin by a chain of lakes, 241

——— volcanic rocks of, 257

——— igneous rocks associated with the lacustrine strata of, 258

——— volcanic breccias of, how formed, 259

——— minor volcanos of, 260, 263

——— long succession of eruptions in, 260

——— ravines excavated through lava in, 264

Auvergne, lavas resting on alluviums of different ages in—see wood-cut No. 61, 266

——— age of the volcanos of, 268, 269

Aventine, Mount, a deposit of calcareous tufa on, 138

Bagneux, alternation of plastic clay and calcaire grossier at, 244

Bagshot sand, its composition, &c., 280

Banos del Pujio, elevated sea-cliff near, 131

Baobab tree, its size, probable age, &c., 99, 272

Baraque, la Petite, section of vertical marls in a ravine near—see wood-cut No. 57, 231

Barcelona, height of the marine tertiary strata of, 193

Barcombe, section from the north escarpment of the South Downs to—see wood-cut No. 71, 296

Barzone, gypsum found in the Subapennine marls near, 159

Basalt, theory of the aqueous origin of, 4

Basalts of the Bay of Trezza, Paternò, &c., their relative age, 82

Basterot, M. de, on the fossil shells of Bordeaux and Dax, 20, 206

Battoch, Mount, granite veins of, 357

Bay of Trezza, sub-Etnean formations exposed in the, 78

——— proofs of ancient submarine eruptions in the, 78

Bayonne, age of the tertiary strata near, 343

——— age of the newest secondary strata near, 343

Bawdesey, inclination of the crag strata near, 174

Beauchamp, remains of a palæotherium and fresh-water shells in calcaire grossier at, 252

Beachy Head, termination of the chalk escarpment at, 291

——— thickness of the upper green-sand at, 292

Beginning of things, supposed proofs of, 383

Belbet, section of white limestone in the quarry of, 237

Belgium, tertiary formations of, 276

——— fossil shells from—see table, Appendix I.

Beliemi, Mount, caves in, 143

Beudant, M., on the volcanic rocks of Hungary, 222

Bingen, gorge of, 152

Binstead, mammiferous remains found in the quarries of, 281, 317

Blaye, limestone of, 200

——— its position—see wood-cut No. 52, 209

Blue marl with shells of the Val di Noto, 67

Boblaye, M., on the successive elevations of the Morea, 113, 132
—— on the formation of osseous breccias in the Morea, 144
—— on the tertiary strata of the Morea, 170

Bolos, Don Francisco, on the volcanos of Olot, in Catalonia, 187, 191, 193
—— on the destruction of Olot by earthquake, in 1421, 191

Bonelli, Signor, on the fossil shells of Savona, 166
—— on the fossil shells of the Superga, 211

Bonn, blocks of quartz containing casts of fresh-water shells found near, 199
—— remains of frogs from the brown coal formation in the museum at, 200

Bordeaux, tertiary strata of, 20, 206
—— Eocene strata in the basin of, 208
—— fossil shells of — see table, Appendix I.

Bormida, tertiary strata of the valley of the, 211

Bosque de Tosca, a mound of lava near Olot, 186

Botley Hill, height of, 288

Boué, M., on the loess of the valley of the Rhine, 151
—— on the value of zoological characters in determining the chronological relations of strata, 208
—— term molasse vaguely employed by, 212
—— on the tertiary formations of Hungary and Transylvania, 213
—— on the fossil shells of Hungary, 223
—— on the volcanic rocks of Transylvania, 223
—— his objections to the theory of M. Elie de Beaumont, 346, 347

Bouillet, M., on the extinct quadrupeds of Mont Perrier, 218
—— on alluviums of different ages in Auvergne, 267

Boulade, position of the alluviums of the —see wood-cut No. 54, 217

Boulon and Ceret, dip of the tertiary strata between, 170

Bourbon, Isle of, a volcanic eruption every two years in the, 363

Bowdich, Mr., fossil shells of recent species brought from Madeira by, 134

Braganza river, brown clay deposited by the, 161

Breaks in the series of superimposed formations, causes of, 26, 33

Breccias in the Val del Bove, 93
—— osseous, in Sicilian caves, 139

Breccias, in Australian caves, 143
—— now in progress in the Morea, 144
—— trachytic, alternations of alluvium and—see wood-cut No. 54, 217
—— volcanic, of Auvergne, 259

Brighton, deposit containing recent shells in the cliffs near, 182

British alluviums, how formed, 147
—— their age, 147, 272

Diocchi on the tertiary strata of the Subapennines, 18, 155
—— on the number of shells common to Italy and the Paris basin, 156
—— on the age of the Italian tertiary strata, 156
—— on the organic remains of the sub-Apennine strata, 163

Bromley, pebble with oysters attached to it found in the plastic clay at, 278

Brongniart, M. Alex., on the formations of the Paris basin, 16
—— on the conglomerate of the hill of the Superga, 211
—— tabular view of his arrangement of the strata of the Paris basin—see wood-cut No. 58, 243, 247

Bronn, M., on the loess of the Rhine, 151, 153, 154

Brown coal formation near the valley of the Rhine, 199
—— organic remains of the, 200

Bruel, quarry of, 237

Buckland, Dr., on the Val del Bove, 83
—— on the grooved summits of the Corstorphine Hills, 147
—— on the effects of the Deluge, 271
—— on the Plastic clay, 278
—— on tertiary outliers on chalk hills, 283
—— on the former continuity of the London and Hampshire basins, 283
—— on valleys of elevation, 305, 307, 308

Budoshagy, rent exhaling sulphureous vapours in the mountain of, 223

Bufadors, jets of air from subterranean caverns called, 190

Bulimus montanus drifted from the Alps into the Mediterranean, 48

Buried cones on Etna, sections of, 88

Burton, Mr. J., his discovery of tertiary strata on the western borders of the Red Sea, 135

Cadibona, section of the fresh-water formations of—see wood-cut No. 55, 221
—— lignites of, remains of an anthracotherium found in, 222

Caernarvonshire, tertiary strata of, 135

Cæsar, volcanos of Auvergne not mentioned by, 269

Cairo, green sand containing shells at, 211

Calabria, recent tertiary strata of, 22

—— effects of the earthquake of 1783, 142, 319

Calais, ripple marks formed by the winds on the dunes near—see wood-cut No. 36, 176

Calanna, lava of Etna turned from its course by the hill of—see wood-cut No. 18, 86

—— description of the valley of, 85, 91

Calcaire grossier, alternation of the Plastic clay and, 244

—— number of fossil shells of the, 245

—— abundance of cerithia in the, 245

—— alternates with fresh-water limestone at Triel, 246

—— manner in which it was deposited, 246

—— in part destroyed when the upper marine strata were formed, 248

—— abundance of microscopic shells in the, 250

—— Palæotherium and fresh-water shells in, 252

Calcaire siliceux of the Paris basin, 246

—— alternates with calcaire grossier at Triel, 246

—— how formed, 246

Calcareous grit and peperino, sections of —see diagrams Nos. 9 and 10, 72

Caltagirone, blue shelly marl of, 66, 67

—— fossil shells from—see list, Appendix II., 55

Caltanisetta, dip of the tertiary strata at, 74

—— list of fossil shells from,—Appendix II., 54

Cambridgeshire, great line of chalk escarpment from, to Dorsetshire, 315

Campagna di Roma, age of the volcanic rocks of the, 183

Campania, tertiary formation of, 118

—— comparison of recorded changes in, with those commemorated by geological monuments, 118

—— age of the volcanic and associated rocks of, 126

—— external configuration of the country how caused, 127

—— affords no signs of diluvial waves, 128

Canadian lakes, changes which would take place in the Gulf of St. Lawrence if they were filled up, 28

Cantal, fresh-water formations of, 230

—— fresh-water limestone and flints resembling chalk in the, 237

—— proofs of the gradual deposition of marl in the, 239

Cape Wrath, granite veins of—see wood-cuts Nos. 85 and 86, 354

Capitol, hill of the, a deposit of calcareous tufa found on the, 138

Capo Santa Croce, shelly limestone resting on lava at, 68

Capra, flowing of the lavas of 1811 and 1819 round the rock of—see wood-cut No. 21, 92

—— traversed by dikes, 92

Carboniferous series, 326

Carcare, tertiary strata of—see wood-cut No. 55, 207, 222

—— fossil shells of, 211

Cardona, rock salt of, its relative age, 333

Casamicciol, shells found in stratified tuff at, 126

Caspian Sea, level of the, 29, 271

Castell de Stolles, ravine excavated in lava opposite the, 189

Castell Follitt, extent of the lava stream of —see map, wood-cut No. 43, 184

—— section of lava cut through by river at—see wood-cut No. 46, 189, 190

Castello d'Aci, 81

Castrogiovanni, section of the Val di Noto series at—see diagram No. 5, 64

—— hill of, its height, 66

—— capped by the Val di Noto limestone, 66

—— fossil fish found in gypseous marls at, 67

—— list of fossil shells from—Appendix II., 55

Castelgomberto, fossil shells of—see Table, Appendix I.

Catalonia, volcanic district of, 183

—— extent of the volcanic region of—see map, wood-cut No. 43, 184

—— volcanic cones and lavas of—(see Frontispiece), 185

—— ravines, excavated through lava in, 188, 189

—— age of the volcanos of, 191

—— superposition of rocks in the volcanic district of—see wood-cut No. 47, 192

Catania, volcanic conglomerates forming on the beach at, 73

—— plain of, 75, 76

—— marine formation near, 78

Catastrophes, remarks on theories respecting, 6, 33

Catcliff, Little, section of part of, showing the inclination of the layers in opposite directions—see wood-cut No. 33, 175

Cavalaccio, Monte, shells procured from the tuffs of, 79

Caves in Sicily, osseous breccias found in, 139

—— perforated in the interior by lithodomi, 140, 141

—— Australian, bones of marsupial animals in, 143

Cavo delle Neve, hollow in Ischia called the, 127
—— ancient sea-beach seen near, 127
Cellent, lava current of—see map, wood-cut, No 43, 184
—— section above the bridge of,—see wood-cut No. 45, 188
Central France, volcanic rocks of, 224, 257
—— fresh-water, formations of, 225
—— analogy of the tertiary deposits of, to those of the Paris basin, 241, 247
—— valleys of, how formed, 319
Cer, valley of the, sections of foliated marls in the, 239
Ceret and Boulon, dip of the tertiary strata between, 170
Cerithia, abundance of in the calcaire grossier, 245
Chabriol, M., on the fossil mammalia of Mont Perrier, 218
Chadrat, pisolitic limestone of, 232
Chalk, protruded masses of in the crag strata—see wood-cuts Nos. 41 and 42, 179, 180
—— English tertiary strata, conformable to the, 282
—— deep indentations filled with sand, &c., on its surface, 282
—— tertiary outliers on, 283
—— fissure in the, filled with sand near Lewes, 283
—— and upper green sand of the Weald valley, 286
—— escarpments of the Weald valley, once sea-cliffs—see wood-cuts Nos. 65 and 66, 289, 290, 291
—— why no ruins of, on the central district of the Weald, 295
—— of the North and South Downs, its former continuity, 303
—— the alternative of the hypothesis that it was once continuous considered, 304
—— valleys and furrows in the, how caused, 311
—— cliffs, rapid waste of on Sussex coast, 311
—— greatest elevation attained by it in England, 314
—— great line of escarpment formed by the, through the central parts of England, 315
—— nearly all the land in Europe has emerged since the deposition of the, 330
—— has been elevated at successive periods, 331
—— converted into marble by trap dike in Antrim, 369
Chalk-flints, analysis of, 238

Chamalieres, near Clermont, section at, 228
Chambon, lake of, formed by the lava of the Puy de Tartaret, 264
Chamouni, glaciers of, 150
Champheix, tertiary red marls of, 229
Champoleon in the Alps, strata altered near, 371
Champradelle, section of vertical marls at, —see wood-cut No. 57, 231
Chili, Newer Pliocene marine strata at great heights in, 130
Christie, Dr. T., his account of the Cave of San Ciro, 140
—— on caverns in Mount Beliemi, Sicily, 143
Cirque of Gavarnie, in the Pyrenees, 88
Cisterna on Etna, formed by a subsidence in 1792, 96, 129
Classification of tertiary formations in chronological order, 45
Clay-slate, lamination of, in the Pyrenees —see wood cut No. 89, 366
—— may be altered into shale, 373
—— convertible into hornblende schist, 373
Clermont, section of littoral deposits near, 228
—— section of vertical marls near—see wood-cut No. 57, 231
—— alternations of volcanic tuff and fresh-water limestone near, 258
Clift, Mr., on the bones of animals from Australian caves, 144
Climate, effects of changes of, on species, 44
Coal reduced to cinder by trap dike, 370
Colle, fresh-water formation of, 137
—— fossil shells of living species in the, 138
Comb Hurst, hills of, 279
Côme, lava current of, 186
Conception Bay, fossil shells of recent species found at great heights in, 130
Conglomerate, tertiary, of Nice, 167
—— now formed by the rivers near Nice, 168, 169
—— time required for the formation of great beds of, 170
Conglomerates, volcanic, of the Val di Noto, 73
—— now forming on the shores of Catania and Ischia, 73
Contemporaneous origin of Rocks, how determined, 37
Contemporaneous, remarks on the term, 52
Continents, position of former, 328, 330
Contortions in the Newer Pliocene strata in the Isle of Cyclops—see wood-cut No. 15, 80
Conybeare, Rev. W. D., on the English crag, 19

Conybeare, Rev. W. D., on the thickness of the London clay, 279
—— on the organic remains of the London clay, 280
—— on indentations in the chalk near Rochester, 282
—— on the transverse valleys of the North and South Downs, 298
—— on the former continuity of the chalk of the North and South Downs, 303
—— his objections to the theory of M. E. de Beaumont, 348
Coomb, view of the ravine called the, near Lewes—see wood-cut No. 75, 301
Coquimbo, parallel roads of, 131
Corals standing erect among igneous and aqueous formations at Galieri, 73
Cornwall, granite veins of—see wood-cut No. 87, 355, 370
—— argillaceous schist, containing organic remains in, 376
Costa de Pujou, structure of the hill of—see frontispiece, 186
Corstorphine hills, parallel grooves on their summits, how formed, 147
Cotentin, tertiary formation of the, 276
Coudes, tertiary red marl and sand-stone of, like 'new red sand-stone,' 229
Couze, river, lake formed by the filling up of its ancient bed by lava, 264
Crag of England, organic remains of the, 19
—— its relative age, 171
—— number of shells found in the, 171
—— its mineral composition, 171
—— relative position of the—see diag. No. 30, 172
—— lacustrine deposits resting on the, 173
—— forms of stratification of the—see wood-cuts 173, 174, 175
—— dip of the strata of the, 174, 175
—— comparison between the Faluns of Touraine and the, 203
—— derangement in the strata of the—see wood-cuts, 177
—— passage of, into alluvium, 181
—— its resemblance to formations now in progress, 177, 182
—— proportion of living species in the fossil shells of the—see Appendix I., p. 47
—— number common to Italy and the, ib. 47
—— number common to Sicily and the, ib. 47
—— number common to Italy, Sicily, and the, ib. 47
—— geographical distribution of the living species found in the, ib. 47, 51
Craters, volcanic, of the Eifel, how formed, 196
Creta, argillaceous deposit called, 67, 76
—— resting on columnar lava in the Isle of Cyclops—see wood-cut No. 14, 79

Crocodile of the Ganges found in both salt and fresh water, 330
Croizet, M., on extinct quadrupeds of Mount Perrier, 218
—— on alluviums of different ages in Auvergne, 267
Cromer, bent strata of loam in the cliffs near—see wood-cut No. 37, 178
Crowborough hill, height of, 288
—— thickness of strata removed from the summit of, 313
Cruckshanks, Mr. A., on distinct lines of ancient sea-cliffs on the coast of Peru, 130
Cuckmere, transverse valley of the, 298, 299
Curtis, Mr. J., on the fossil insects of Aix, in Provence, 277
Cussac, bones of extinct quadrupeds in alluvium under lava at, 219
Cutch, changes caused by the earthquake of 1819 in, 104, 249, 318
Cuvier, M., on the mammiferous remains of the Upper Val d'Arno, 221
—— on the tertiary strata of the Paris basin, 16, 247, 243
—— on the fossil organic remains of the Paris basin, 253
Cyclops, view of the island of, in the Bay of Trezza—see wood-cut No. 14, 79
—— its height, &c., 79
—— stratified marl resting on columnar lava in the—see wood-cut No. 14, 79
Cypris, abundance of the remains of, in the fresh-water strata of Auvergne, 230
—— habits of the living species of, 230

Darent, transverse valley of the, 298, 299
Daubeny, Dr., on the Val di Noto limestone, 66
—— on the volcanic region of Olot, in Catalonia, 184
—— on the volcanic district of the Lower Rhine and Eifel, 201
—— on the age of the Auvergne volcanos, 269
D'Aubuisson, on the appearance of some of the Auvergne lavas, 94
Daun, lake-craters of the Eifel seen near, 195
Dax, tertiary formations of, 20, 206
—— section of tertiary strata overlying the chalk near,—see diag. No. 51, 207
—— section of inland cliff near—see wood-cut No. 53, 210
—— fossil shells of—see tables Appendix I.
De Beaumont, M. Elie, on the cause of the historical deluge, 148, 272
—— his theory of the contemporaneous origin of parallel mountain chains considered, 337
—— his proofs that different chains were raised at different epochs, 340

De Beaumont, M. Elie, objections to the theory of, 341
—— on modern granite of the Alps, 358
De Candolle on the longevity of trees, 99
De la Beche, Mr., on M. Elie de Beaumont's theory, 347
Delta, of the Niger, size of the, 329
—— of the Nile, preyed on by currents, 28
—— of Rhone, in lake of Geneva, 27
De Luc, on the deluge, 271
Deluge, on the changes caused by the, 270
—— M. Elie Beaumont, on the cause of the historical, 148
Denudation, effects of, 30, 32
—— of the Valley of the Weald, 285
Deposition, sedimentary, shifting of the areas of, 26
Descartes, 97
Deshayes, M., his comparison of the fossil shells of Touraine, S. E. of France, Piedmont and Vienna, 21
—— his tables of fossil shells, 49—see Appendix I.
—— on the shells of the Val di Noto, series, 65, 67
—— on shells of the sub-Etnean beds, 79
—— on the fossil shells of Ischia, 126
—— on the fossil shells of the Antilles, 133
—— on the fresh-water shells of Colle, 138
—— on the fossil shells of the Crag, 171
—— on the limestone of Blaye, 208
—— on the fossil shells of Volhynia and Podolia, 215
—— on the fossil shells of Hungary, 223
—— on the abundance of Cerithia in the Paris basin, 245
—— on the changes which the Cardium porulosum underwent during its existence in the Paris basin, 250
—— on the microscopic shells of the Paris basin, 251
—— on the fossil shells of the Netherlands, 276
—— on the number of shells common to the Maestricht beds, chalk, and upper green sand, 325
—— on the distinctness of the secondary and tertiary fossil shells, 327
—— on the secondary fossil shells of the Pyrenees, 343
Desmoulins, M. Ch., on the Eocene deposits of the environs of Bordeaux, 209
Desnoyers, M., on the organic remains of the Faluns, 205
—— on the tertiary formations of Touraine, 20, 203
—— on the resemblance of the English Crag and the Faluns of the Loire, 204

Desnoyers, M., on the fossil organic remains of the Orleanais, 219
—— on the alternation of the plastic clay and calcaire grossier in the Paris basin, 244
—— on the tertiary formations of the Cotentin, 276
—— on the marine tertiary strata near Rennes, 276
Devil's-dyke, view of the chalk escarpment of the South Downs, taken from the—see wood-cut No. 65, 290
Diagonal stratification of the Crag strata—see wood-cuts, 174, 175
—— cause of this arrangement, 176
Dikes, intersecting limestone, 69
—— traversing peperino near Palagonia, —see diagrams Nos. 6 and 7, 69
—— on the summit of the lime-stone platform, Val di Noto, 70
—— off tuff or peperino, how formed, 70
—— changes caused in argillaceous strata by, 70
—— on Etna, their form, origin, and composition, 90
—— at the base of the Serra del Solfizio —see wood-cut, No. 19, 90
—— changes caused by, in the escarpment of Somma, 91
—— in the Val del Bove, as seen from the summit of Etna—see wood-cut No. 22, 93
—— some caused by the filling up of fissures by lava, 122, 123
—— of Somma—see wood-cut No. 25, 122
—— cause of the parallelism of their opposite sides, 122
—— varieties in their texture, 124
—— volcanic, in Madeira, 134
—— strata altered by, 368
Diluvial theories, 270
Diluvial waves, whether there are signs of their occurrence on Etna, 101
—— no signs of, in Campania, 128
Dip and direction of the tertiary strata of Sicily, 73
—— of the marine strata at the foot of Etna, 78
Dominica, alternations of coral and lava in, 133
Dorsetshire, valleys of elevation in, 308
Dorsetshire and Cambridgeshire, great line of chalk escarpment between, 315
Doue, M. Bertrand de, on the fossil mammiferous remains of Velay, 219
—— on the lacustrine deposits of Velay, 235, 236
—— on the igneous rocks of Velay, 262

Doue, M. Bertrand de, on Auvergne alluviums under lava, 267

Du Bois, M., on the tertiary strata of Volhynia and Podolia, 215

Dufrénoy, M., on the limestone of Blaye, near Bordeaux, 209

—— on the hill of Gergovia, 258

—— on the age of the red marl and rocksalt of Cardona, 333

Durance, river, land-shells brought from the Alps into the Rhone by the, 48

Dunwich, thickness of the crag strata in the cliffs near, 172

Dunwich, dip of the crag strata in a cliff between Mismer and—see wood-cut No. 33, 175

Dunes, near Calais, ripple marks formed by the winds on the—see wood-cut, No. 36, 176

Earthquake, Olot destroyed by, in 1421, 191

—— of Cutch, effects of the, 104, 249, 318

Earthquakes, their effects on the excavation of valleys, 113

—— during the Eocene period, 312

Earth's crust, signs of a succession of former changes recognizable in, 1

—— arrangement of the materials composing the, 8

Earth's surface may be greatly changed in one part while an adjoining tract remains stationary, 128

East Indian Archipelago, tertiary formations of the, 133

Ehrenhausen, coralline limestone of the hills of, 214

Eichwald, M., on the tertiary deposits of Volhynia and Podolia, 215

Eifel, volcanos of the, 193

—— map of the volcanic district of the—see wood-cut No. 48, 194

—— lake-craters of the—see wood-cut, No. 49, 195

—— trass of the, and its origin, 197

—— age of the volcanic rocks of the, 199

Elevation of land, how caused, 105

Elevation, proofs of successive, 111

Elsa, valley of the, fresh-water formations of, 137

England, tertiary strata of, 19, 135, 171, 284

—— comparison between the tertiary strata of Paris and those of, 282

—— tertiary strata of, conformable to the chalk, 282

—— origin of the tertiary strata of, 284

—— great line of chalk escarpment through the central parts of, 315

—— elevation of land on the east coast of, since the Older Pliocene period, 316

England, elevation of land gradual in the S.E. of, 318

—— on the excavation of valleys in the S.E. of, 319

Enza, river, nature of the sediment deposited by the, 161

Eocene period, derivation of the term, 55

—— proportion of living species in the fossil shells of the, 55

—— position of the beds referrible to this era—see diagrams Nos. 3 and 4, 20, 21

—— geographical distribution of the recent species found in the, 55

—— mammiferous remains of the, 59

—— fresh-water formations of the, 225

—— marine formations of the, 241

—— our knowledge of the physical geography, fauna and flora of the, considerable, 254

—— volcanic rocks of the, 257

—— map of the principal tertiary basins of the—see wood-cut No. 62, 275

—— earthquakes during the, 312

—— alluviums of the, 317

—— chasm between the newest secondary formations and those of the, 328

—— great volume of hypogene rocks formed since, 381

—— number of species of fossil shells common to different formations referrible to the, Appendix I., p. 49

—— number of living species in the fossil shells of the, ib., 50

—— number common to the Pliocene, Miocene, and, ib., 50

—— geographical distribution of the living species found in the, ib., 51

Eocene strata in the Bordeaux basin, 208

—— its relative position—see wood-cut No. 52, 209

Epomeo, shells found in volcanic tuff near the summit of, 126

Erratic blocks of the Alps, 148

—— transported by ice, 149

Escarpments, manner in which the sea destroys successive lines of, 111, 292

Escarpments of the chalk in the Weald valley, once sea-cliffs—see wood-cuts, Nos. 65 and 66, 289, 291

Estuary deposits, arrangement of, 9

Eternity of the earth, or of present system of changes not assumed in this work, 383

Etna, marine and volcanic formations at its base, 75

—— view of, from the limestone platform of Primosole—see diagram No. 11, 75

—— connexion of the strata at its base with those of the Val di Noto—see diagram No. 12, 76

Etna, southern base of, 77
—— recent shells in clay at the foot of, 77
—— dip of the marine strata at the base of, 78
—— eastern side of, 78
—— shells in tuffs and marls on the east side of, 79
—— lavas of the Cyclopian isles, not currents from, 81
—— internal structure of the cone of, 83
—— great valley on the east side of—see wood-cut No. 17, 83
—— lateral eruptions of, 84
—— manner of increase of the principal cone of, 84
—— sections of buried cones on, 88
—— form, composition, and origin of the dikes on, 90
—— veins of lava on—see wood-cut No. 20, 91
—— view from the summit of, into the Val del Bove—see wood-cut No. 22, 93
—— subsidences on, 96
—— antiquity of the cone of, 97
—— whether signs of diluvial waves are observable on, 101
—— list of fossil shells from the flanks of—Appendix II., p. 53,
Europe, newest tertiary strata of, 22
—— large portions of, submerged when the secondary strata were formed, 23
—— almost all the land in, has emerged since the deposition of the chalk, 330
European tertiary strata, successive origin of the, 18
European alluviums in great part tertiary, 150
Excavation of valleys, 319

Faluns of Touraine, 203
—— comparison between the English crag and the, 203, 204
—— were formed in a shallow sea, 204
—— organic remains of the, 204, 206
Fasano, escarpment of marine strata seen near, 78
Fault in the cliff-hills near Lewes—see section, wood-cut No. 76, 301
Finochio, view of the rock of, with the lavas of 1811 and 1819 flowing round it—see wood-cut No. 21, 92
Firestone of the Weald Valley, 286
—— terrace formed by the harder beds of—see wood-cut No. 67, 291, 292
Fish, skeletons of, by no means frequent in a fossil state, 47
—— fossil, of Castrogiovanni, 67
Fitton, Dr., on the secondary rocks of the Valley of the Weald, 286
—— on the denudation of the Weald Valley, 289

Fitton, Dr., on faults in the strata of the Forest ridge, 293
—— on a line of vertical and inclined strata from the Isle of Wight to Dieppe, 315
—— an ammonite found in the Maestricht beds by, 325
—— on the extent and thickness of the Wealden, 329
—— on the delta of the Niger, 329
Fiume Salso, in Sicily, 252
Fleming, Dr., on the effects of the deluge, 271
Flinty slate, slate-clay of the lias, converted into, by trap dike, 370
Flood, supposed effects of the, 270
—— hypothesis of a partial, 270
Floridia, schistose and arenaceous limestone of, 66
Fluvia, river, ravines in lava excavated by, 186, 189
Forest ridge of the Weald Valley, 293
—— faults in the strata of the, 293
—— thickness of masses removed from the, 313
Formations, causes of the superposition of successive, 26
—— universal, remarks on the theory of, 38
—— new subdivisions of the tertiary, 52
Fossa Grande, section of Vesuvius seen in, 84
Fossilization of plants and animals partial, 31
Fossils, distinctness of the secondary and tertiary, 327
Fresh-water deposits, secondary, why rare, 330
Fuveau, in Provence, tertiary strata of, 276

Gabel Tor, volcano of, 136
Galieri, a bed of corals found standing erect among igneous and aqueous formations at, 73
Ganges, the crocodile of the, found both in fresh and salt water, 330
Gannat, fresh-water limestone of, 232
Garnets, in altered shale, 369
Garrinada, hill of, described—(see frontispiece,) 187
Gavarnie, cirque of, 88
—— lamination of clay-slate near—see wood-cut No. 89, 366
Gault of the Valley of the Weald, 286
—— valley formed at its out-crop, 292
—— forms an escarpment towards the Weald clay, 293
Gemunden Maar, view of the—see wood cut No. 49, 195

Geneva, lake of, advance of the delta of the Rhone in, 27
—— change which will take place in the distribution of sediment when it is filled up, 27
Genoa, height of the tertiary strata above the sea at, 165, 166
—— position of the strata—see diagram No. 28, 166
Geological periods, their distinctness may arise from our imperfect information, 56
Gergovia, hill of, alternation of volcanic tuff and fresh-water marls in the, 258
—— section of, 259
—— intersected by a dike of basalt—see wood-cut, No. 60, 259
Giacomo, St., valley of, described, 84, 85, 91
Gillenfeld, description of the Pulvermaar of, 197
Girgenti, section at—see diagram No. 5, 64
—— shells found in the limestone of, 65
—— dip of the tertiary strata at, 74
—— list of fossil shells from—Appendix II., p. 54
Gironde, tertiary strata of the basin of the, 206
Glaciers of Savoy, great quantities of rock brought down by the, 149
Glen Roy, parallel roads of, 131
Glen Tilt, junction of limestone and granite in—see wood-cut No. 88, 356
Gly, river, tertiary strata in the valley of the, 170
Gneiss, mineral composition of, 365, 367
—— passage of, into granite, 367, 372
—— was originally deposited from water, 367
—— whence derived, 373
Gozzo degli Martiri, dikes intersecting limestone at, 69
—— view of the valley of—see wood-cut No. 23, 110
Grammichele, beds of incoherent yellow sand with shells found near, 66
—— bones of the mammoth found in alluvium at, 151
Grampians, granite veins of the, 357
Granada, tertiary strata of, 170
Granite, junction of limestone and, in Glen Tilt—see wood-cut No. 88, 356
—— formed at different periods, 13, 357
—— passage from trap into, 361
—— origin of, 12, 363
—— passage of gneiss into, 367, 372
—— changes produced by its contact with strata of lias and oolite in the Alps, 371
Granite veins, their various forms and mineral composition—see wood-cuts, Nos. 85, 86, 87, and 88, 353, 356, 370

Gravesend, deep indentations in the chalk filled with sand, gravel, &c., near, 282
Greywacke, 377
—— of the Eisel, 194
—— age of the rocks termed, 327
Greenough, Mr., on fossil shells from the borders of the Red Sea, 136
Grifone, Monte, caves containing the remains of extinct animals in, 141
Grit, calcareous, and peperino, sections of —see diagrams Nos. 9 and 10, 72
Grooved surface of rocks, how formed, 147
Grosœil, near Nice, tertiary strata found at, 135
Guadaloupe, active volcanos in, 133
Guidotti, Signor, on the shells of the gypsum of Monte Cerio, 159
Gypseous marls containing fish found at Castrogiovanni, 63, 67
Gypsum, and marls, of the Paris basin, 247
—— bones of quadrupeds, &c., in, 251
—— on the entire absence of marine remains in the, 252
—— of St. Romain on the Allier, 233
—— beds of, interstratified with the sub-Apennine marls, 159
—— unaltered shells in the, 159
Gyrogonites, abundant in the fresh-water formations of the Paris basin, 250

Hall, Sir James, his experiments on rocks, 124
—— on the grooved summits of the Corstorphine hills, 147
Hall, Capt. B., on the parallel roads of Coquimbo, 131
—— on vertical dikes of lava in Madeira, 134
—— on the veins traversing the Table Mountain, Cape of Good Hope, 354
Hamilton, Sir W., his account of the eruption of Vesuvius in 1779, 122
Hampshire basin, tertiary formations of the 18, 280
—— mammiferous remains of the, 280, 281
—— on the former continuity of the London and, 283
Happisborough, diagonal stratification of the crag strata near—see wood-cut No. 32, 174
Hartz mountains, geological and geographical axes of the, 346
Hastings sands, their composition, 280
—— anticlinal axis formed by the, 287
Haute Loire, fresh-water formation of the, 235
Headen Hill, section of, 281
Heat, its influence on the consolidation of strata, 334
Hebrides, age of the volcanic rocks of the, 336

Heidelberg, shells found in the loess at, 152

—— loess and gravel alternating at, 153

—— granites of different ages near, 357

Henslow, Professor, on the changes caused by a volcanic dike in Anglesea, 368

Hibbert, Dr., on the extinct volcanos of the Rhine, 197, 201

—— on the loess of the valley of the Rhine, 151

—— on the mammiferous remains of Velay, 219

Highbeach, in Essex, height of the London clay at, 312

Hoffmann, Professor, his examination of Sicily, 63

—— on the limestone of Capo Santa Croce, 68

—— on the new island of Sciacca, 71

—— on the Val del Bove, 88

—— on cave deposits in Sicily, 139, 140, 141

Honduras, recent strata of the, 133

Hornblende schist, altered clay or shale, 373

Horner, Mr. Leonard, his map of the volcanic district of the Eifel and Lower Rhine—see wood-cut No. 48, 194

—— on the geology of the Lower Rhine and Eifel, 201

Hugi, M., on secondary strata altered into gneiss in the Alps, 372

—— on modern granite in the Alps, 358

Human remains now becoming imbedded in osseous breccias in the Morea, 144

Humboldt, on the depression of a large part of Asia, below the level of the sea, 29

Hundsruck, beds and veins of quartz found in the mountains of the, 201

Hungary, tertiary formations of, 212

—— age of the tertiary strata of, 215

—— volcanic rocks of, 222

—— age of the igneous rocks of, 223

Hutton, his opinion as to altered sedimentary rocks, 382

Huttonian hypothesis of the origin of gneiss, 366

Hypogene, term proposed as a substitute for primary, 374

—— formations, no order of succession in, 375

—— rocks, their identity of character in distant regions, 376

—— produced in all ages in equal quantities, 377

—— their relative age, 377

—— volume of, formed since the Eocene period, 381

Icebergs, rocks transported by, deposited wherever they are dissolved, 149, 150

Idienne, volcanic mountain of, 252

Indusial limestone of Auvergne, 232

Inkpen Hill, the highest point of the chalk in England, 314

Inland cliff near Dax—see wood-cut No. 53, 209

Inland cliffs on East side of Val di Noto, 111

Insects, fossil, of Aix, 277

Ischia, volcanic conglomerates now in progress on the shores of, 73

—— fossil shells of recent species found at great heights in, 126

—— external configuration of, how caused, 127

—— list of fossil shells from—Appendix II., 57

Isle of Bourbon, a volcanic eruption every two years in, 363

Isle of Cyclops, in the bay of Trezza, view of—see wood-cut No. 14, 79

—— its height, &c., 79

—— stratified marl resting on columnar lava in the—see wood-cut No. 14, 79

—— contortions in the newer Pliocene, strata of—see wood-cut No. 15, 80

—— divided into two parts by a great fissure, 80

—— newer Pliocene strata invaded by lava in—see wood-cut No. 16, 81

—— lavas of, not currents from Etna, 81

Isle of Purbeck, traversed by a line of vertical or inclined strata, 315

Isle of Wight, geology of the, 18

—— fall of one of the Needles of the, into the sea in 1772, 181

—— fresh-water strata of the, 280

—— mammiferous remains of the, 281, 317

—— vertical strata of the, 315

Italy, tertiary strata of, 18

—— age of the volcanic rocks of, 183

—— number of living species in the fossil shells of—see Appendix I., 47

—— number of those common to Sicily and, ib. 47

—— number common to the Crag and, ib. 47

—— number common to Sicily, the and, ib. 47

Jack, Dr., on the geology of the island of Pulo Nias, 134

Jamaica, fossil shells of recent species from, in the British Museum, 133

Java, subsidence of the volcano of Papandayang, in the island of, 96

—— vegetation destroyed by hot sulphuric water from a mountain in, 252

Jobert, M., on the extinct quadrupeds of Mont Perrier, 218

—— on the hill of Gergovia, 258

Jobert, M., on the different ages of Auvergne alluviums, 267
Jorullo, time for which the lava of, retained its heat, 363
Jura, erratic blocks of the, 148

Kaiserstuhl, volcanic hills in the plains of the Rhine, 152
—— covered nearly to their summits with loess, 152
Katavothrons of the plain of Tripolitza now filling up with osseous breccias, 144
Kater, Capt., on recent deposits near Ramsgate, 182
Keferstein, M., his objections to M. de Beaumont's theory, 347
Kingsclere, valley of, ground plan of the —see wood-cut No. 78, 305
—— section across the, from North to South—see wood-cut No. 79, 305
—— section of the, with the heights on a true scale—see wood-cut No. 80, 306
—— anticlinal axis of the, 306
—— proofs of denudation in the, 307
Killas of Cornwall, 370

Laach, lake-crater of, 197
Lacustrine deposits overlying the crag—see diagram No. 30, 173
Lake Aidat, formed by the damming up of a river by lava, 269
Lake-craters of the Eifel—see wood-cuts Nos. 49 and 50, 195, 196
—— how formed, 196
Lakes, arrangement of deposits in, 8
Lake Superior, recent deposits in, analogous to those of the Eocene lakes in Auvergne, 230
—— nature of the recent deposits in, 334
—— the bursting of its barrier would cause an extensive deluge, 270
Lamarck, his list of the fossil shells of the Paris basin, 156
La Motta, valleys excavated through blue marl capped with columnar basalt at, 77
—— volcanic conglomerate of—see diagram No. 13, 77
—— relative age of the basalts of, 82
Lancashire, tertiary strata of, 135
Land, elevation of, caused by subterranean lava, 105
Land-shells drifted from the Alps into the Mediterranean, 48
Landers, on the delta of the Niger, 330
Landes, tertiary strata of the, 206
La Roche, section of the hill of, 190
Las Planas, lava current of, 189
La Trinità, near Nice, fossil shells of, 168
Lauder, Sir T. D., on the parallel roads of Glen Roy, 131
Lava, a bed of oysters between two currents of, at Vizzini, 73

Lava, columnar, stratified marl resting on, in the Isle of Cyclops—see wood-cut No. 14, 79
—— minerals in cavities of, 81
—— veins of, on Etna, 91
—— great length of time which it requires to cool, 363
Lava streams solid externally while in motion, 86
Lavas of the Cyclopian isles not currents from Etna, 81
Lavas and breccias of the Val del Bove, 93
Lavas excavated by rivers in Catalonia, 186, 189
Lavas and alluviums of different ages in Auvergne—see wood-cut No. 61, 266
La Vissiere, fresh-water limestone covered by volcanic rocks at, 263
—— faults in the limestone at, 263
Leeward Islands, geology of the, 132
Le Grand d'Aussi, M., on alluviums under lava in Auvergne, 267
Leith Hill, height of, 293
Lentini, volcanic pebbles covered with serpulæ in the limestone near, 73
—— dip of the strata at, 74
—— valleys near, their origin, 111
Leonhard, M., on the loess of the valley of the Rhine, 151
—— on the volcanic district of the Lower Rhine, 201
—— on granites of different ages near Heidelberg, 357
Lewes, fissures in the chalk filled with sand near, 283
—— view of the ravine called the Coomb near—see wood-cut No. 75, 301
—— fault in the cliff-hills near—see wood-cut No. 76, 301
Leybros, fresh-water limestone of, 237
Lias, strata of the, 326
—— strata between the Carboniferous group and the, 326
—— converted into flinty slate by trap dike in Antrim, 370
—— altered in the Alps, 372
—— altered in Hebrides, 378
Licodia, relative age of the basalts of, 82
Lignite interstratified with the sub-Apennine marls, 159
Lima, valley of, proofs of the successive rise of the, 130
Limagne d'Auvergne, lacustrine deposits and volcanic rocks of the—see map, wood-cut No. 60, 226
Limestone formation of the Val di Noto described—see diagram No. 5, 64
—— its organic remains, 65
Limestone, resting on lava at Capo Santa Croce, 68
Lithological character of the sub-Apennine beds, 157, 162

Lockart, M., on the fossil remains of the Orleanais, 219

Loess of the valley of the Rhine, 151
—— mineral, composition of the, 151
—— its thickness and origin, 152
—— gravel, &c. alternating with, 153
—— list of shells from the—see Appendix II., 58

Loire, tertiary strata of the basin of the, 20
—— relative age of the strata of the—see diagram No. 3, 20
—— 'faluns' of the, 203

London basin, tertiary deposits of the, 18, 277
—— on the former continuity of the Hampshire and, 283
—— fossil shells of the—see Tables, Appendix I.
—— proportion of living species in the fossil shells of the—Appendix I., 50

London clay, its composition, thickness, &c., 279
—— septaria of the, 279
—— the fossil shells identifiable with those of the Paris basin, 280
—— organic remains of the, 280

Lower green-sand described, 286
Lower Rhine, see Rhine.

Lucina divaricata, wide geographical range of the, 254

Luy, section of tertiary strata in the valley of the—see diagram No. 51, 207

Maars, or lake-craters of the Eifel—see wood-cuts Nos. 49 and 50, 195, 196
—— how formed, 196

Macculloch, Dr., on the parallel roads of Glen Roy, 131
—— sub-Apennine strata termed marine alluvia by, 157
—— on the granite veins of Cape Wrath, in Scotland, 354
—— on the junction of granite and limestone in Glen Tilt, 356
—— on the granitic rocks of Shetland, 357
—— on the granite of Sky, 358
—— on the trap rocks of Scotland, 360
—— on the granite of Aberdeenshire, 361
—— on the passage of gneiss into granite, 372

Macigno of the Italians the greywacke of the Germans, 162

Maclure, Dr., on the geology of the Leeward Islands, 132
—— on the volcanic district of Olot in Catalonia, 184
—— his observations preceded by those of Don Bolos, 193

Madeira, fossil shells of recent species brought from, 134

Madeira, vertical dikes of compact lava seen in, 134
—— violently shaken by earthquakes during the last century, 134

Maestricht beds, fossils of the, 324
—— chasm between the Eocene and, 325
—— number of fossil shells common to the chalk and, 325
—— number common to the upper greensand and, 325

Magnan, river, 167
—— section from Monte Calvo to the sea by the valley of—see diagram, No. 29, 167

Malaga, tertiary strata of, 170

Mammalia, fossil, importance of the remains of, 47
—— duration of species in, more limited than in testacea, 140
—— shells of living species found with extinct, 140

Mammiferous remains of the successive tertiary eras, 59

Mammoth, tusk of, found in calcareous tufa near Rome, 139

Man, remains of, now becoming imbedded in osseous breccias in the Morea, 144

Mantell, Mr., on the fossil shells of the crag, 171
—— on deposits containing recent shells in the cliffs near Brighton, 182
——— on tertiary outliers on the chalk, 283
—— on the secondary rocks of the Weald valley, 286
—— his section of the valley of the Weald, with the heights on a true scale—see wood-cut No. 64, 288
—— his section from the North escarpment of the South Downs to Barcombe —see wood-cut No. 71, 296
—— on the absence of chalk detritus on the central ridge of the Weald, 296
—— his section of a fault in the cliff-hills near Lewes—see wood-cut No. 76, 301
—— his discovery of the Mososaurus of Maestricht in the English chalk, 325

Map of the volcanic district of Catalonia —see wood-cut, No. 43, 184
—— of the volcanic region of the Eifel— see wood-cut No. 48, 194
—— of Auvergne, showing its geographical connexion with the Paris basin—see wood-cut No. 56, 226

Marculot, fresh-water limestone of, 232

Mardolce, grotto of, bones of extinct quadrupeds found in the, 140
—— pierced in the interior by boring testacea, 141
—— breccia in, how formed, 141

Marienforst, blocks of quartz containing casts of fresh-water shells found near, 199

Marine alluviums, 145

Marine testacea, wide range of, 44, 48

Marls, sub-Apennine, localities of the, 158, 159

—— sometimes thinly laminated, 158

—— interstratified with lignite and gypsum, 159

—— capped by basalt at some places, 159

Martin, Mr., on the Valley of the Weald, 293

—— on the transverse valleys of the North and South Downs, 299

—— his supposed section of a transverse valley—see wood-cut No. 74, 300

—— his estimate of the thickness of strata removed from the summit of the Forest ridge, 313

Maritime Alps, tertiary strata at the base of the, 164

Marsupial animals, their remains found in breccias in Australian caves, 143

Mascalucia, subsidence on Etna near the town of, 96

Medesano, lignite in the sub-Apennine marls at, 159

Mediterranean, organic remains of the, 40

—— distinct from those of the Red Sea, 41, 205

—— shells drifted from the Alps into the, 48

Medway, transverse valley of the, 298, 299

Meerfelder Maar described, 197

Melilli, view of a circular valley near—see wood-cut No. 23, 110

—— inland cliffs seen near, 111

Merdogne, fresh-water marls intersected by a dike of basalt above the village of, 259

Metamorphic, the term proposed and defined, 374

—— rocks of the Alps, altered lias and oolite, 371

—— sometimes pass into sedimentary, 376

—— sometimes divided by strong line of demarcation, 376

—— in what manner their age should be determined, 378

—— why those visible to us are for the most part ancient, 380

—— why they appear the oldest, 379

Micaceous schist, whence derived, 373

Microscopic fossil shells abundant near Sienna, 163

—— shells of the Paris basin,—see Plate IV., 250

Militello, list of fossil shells from—see Appendix II., p. 54

Mineral character, persistency of, why apparently greatest in older rocks, 331

—— characters, proofs of contemporaneous origin derived from, 37

Mineral composition of the sub-Apennine strata, 157

—— of rocks no proof of contemporaneous origin, 161

Minerals in the cavities of lava, Isle of Cyclops, 81

Miocene period, term whence derived, 54

—— proportion of living species in the fossil shells of the, 54

—— position of the beds referrible to the —see diagrams, Nos. 3 and 4, 20, 21

—— mammiferous remains of the, 59

—— Marine formations of the, 202

—— fresh-water formations of the, 219

—— volcanic rocks of the, 222

—— alluviums, localities of, 217

—— fossil shells of the—see tables Appendix I.

—— general results derived from the fossil shells of the—Appendix I., p. 47

—— number of fossil species of shells common to different formations referrible to the, ib., p. 47.

—— number of living species in the fossil shells of the, ib., p. 48

—— number of species common to the pliocene and, ib., p. 49

—— geographical distribution of the living species of the, ib., 51

Mirambeau, red clay and sand of, 208

Mismer, 'dip of the crag strata in a cliff between Dunwich and—see wood-cut, No. 33, 175

Misterbianco, valleys excavated through blue marl at, 77

Mitchell, Major, on breccias in Australian caves, 143

Mitscherlich, M., on the minerals found in Somma, 121

Modern causes, remarks on the term, 319

Molasse, thickness of, at Stein, 153

—— of Switzerland, 212

—— its place in the series of tertiary formations not yet known, 212

Mole, transverse valley of the, 298

Molluscous animals, superior longevity of the species of, 48

Mont Dor, age of the volcano of, 260, 262

—— its height, form, and composition, 261

Mont Ferrat, tertiary strata of the hills of, 21

—— hills of, geological structure of the, 211

Monte Calvo, section from to the sea—see diagram No. 29, 167

Monte Cerio, unaltered shells found in the gypsum of, 160

Monte Grifone, caves containing osseous breccias in, 141

Montlosier, M., on alluviums of different ages in Auvergne, 267

Monte Mario, marine strata of, 138

—— shells changed into calcareous spar in, 160

Montmartre, gypsum of, 247
—— bones of quadrupeds, &c. in the gypsum of, 251
—— entire absence of marine remains in the gypsum of, 252
Mont Mezen, age of the, 260
Monte Nuovo, formation of, 104, 128, 125
Montpellier, tertiary strata of, 215
Mont Perrier, position of the Miocene alluviums of—see wood-cut, No. 54, 217
—— remains of extinct quadrupeds in the alluviums of, 218
—— age of the trachytic breccias of, 262
Montsacopa, volcanic cone of—(see Frontispiece,) 186
Mountain chains formed of successive igneous and aqueous groups superimposed on each other, 240
—— on the relative antiquity of, 337
—— difficulty of determining the relative ages of, 350
Moravia, fossil shells of— see tables, Appendix I.
Morea, osseous breccias now forming in the, 144
—— tertiary strata of the, 170
—— distinct ranges of sea cliffs at various elevations in the, 113, 132
—— fossil shells of the—see tables, Appendix I.
Moropano, fossil shells found in tuff near the town of, 126
Mosenberg, a mountain with a triple volcanic cone, 197
Mososaurus of Maestricht found also in the English chalk, 325
Mundesley, protuberances of chalk in the crag strata near, 180
Murat, fresh-water deposits covered by volcanic rocks near, 263
Murchison, Mr., on the tertiary strata of Grosœil, near Nice, 135
—— on tertiary strata at the base of the Maritime Alps, 166, 168
—— his section of the manner in which the crag rests on the chalk—see diagram No. 30, 173
—— on the Superga, 211
—— on the tertiary formations of Styria, gaam 213, 214
—— on the fresh-water formation of Cadibona, 222
—— on the volcanic rocks of Styria, 224
—— on central France, 227
—— on the lacustrine strata of the Cantal, 239
—— on Auvergne, 258
—— on the Plomb du Cantal, 263
—— on the excavation of valleys, 265
—— on the tertiary formations of Aix, in Provence, 277
—— on the terrace formed by the hard beds of the upper green-sand, 292

Murphy, Lieut. H., on the height of the North Downs, 288
Musara, sections of buried cones seen near the rock of, 88
—— flowing of the lava of 1811 and 1819 round—see wood-cut, No. 21, 92
—— traversed by dikes, 92
Nadder, valley of the, 308
Nantes, tertiary strata near, resting on primary rocks, 204
Naples, recent tertiary strata in the district around, 22
—— volcanic region of, changes which it has undergone in the last 2000 years, 118
—— recent shells in volcanic tuffs near, 126
Necker, M. L. A., on the dikes of Somma, 121
—— on the cause of the parallelism of their opposite sides, 122
—— on the varieties in texture of the dikes of Somma, 124
Needles of the Isle of Wight, fall of one of them into the sea in 1772, 181
Nesti, M., on the fossil elephant of the upper Val d'Arno, 221
Netherlands, tertiary formations of the, 276
Newer Pliocene period—see Pliocene period, newer
Newhaven, patches of tertiary strata found on the chalk near, 286
Nice, height of the tertiary strata above the sea at, 165, 167
—— section from Monte Calvo to the sea, by the valley of Magnan near—see diagram No. 29, 167
—— great beds of conglomerate near, 167
—— dip of the strata, 168
Niger, delta of the, area covered by the, 329
Nile, its delta now preyed on by currents, 28
Noeggerath, M., his map of the Eifel district, 193
—— on volcanic district of the Rhine, 201
Norfolk, crag strata of, 171
—— rapid waste of the cliffs on the coast of, 297
Northampton, Lord, fossil fish found near Castrogiovanni by, 67
North Downs, chalk ridge called the, 287
—— section across the valley of the Weald from the south to the—see woodcuts, No. 63 and 64, 288
—— highest point of the, 288
—— on the former continuity of the chalk of the, with that of the South Downs, 303
Noto, Val di, formations of the, 63
—— volcanic rocks of the, 63, 67
Novera, hill of, in Sicily, junction of tuff and limestone in the—see diagram No. 8, 70

Odoardi, on the recent origin of the tertiary strata of Italy, 19

Œiningen, fossil reptile found at, 7

Older Pliocene period—see Pliocene period, *older*

Olivet, volcanic cone of—(see frontispiece,) 187

Olot, volcanic district of, 183

—— its extent—see map, wood-cut No. 43, 184

—— number of volcanic cones in—(see frontispiece, 185)

—— geological structure of the district around, 185

—— age of the volcanos of, 191

—— town of, destroyed by an earthquake in 1421, 191

—— country between Perpignan and, occasionally shaken by earthquakes, 191

Omalius d'Halloy, on the former connexion of Auvergne and the Paris basin by lakes, 241

Oolite, or jura limestone formation, 326

—— converted into hypogene rock in the Alps, 371

Organic remains, controversy as to the real nature of, 3

—— theories to account for their occurrence in high mountains, 4

—— contemporaneous origin of rocks proved by, 39

—— comparative value of different classes of, 46

Origin of the globe, no geological proofs of, 384

Orleanais, fossil remains of the, 219

Orthès, tertiary strata of, 207

Osseous breccias, in Sicilian caves, 139

—— in Australian caves, 143

—— now forming in the Morea, 144

Otranto, tertiary strata of, 22

Ouse, transverse valley of the, 298, 299

—— has filled up an arm of the sea, 300

Outlying patches of tertiary strata on chalk hills, 283

Pachydermata, great abundance of this order in the Eocene period, 59

Pacific, lines of ancient sea cliffs on the shores of the, 130

Palæotherium found in the fresh-water strata of the Isle of Wight, 281, 317

Palagonia, dikes traversing peperino at—see diagrams Nos. 6 and 7, 69

—— section to Paterno from—see diagram No. 12, 76

Palermo, caves containing osseous breccias near, 140

—— fossil shells from—see list Appendix II., p. 55, 56

Panella, in Ischia, ancient sea-beach seen near, 127

Papandayang, subsidence of the volcanic cone of, 96

Paraliel roads of Coquimbo, 131

—— of Glen Roy, 131

Paris, comparison between the tertiary strata of, and those of England, 282

Paris basin, formations of the, 16

—— organic remains of the, 16

—— all tertiary formations at first referred to the age of those of the, 17,

—— analogy of the deposits of central France to those of the, 241

—— geographical connexion of Auvergne and the, 241

—— subdivisions of strata in the, 242

—— diagrams showing the relation which the strata bear to each other—see wood-cuts, Nos. 58 and 59, 243

—— superposition of different formations in the, 244

—— plastic clay and sands of the, 244

—— calcaire grossier, 245

—— calcaire siliceux described, 246

—— gypsum and marls of the, 246

—— second or upper marine group, 248

—— third fresh-water formation, 249

—— age of the deposits of the, 20, 250

—— abundance of microscopic shells in the, 20, 250

—— bones of quadrupeds in gypsum, 251

—— alternation of strata with and without organic remains in the, 254

—— number of living species in the fossil testacea of the, 55, 253

—— concluding remarks on the tertiary strata of the, 254

—— fossil shells of the—see tables, Appendix I.

—— number of living species in the fossil shells of the—Appendix I., p. 50

Parkinson, Mr., on the crag, 19, 156

Parma, sub-Apennine marls thinly laminated near, 158

—— these marls interstratified with lignite in the territory of, 159

—— silicified shells found in the marls near, 160

—— blue marl of, a fresh-water univalve filled with marine shells found in the, 163

—— river, brown clay deposited by the, 161

Paroxysmal elevations, theory of, 128

Partsch, M., on the tertiary strata of the basin of Vienna, 213

Paterno, section from, to Palagonia—see diagram, No. 12, 76

—— valleys excavated through blue marl at, 77

—— relative age of the basalts of, 82

Pauliac, lime-tone of, 208

Pegwell bay, recent deposits in, 182

Pentalica, great limestone of the Val di Noto seen in the valley of, 64

Pentland, Mr., on the bones of animals from Australian caves, 144

—— on the mammiferous remains of the Upper Val d'Arno, 220

Peperino, traversed by dikes near Palagonia—see diagrams, Nos. 6 and 7, 69

—— dikes of, how formed, 70

—— sections of calcareous grit and—see diagrams Nos. 9 and 10, 72

Peperinos, of the Val di Noto, 71

—— how formed, 71

Perpignan, the country between Olot and, occasionally shaken by earthquakes, 191

—— fossil shells of—see Tables, Appendix I.

Peru, proofs of successive elevation of the coast of, 130

Pewsey, Vale of, 308

Phillips, Mr., his analysis of chalk flints, 230

Philosopher's Tower on Etna, 128

Phlegræan Fields, minor cones of the, 125

Piana, conglomerate of, 211

Piazza, dip of the tertiary strata at, 74

Piedmont, tertiary strata of, 20, 211

—— their relative age—see diagram, No. 4, 21

Pitchstone, a thin band of, formed at the contact of the dikes of Somma and intersected beds, 124

Placentia, character of the sediment transported by rivers in the territory of, 161

Plants, their fossilization partial, 31

—— fossil, importance of, in geology, 47

Plas Newydd, changes caused in sedimentary strata by a volcanic dike near, 368

Plastic clay and sand of the London basin, 278

—— its thickness, composition, &c., 278

—— organic remains rare in the, 279

—— clay and sand of the Paris basin, 244

—— alternates with calcaire grossier, 244

Pliny does not mention the Auvergne volcanos in his Natural History, 269

Pliocene period, newer, derivation of the term, 53

—— proportion of living species in the fossil shells of the, 53

—— marine formations of the, 62

—— contortions in strata of the, in the Isle of Cyclops—see wood-cut, No. 15, 80

—— strata of the, invaded by lava—see wood-cut No. 16, 81

—— subterranean rocks of fusion, formed during the, 107

—— fresh-water formations of the, 137

Pliocene period, newer, osseous breccias and cave deposits of the, 139

—— alluviums of the, 145

—— extinct animals in breccias of the, 140

Pliocene period, older, proportion of living species in the fossil shells of the, 54

—— position of the beds referrible to this era—see diagrams Nos. 3 and 4, 20, 21

—— mammiferous remains of the, 59

—— tertiary formations referrible to the, 155

—— volcanic rocks of the, 183

—— elevation of land on the East coast of England since the commencement of the, 316

Pliocene period, fossil shells of the—see Table, Appendix I.

—— general results derived from the fossil shells of the—Appendix I., p. 47

—— number of species of fossil shells common to different formations of the—Appendix I., p. 47

—— number of living species in the fossil shells of the—Appendix I., p. 47

—— number of species common to the Miocene and—Appendix I., p. 49

—— geographical distribution of the living species of the—Appendix I., p. 51

—— strata of Sicily, their dip and direction, 73

Pliocene strata of Sicily, origin of the, 103

—— changes of the surface during and since their emergence, 109

—— strata, newer, only visible in countries of earthquakes, 129

Plomb du Cantal, successive accumulation of the, 240

—— age of the volcanic rocks of the, 260, 262

—— its height, form, structure, &c., 263

—— fresh-water limestone covered by volcanic rocks on the northern side of the, 263

Plutonic rocks, 353

—— distinction between volcanic and, 359

—— their relative age, 364, 377

—— changes produced by, 370

—— why those now visible are for the most part very ancient, 379

Podolia, tertiary formations of, 215

Poggibonsi, conglomerate of, 160

Pont du Chateau, alternation of volcanic tuff and fresh-water limestone at, 258

Portella di Calanna, furrows in the defile called, how formed, 147

Pratt, Mr., on the mammiferous remains of the Isle of Wight, 281

Pressure, effects of, on the consolidation of strata, 334

Prevost, M. Constant, on the tertiary strata of Vienna, 21, 212
—— tabular view of his arrangement of the strata of the Paris basin—see wood-cut No. 59, 243
—— on the alternation of the calcaire grossier, and siliceous limestone, 246, 248
—— on the manner in which the mammiferous remains may have been preserved in the Paris gypsum, 252
—— on the alternation of strata with and without organic remains, 254
Primary, on the rocks usually termed, 10, 352
—— their relation to volcanic and sedimentary formations, 352
—— divisible into two groups, the stratified and unstratified, 353
—— on the stratified rocks called, 12, 365
—— the term why faulty, 374
—— strata, how far entitled to the appellation, 377
Primitive, term now abandoned, 13
Primosole, termination of the Val di Noto, limestone at, 75
—— view of Etna from—see diagram No. 11, 75
Procida, island of, would resemble Ischia if raised, 127
Pulo Nias, fossil shells of recent species found in the island of, 134
Pulvermaar, description of the, 197
Punto del Nasone on Somma, dikes or veins of lava seen at—see wood-cut No. 25, 122
Punto di Guimento, veins of lava at—see wood-cut No. 20, 91
Puracé, extinct volcano of, 252
Pusanibio river, sulphuric and muriatic acids, and oxide of iron in the waters of the, 252
Puy Arzet, chalk with conformable beds of tuff in the hill called, 207
Puy de Come, ravine excavated through the lava of the, 264
Puy de Jussat, quartzose grits of, 229
Puy de Marmont, alternation of volcanic tuff and fresh-water marl in the, 258
Puy de Pariou, 268
Puy Rouge, ravine cut through the lava of the, 265
Puy de Tartaret, 264
Puy en Velay, bones of extinct quadrupeds in alluvium under lava near, 219
—— fresh-water formation of, 236
Puzzuoli, inland cliff near, will be destroyed, 112
—— no great wave caused by the rise of the coast near, in 1538, 128
Pyrenees, tertiary strata at the eastern extremity of the, 170

Pyrenees, tertiary formations between the basin of the Gironde and the, 206
—— their relative age, 341
—— tertiary strata abutting against vertical mica-schist at the eastern end of the, 348
—— lamination of clay-slate in the—see wood-cut No. 89, 366

Quartz, compact, whence derived, 373
Quorra, or Niger, delta of the, 329

Radicofani, sub-Apennine marls capped by basalt at, 159
—— age of the volcanic rocks of, 183
Radusa, fossil fish found in great abundance at, 67
Ramond, M., on alluviums of Auvergne, 267
Ramsgate, recent deposits in the cliffs near, 182
Ravines excavated through the lavas of Auvergne, 264, 265
Recent formations, description of, 52
—— form a common point of departure in all countries, 58
—— why first considered, 62
Recent and Tertiary formations, synoptical table of, 61
Red marl and sandstone of Auvergne like 'new red sandstone,' 229, 333
Red marl, supposed universality of, 333
Red Sea, and Mediterranean, distinct assemblages of species found in the, 41, 205
—— tertiary strata found on its western borders, 135
—— list of fossil shells from—see Appendix II., 57
Rennes, tertiary strata near, 276
Rhine, lower, volcanos of the, 193
—— map of the volcanic district of the, 194
—— age of the volcanic rocks of the, uncertain, 199
—— origin of the trass of the, 197
—— ancient alluviums of the, 200
Rhone, delta of, in lake of Geneva, 27
—— shells drifted from the Alps to the Mediterranean by the, 48
Riccioli, Signior, tusk of the mammoth from the Roman travertin shown to the author by, 138
Rimao, valley of, lines of ancient sea-cliffs in, 130
Ripple marks formed by the wind on the dunes near Calais—see wood-cut No. 36, 176
Risso, M., on the fossil shells of Groseil, near Nice, 135
—— on the fossil shells of St. Madeleine, near Nice, 168
Rivers, difference in the sediment of, 40

Robert, M., on extinct quadrupeds of Cussac, 219

Rocca di Ferro, shells in the tuffs of, 79

Rochester, indentations in the chalk filled with sand, &c., near, 282

Rocks, distinction between sedimentary and volcanic, 10, 352

—— primary, 10

—— origin of the primary, 11, 363

—— distinction between primary, secondary and tertiary, 10

—— persistency of mineral character, why apparently greatest in the older, 331

—— older, why most consolidated, 334

—— older, why most disturbed, 335

—— secondary volcanic, of many different ages, 335

—— relative age of, how determined, 35

—— proofs of, by superposition, 35

—— proofs by included fragments of older rocks, 36

—— proofs of their contemporaneous origin derived from mineral characters, 37

—— proofs from organic remains, 39

—— volcanic of the Val di Noto, 63, 67

—— grooved surface of, 147

—— transportation of, by ice, 149

—— identity of their mineral composition no proof of contemporaneous origin, 161

Roderberg, crater of the, described, 198

Rome, travertins of, 138

—— hills of, capped by calcareous tufa, 138

Ronca, fossil shells found at—see Table, Appendix I.

Royat, ruins of Roman bridges and baths at, prove that no great changes have taken place since their erection, 269

Rozet, M., on the loess of the valley of the Rhine, 151

Runton, folding of the crag strata in the cliffs near—see wood-cut No. 38, 178

St. Christopher's, alternations of coral and volcanic substances in, 133

St. Eustatia, tertiary formations in, 133

St. Hospice, tertiary strata in the peninsula of, 135

St. Lawrence, Gulf of, changes which would result in, on the filling up of the Canadian lakes, 28

St. Madeleine, near Nice, shells abundant in the loamy strata of, 168

St. Michael's Mount, Cornwall, 371

St. Peter's Mount, Maestricht, fossils of, 325

St. Romain, gypsum worked at, 233

St. Vincents, active volcanos in, 133

Salisbury Craig, altered strata in, 369

San Ciro, cave of, breccia containing bone of extinct quadrupeds in, 141

San Ciro, position of the cave of,—see diagram No. 27, 141

San Feliu de Pallerols, deep ravine cut through lava near the town of, 189

San Quirico, hills of, their composition, 159

Sand and conglomerate of the sub-Apennine strata described, 159

Santa Croce, Cape of, limestone resting on lava at, 68

Santa Madalena, section at the bridge of, 186

Santa Margarita, size of the volcanic crater of, 187

Sardinian volcanos, their age uncertain, 193

—— rest on a tertiary formation, 193

Sasso, Dr., on the tertiary strata of Genoa, 166

—— on the fossil shells of Albenga, 167

Saucats, fresh-water limestone of, 207

Savona, tertiary strata of—see wood-cut No. 55, 166, 222

Sciacca, volcanic island of, 69, 71

Scoresby, Capt., on the transportation of rocks by icebergs, 150

Scotland, parallel grooves formed in the beds of torrents in, 147

—— granite veins of—see wood-cuts Nos. 85 and 86, 354

Scrope, Mr. G. P., on the volcanic district of Naples, 125

—— on the extinct volcanos of the Rhine, 197, 201

—— on the hill of Gergovia, 258

—— on Mont Dor, 261

—— on the excavation of lava by the river Sioule, 265

—— on alluviums under lava at different elevations in Auvergne, 267

Sea-cliffs, successive elevations proved by —see wood-cut No. 24, 111

—— manner in which the sea destroys successive ranges of, 111, 292

—— distinct ranges of ancient, in the Morea, 113

—— found elevated to great heights in Peru, 130

Seaford, waste of the chalk cliffs at, 311

Secondary rocks, 14

—— of the Weald valley divisible into five groups, 286

—— their rise and degradation gradual, 308

—— enumeration of the principal groups of, 324

—— no species common to the tertiary and, 327

—— circumstances under which they originated, 23, 329

—— why more consolidated, 334

—— why more disturbed, 335

Secondary rocks, volcanic, of many different ages, 335

Secondary fresh-water deposits why rare, 330

Secondary periods, duration of, 328

Sedgwick, Professor, on diluvial waves, 101, 272

—— on the tertiary formations of Styria, 213, 214

—— on the volcanic rocks of Styria, 224

—— on the Isle of Wight, 281, 315

—— on synclinal lines, 293

—— on the theory of M. Elie de Beaumont, 347

—— on the Cornish granite veins, 355

—— on garnets in altered shale, 369

Sediment, changes in the distribution of, which would take place on the filling up of large lakes, 27

Sedimentary deposition, causes which occasion the shifting of the areas of, 26

Sedimentary rocks, distinction between volcanic and, 10

Seguinat, Montagne de, lamination of clay-slate in the—see wood-cut, No. 89, 366

Selenite found in clay at the foot of Etna, 77

Septaria of the London clay described, 279

Serre del Solfizio, sections of buried cones in the cliffs of, 88

—— dikes at the base of—see wood-cut No. 19, 90

Serres, M. Marcel de, on the drifting of land shells to the sea by the Rhone, 48

—— on the tertiary strata of Montpellier, 215

—— on the fossil insects of Aix, 277

Sicily, geological structure of, 22, 63

—— dip and direction of the newer Pliocene strata of, 73

—— origin of the newer Pliocene strata of, 103

—— form of the valleys of, 109

—— no peculiar indigenous species found in, 115

—— breccias containing bones of extinct animals in caves in, 139

—— alluviums of the newer Pliocene period in, 151

—— fossil shells of—see Tables, Appendix I.

—— number of living species in the fossil shells of—see Appendix I., 47

—— number common to Italy and, ib. 47

—— number common to Italy, the Crag and, ib. 47

—— number of species proper to, ib. 47

Shells, tables of fossil—(see Appendix,) 49

—— characteristic tertiary—(see Plates,) 50

—— necessity of accurately determining the species of, 50

Shells, recent, numerical proportion of in the different tertiary periods, 58

—— number of species of, found both living and fossil, 394

—— fossil tertiary, number examined to construct the tables, 394

—— fossil, number common to all the tertiary periods, Appendix I., 50

—— living, number of those found in a fossil state in all the tertiary periods, ib. 50

—— geographical distribution of those species which have their fossil analogues, ib. 51

Sherringham, sections in the cliffs east of —see wood-cuts, Nos. 39 and 40, 178, 179

—— rapid waste of the cliffs at—see section, wood-cut No. 72, 297

Shetland, action of the sea on the coast of, 146

—— granites of different ages in, 357

—— passage of trap into granite in, 362

Siebengebirge, volcanic phenomena of the, 198

Sienna, Subapennine strata near the town of, 160

—— microscopic fossil shells very abundant near, 163

—— list of fossil shells from—Appendix II., 59

Siliceous schist, clay converted into by lava, 70, 81

Silvertop, Col., on the tertiary strata of Spain, 170

Simeto, plain of the, 76

Sioule, river, ravines cut through lava-currents by the, 265

Sky, age of the granite of, 358

Smyth, Capt. W. H., his drawing of the Isle of Cyclops—see wood-cut No. 14, 79

—— on the extinct volcanos of Sardinia, 193

Somma, escarpment of, 84, 85, 87, 96

—— changes caused by dikes in the, 91

—— dikes of, 121

—— minerals found in, 121

—— parallelism of the opposite sides of the dikes of, 122

—— varieties in the texture of the dikes of, 124

Somma and Vesuvius, differences in the composition of, 120

Sortino, great limestone formation seen in the valleys of, 64

—— bones of extinct animals in caves near, 139

South Downs, chalk ridge called the, 287

—— section from to the North Downs across the Weald Valley—see wood-cuts No. 63 and 64, 288

—— highest point of the, 288

South Downs, view of the escarpment of the—see wood-cut No. 65, 290
—— section from their northern escarpment to Barcombe—see wood-cut No. 71, 296
—— on the former continuity of the chalk of the North and, 303
Spaccaforno limestone, 65
Spain, tertiary formations of, 170
—— extinct volcanos of the north of, 183
—— lavas excavated by rivers in, 186, 189
Species, changes of, everywhere in progress, 30
—— effects of changes of climate on, 44
—— superior longevity of molluscous, 48
—— necessity of accurately determining, 49
—— living, proportion of in the fossils of the newer Pliocene period, 53
—— in the older Pliocene period, 54
—— in the Miocene period, 54
in the Eocene period, 55
—— their geographical distribution, 55
—— in Sicily older than the country they inhabit, 115
—— outlive great revolutions in physical geography, 115
—— none common to the secondary and tertiary formations, 327
Spinto, fossil shells in green sand at, 211
Steininger, M., on the loess of the Rhine, 151
—— on the volcanic district of the Eifel, 201
Steyning, chalk escarpment as seen from the hill above—see wood-cut No. 66, 291
Stirling Castle, altered strata in the rock of, 369
Stour, transverse valley of the, 298
Strata, cause of the limited continuity of, 9
— — order of succession of—see diagram No. 1, 14
—— origin of the European tertiary, at successive periods, 18
—— Recent, form a common point of departure in all countries, 58
—— with and without organic remains alternating in the Paris basin, 254
—— on the consolidation of, 334
Stratification, unconformable, remarks on, 30, 33
—— of the Crag—see wood-cuts, 174, 175
—— of primary rocks—see wood-cut No. 89, 365, 366
Strike of beds, explanation of the term, 346
Stromboli, lava of, has been in constant ebullition for 2000 years, 363
Studer, M., on the loess of the valley of the Rhine, 152
—— on the molasse of Switzerland, 212

Styria, tertiary formations of, 212
—— age of the tertiary strata of, 214
—— volcanic rocks of, 223
Sub-Apennine strata, 18, 155
—— opinions of Brocchi on the, 155
—— lithological characters of the, 157, 162
—— not all of the same age, 157
—— termed marine alluvia by Dr. Macculloch, 157
—— subdivisions of the, described, 158
—— how formed, 160
—— organic remains of the, 163
—— fossil shells of the—see Tables, Appendix I.
Subaqueous deposits, our continents chiefly composed of, 9
—— how raised, 104
—— distinction between alluvium and, 145
Submarine eruptions, proofs of ancient, in the Bay of Trezza, 78, 81
Subsidence on Papandayang, in Java, 96
—— on Etna, 96
Subterranean lava the cause of the elevation of land, 105
Subterranean rocks of fusion, probable structure of the recent, 107
Suffolk, relative age of the tertiary strata of—see diagram No. 4, 21
—— crag strata of, 171
—— cliffs, thickness of the crag in the, 172
Superga, strata composing the hill of the, highly inclined, 211
—— fossil shells of the, 211
Superior, Lake. See Lake Superior.
Superposition, of successive formations, causes of the, 26
—— proof of more recent origin, 35
—— exceptions in regard to volcanic rocks, 36
—— no invariable order of, in Hypogene formations, 375
Surface, different states of the, when the secondary and tertiary strata were formed, 23
Switzerland, 'molasse' of, 212
Synclinal and anticlinal lines described—see wood-cut No. 68, 293
Syenites not distinguishable from granites, 358
Synoptical Table of Recent and Tertiary Formations, 61
Syracuse, section at—see diagram No. 5, 64
—— shells found in the limestone of, 65
—— range of inland cliffs seen to the north of, 111
—— bones of extinct animals in caves near, 140
—— list of fossil shells from—Appendix II., p. 54

Table-Mountain, Cape of Good Hope, intersected by veins—see wood-cut No. 85, 354

Tanaro, plains of the, 211

Taro, river, nature of the sediment deposited by the, 161

Taunus, beds and large quartz veins found in the mountains of the, 201

Tech, tertiary strata in the valley of the, 170

Ter, valley of the, 185

Teronel, river, lava excavated by the, 189

Terraces, manner in which the sea destroys successive lines of—see wood-cut No. 24, 111, 292

Terranuova, dip of the tertiary strata at, 74

Tertiary formations, general remarks on the, 15

—— of the Paris basin, 16, 241

—— at first all referred to the age of those of the Paris basin, 17, 19

—— origin of the European, at successive periods, 18

—— of the sub-Apennine hills, 18

—— of Touraine, 20

—— of Bordeaux and Dax, 20

—— of Piedmont, 20

—— of the Valley of the Bormida, 21

—— of the Superga, near Turin, 21

—— of the basin of Vienna, 21

—— newer than the sub-Apennines, 21

—— the newest often blend with those of the historical era, 22

—— different circumstances under which these and the secondary formations may have originated, 23, 329

—— state of the surface when they were formed, 24

—— classification of, in chronological order, 45

—— new subdivisions of the, 52

—— numerical proportion of recent shells in different, 53, 54, 55, 58

—— mammiferous remains of successive, 59

—— Synoptical Table of Recent and, 61

—— of Sicily, 63

—— of Campania, 118

—— of Chili and Peru, 130

—— of the West India Archipelago, 132

—— of the East India Archipelago, 133

—— of Norway and Sweden, 135

—— on the western borders of the Red Sea, 135

—— identity of their mineral composition no proof of contemporaneous origin, 161

—— of the Po, Arno, and Tiber, their resemblance, 161

—— at the base of the Maritime Alps, 164

—— at the eastern extremity of the Pyrenees, 170

—— in Spain, 170

—— in the Morea, 170

—— of England, 18, 19, 135, 171, 277

Tertiary formations, of Touraine, 20, 203

—— of the basin of the Gironde and the district of the Landes, 206

—— of Piedmont, 211

—— of Switzerland, 212

—— of Styria, Vienna, Hungary, &c., 212

—— of Volhynia and Podolia, 215

—— of Montpellier, 215

—— of Auvergne, 217, 226

—— of Velay, 219, 235

—— of the Orleanais, 219

—— of the Upper Val d'Arno, 219

—— of Cadibona, 221

—— of the Cantal, 236

—— of the Cotentin, or Valognes, 276

—— of Rennes, 276

—— of the Netherlands, 276

—— of Aix in Provence, 276

—— no species common to the secondary and, 327

Testacea, fossil, of chief importance, 47

—— marine, wide range of, 44, 48

—— longevity of the species of, 48, 56

Tet, valley of the, tertiary strata found in, 170

Thames, basin of the, 18

Theorizing in geology, different methods of, 1

Tiber, river, has flowed in its present channel since the building of Rome, 138

—— yellow sand deposited by the, 161

—— valley of the, 139

Time, effects of prepossessions in regard to the duration of past, 97

Touraine, tertiary strata of, 20, 203

—— and Paris, relative age of the tertiary strata of—see diagram No. 3, 20

—— fossil shells of—see Tables, Appendix I.

Trachytic breccias and alluviums alternating in Auvergne, 217

Transition formations, remarks on, 13

Transverse valleys in the North and South Downs—see wood-cut No. 73, 298

—— remarks on their formation, 299

—— supposed section of one of them—see wood-cut No. 74, 300

Transylvania, tertiary formations of, 213

—— age of the tertiary strata of, 215

—— volcanic rocks of, 223

—— fossil shells of—see Tables, Appendix I.

Trap rocks, origin of the term, 360

—— of Scotland, how formed, 360

—— passage of, into granite, 361

Trass of the Rhine volcanos, 197

—— its origin, 198

Travertins of the valley of the Elsa, 137

—— of Rome, recent shells with the tusk of the mammoth found in, 138

Trees, longevity of, 99, 272

Trezza, bay of, sub-Etnean formations exposed in the, 78

—— proofs of sub-marine eruptions in the, 78, 81

Trimmingham, manner in which the crag strata rest on the chalk near—see diagram No. 30, 173

—— view of a promontory of chalk and crag near—see wood-cut No. 41, 179

—— section of the northern protuberance of chalk at—see wood-cut No. 42, 180

Tripolitza, plain of, breccias now forming in the, 144

Tufa, calcareous, the hills of Rome capped by, 138

—— tusk of the mammoth found in, near Rome, 138

Tuff, dikes of, how formed, 70

—— in the hill of Novera—see diagram No. 8, 70

—— volcanic, recent shells in, near Naples, 126

—— shells found in, at great heights in Ischia, 126

Turin, tertiary formations of, 211

—— fossil shells of—see Tables, Appendix I.

Tuscany, fresh-water formations of, 137

—— age of the volcanic rocks of, 183

Uddevalla, elevated beaches of, 135

Unconformability of strata, remarks on the, 30, 33

Universal formations, remarks on the theory of, 38

Universality of red marl, remarks on the supposed, 333

—— of certain hypogene rocks, 376

Upper marine formations of the Paris basin how formed, 248

Val d'Arno, Upper, mineral character of the lacustrine strata of the, 161

—— fresh-water formations of the, 219

—— mammiferous remains of the, 220

Val del Bove, great valley on east side of Etna—see wood-cut No. 17, 83

—— its length, depth, &c, 84

—— description of the, 87

—— its circular form, 87

—— dikes numerous in the, 87

—— dip of the beds in the, 87

—— section of buried cones seen in, 88

—— difference in the dip of the beds where these occur, 88

—— scenery of the, 88

—— form, composition, and origin of the dikes in, 90

—— view of, from the summit of Etna—see wood-cut No. 22, 93

—— lavas and breccias of the, 93

—— origin of the, 95

—— floods in, caused by melting of snow by lava, 96

Valdemone, formations of, 75

Val di Calanna, its crateriform shape, 85

—— dip of the beds in the, 85

Val di Calanna, its origin, 85

—— began to be filled up by lava in 1811 and 1819—see wood-cut No. 18, 86

Val di Noto, formations of the, 63

—— divisible into three groups—see diagram No. 5, 64

—— volcanic rocks of the, 63, 67

—— volcanic conglomerates of the, 73

—— proofs of the gradual accumulation of the formations of the, 73

—— connexion of the formations of the, with those at the base of Etna—see diagram No. 12, 76

—— form of the valleys in the limestone districts of the, 110

—— inland cliffs seen on the east side of, 111

—— igneous rocks of the, 361

—— fossil shells from the—see Appendix II., p. 53

Vale of Pewsey, 308

Valley of the Nadder, 308

—— of the Weald—see Weald

Valleys, of elevation, 305

—— on Etna, account of, 83

—— of Sicily, their form—see wood-cut No. 23, 110

—— most rapidly excavated where earthquakes prevail, 113, 148

—— and parallel troughs between the North and South Downs, how formed, 294

—— transverse, of the North and South Downs, 298

—— how formed—see wood-cut No. 74, 300

—— and furrows on the chalk, how caused, 311

—— of the South-east of England, how formed, 319

Valmondois, rolled blocks of calcaire grossier in the upper marine sandstone of, 249

Valognes, tertiary strata of the environs of, 276

—— fossil shells of—see Tables, Appendix I.

Van der Wyck, M., on the Eifel district, 201

Var, river, large quantities of gravel swept into the sea by the, 168, 169

Vatican, hill of the, calcareous tufa on the, 138

Veaugirard, alternation of calcaire grossier and plastic clay at, 244

Veins of lava on Etna—see wood-cut No. 20, 91

Velay, bones of extinct quadrupeds in volcanic scoriæ in, 219

—— fresh-water formations of, 235

—— age of the volcanic rocks of, 224, 260, 262

Velay, ancient alluviums covered by lava at different heights in, 262

Vertical and inclined strata, great line of, from the Isle of Wight to Dieppe, 315.

Vesuvius, dikes of, 121

—— channels formed by the flowing of lava from, in 1779, 122

—— and Somma, difference in their composition, 120

Vichy, tertiary oolitic limestone of, 232

—— dip of the lacustrine strata at, 233

Vienna, tertiary formations of, 21, 212

—— age of the tertiary strata of, 214

—— basin, fossil shells of the—see Table, Appendix I.

Vigolano, gypsum interstratified with sub-Apennine marls at, 159

Villasmonde, shells found in limestone at, 65

—— list of fossil shells from—Appendix II., 54

Villefranche, bay of, tertiary strata found near the, 135

Vinegar river, sulphuric and muriatic acids, and oxide of iron, in the waters of the, 252

Virlet, M., on the tertiary strata of the Morea, 170

Viterbo, volcanic tuffs and sub-Apennine marls alternating at, 159

—— age of the volcanic rocks of, 183

Viviani, Professor, on the character of the Sicilian flora, 115

—— on the tertiary strata of Genoa, 166

Vizzini, junction of inclined tuff and horizontal limestone near—see diagram No. 8, 70

—— changes caused by a dike of lava in argillaceous strata at, 70

—— a bed of oysters between two lava-currents at, 73

Volcanic breccias in Auvergne, how formed, 259

Volcanic conglomerates of the Val di Noto, 73

—— now forming on the shores of Catania and Ischia, 73

Volcanic dikes, strata altered by, 70, 368

Volcanic district of Catalonia, superposition of rocks in the—see wood-cut No. 47, 192

Volcanic lines, modern, not parallel, 349

Volcanic region of Naples, changes which it has undergone during the last 2000 years, 118

Volcanic rocks, distinction between sedimentary and, 10

—— relative age of, how determined, 36

—— of the Val di Noto, 68

—— of Campania, their age, 126

—— of central France, 257

—— secondary, of many different ages, 335

Volcanic rocks, distinction between plutonic and, 359

Volcanos, mode of computing the age of, 97

—— sometimes inactive for centuries, 98

—— of Olot, in Catalonia, described—see Frontispiece, 183

—— extinct, of Sardinia, 193

—— of the Lower Rhine and the Eifel, 193

—— the result of successive accumulation, 240

—— attempt to divide them into antediluvian and post-diluvian, 268

Volhynia, tertiary formations of, 215

Voltz, M., on the loess of the Rhine, 151

Von Buch, M., on the Eifel, 201

—— on the tertiary formations of Volhynia and Podolia, 215

—— on the general range of volcanic lines over the globe, 349

Von Dechen, M., on the volcanic district of the Lower Rhine, 201

—— on the Hartz mountains, 346

—— his objections to the theory of M. de Beaumont, 346, 347

—— on the Cornish granite veins—see wood-cut No. 87, 355

Von Oeynhausen, his map of the Eifel district, 193

—— on the volcanic district of the Rhine, 201

—— on the granite veins of Cornwall—see wood-cut No. 87, 355

Walton, section of shelly crag near—see wood-cut No. 31, 174

—— lamination of sand and loam near—see wood-cut No. 34, 175

Warburton, Mr., on the Bagshot sand, 280

Watt, Gregory, his experiments on melted rocks, 124, 372

Weald, denudation of the valley of the, 285

—— secondary rocks of the, divisible into five groups, 286

—— section of the valley of the—see wood-cuts Nos. 63 and 64, 287, 288

—— clay, its composition, 286

—— gradual rise and degradation of the rocks of the, 289

—— alluvium of the valley of the, 295

—— extent of denudation in the valley of the, 303, 313

Wealden, secondary group, called the, 325

—— organic remains of the, 325

—— its great extent and thickness, 329

—— how deposited, 329

Webster, Mr., on the geology of the Isle of Wight, 18, 316

—— on the tertiary formations of the London and Hampshire basins, 278, 280

Wellington Valley, Australia, breccias containing remains of marsupial animals found in, 143

West Indian Archipelago, tertiary formations of the, 132

Wey, transverse valley of the, 298, 299

Whewell, Rev. W., 53

Wildon, thickness of the coralline limestone of, 214

Wiltshire, valleys of elevation in, 308

Wily, valley of the, 308

Winds, ripple-marks caused by, on the dunes near Calais—see wood-cut No. 36, 176

Wrotham Hill, height of, 288

Yarmouth, thickness of crag in the cliffs near, 172

Ytrac, fresh-water flints strewed over the surface near, 237

Zaffarana, valleys extending from the summit of Etna to the neighbourhood of, 83

Zocolaro, hill of, lava of Etna deflected from its course by the—see wood-cut No. 18, 86

Zoological provinces, great extent of, 40

Zoophytes, recent species of, but little known, 47

THE END.

WORKS

ON

SCIENCE AND NATURAL HISTORY,

PUBLISHED BY MR. MURRAY.

I.

VIEW OF THE MOTIONS OF THE HEAVENLY BODIES.

With a popular INTRODUCTION.

By Mrs. SOMERVILLE. 8vo. 30s.

II.

ON THE RECIPROCAL CONNEXION OF THE PHYSICAL SCIENCES.

By Mrs. SOMERVILLE.

Being the substance of the Preliminary Essay prefixed to the " Mechanism of the Heavens," enlarged and adapted for the general and unscientific reader, in one small volume.

" We have, indeed, no hesitation in saying, that we consider the Preliminary Dissertation by far the best condensed view of the Newtonian philosophy which has yet appeared."—*Quarterly Review.*

III.

FIRST AND SECOND REPORT OF THE
PROCEEDINGS OF THE

BRITISH ASSOCIATION FOR THE ADVANCEMENT OF SCIENCE,

At YORK, in the year 1831, and at OXFORD in 1832. In one vol. 8vo.

CONTENTS :—PART I.

1. Rev. WILLIAM VERNON HARCOURT. Exposition of the OBJECT and PLAN of the ASSOCIATION.
2. DETAILED ACCOUNT of the PROCEEDINGS at YORK, in September, 1831.

PART II.

REPORTS read to the SOCIETY at OXFORD, June, 1832, viz.—

1. PROFESSOR AIRY, F.R.S., on the State and Progress of PHYSICAL ASTRONOMY.
2. J. W. LUBBOCK, Esq., on the TIDES.
3. J. D. FORBES, Esq., F.R.S., on the Present State of METEOROLOGICAL SCIENCE.
4. Sir DAVID BREWSTER, F.R.S. L.& E., on the Progress of OPTICAL SCIENCE.
5. Rev. R. WILLIS, on the PHENOMENA of SOUND.
6. Rev. PROFESSOR POWELL, on the PHENOMENA of HEAT.
7. Rev. PROFESSOR CUMMING, on THERMO-ELECTRICITY.
8. F. W. JOHNSTONE, Esq., on the Recent Progress of CHEMICAL SCIENCE.
9. Rev. PROFESSOR WHEWELL, M.A., F.R.S., on the State and Progress of MINERALOGICAL SCIENCE.
10. Rev. W. CONYBEARE, M.A., F.R.S., F.G.S., on the Recent Progress, Present State, and Ulterior Development of GEOLOGY.

Together with an Account of the Public Proceedings of the Society, and of the Daily Transactions of the Sub-Committees during the period of the meeting.

Will be published in a few days.

IV.

CONSOLATIONS in TRAVEL; or the LAST DAYS of PHILOSOPHER.

By SIR HUMPHRY DAVY, late President of the Royal Society.
A New Edition. Small 8vo. 6s.

V.

Also by the same Author,

SALMONIA; or DAYS of FLY-FISHING.

Third Edition, with Plates and Wood-cuts. Small 8vo. 12s.

VI.

The JOURNAL of a NATURALIST.

Third Edition. Plates and Wood-cuts Post 8vo. 15s.

———— Plants, trees, and stones, we note,
Birds, insects, beasts, and many rural things.

VII.

GLEANINGS in NATURAL HISTORY. With Local Recollections.

By EDWARD JESSE, Esq., Surveyor of His Majesty's Parks and Palaces.

To which are added MAXIMS and HINTS for an ANGLER. Being a Companion to the JOURNAL OF A NATURALIST. A New Edition, Post 8vo. 10s. 6d.

VIII.

The ZOOLOGY of NORTH AMERICA.

Part I., containing the QUADRUPEDS.

By JOHN RICHARDSON, M.D., Surgeon of the late Expedition under Captain FRANKLIN. Illustrated by Twenty-eight spirited Etchings, by Thomas Landseer. 4to. 1l. 11s. 6d.

*** Published under the Authority and Patronage of His Majesty's Government.

IX.

The ZOOLOGY of NORTH AMERICA.

Part II., containing the BIRDS.

By JOHN RICHARDSON, M.D., Surgeon of the late Expedition under Captain FRANKLIN, and WILLIAM SWAINSON, Esq., F.R.S., F.L.S. Illustrated with Fifty beautifully coloured Engravings, drawn by W. SWAINSON, Esq., and numerous Wood-cuts. 4to. 4l. 4s. Printed uniformly with the Narratives of Captains Franklin and Parry's Expeditions, to which it is an indispensable Appendix.

X.

BOTANICAL MISCELLANY.

Containing Figures and Descriptions of new, rare, or little-known Plants, from various parts of the World, particularly of such as are useful in Commerce, in the Arts, in Medicine, or in Domestic Economy. By PROFESSOR HOOKER. Now complete in Three volumes, with upwards of One Hundred and Twenty Plates.

*** The NINTH and LAST NUMBER is just published, with 12 plates, 8vo. 10s. 6d.

XI.

TRANSACTIONS of the ROYAL GEOGRAPHICAL SOCIETY of LONDON.

Nos. I. and II. With Maps and Plans. 8vo. 7s. 6d. each.

Printed in the United States
By Bookmasters